Welding and Joining Engineering

Zainul Huda

Welding and Joining Engineering

Design, Processes, Calculations

Springer

Zainul Huda
Department of Mechanical Engineering
King Abdulaziz University
Jeddah, Saudi Arabia

ISBN 978-3-032-20461-5 ISBN 978-3-032-20462-2 (eBook)
https://doi.org/10.1007/978-3-032-20462-2

This Springer imprint is published by the registered company Springer Nature Switzerland AG
The registered company address is: Gewerbestrasse 11, 6330 Cham, Switzerland

Preface

Welding and joining technologies form an integral part of the modern manufacturing industry. This is why all manufacturing/mechanical/metallurgical engineering students (both at undergraduate and graduate levels) are required to study "Welding and Joining" either as a core course or as the topics embedded in the *Manufacturing Technology* course. Despite the industrial/technological importance of the topic, no welding/joining book covering combined qualitative and quantitative aspects of design, processes, and calculations is available in the market. This gap in the scientific literature motivated the author to write the present book.

This unique text covers both qualitative and quantitative aspects of welding and joining technologies. The book includes 200 worked numerical examples, 130 diagrams/illustrations, 100 MCQs with answers, 180 formulas/equations, and several case studies on the failures of weld joints in machine components. This volume would enable BS (Mechanical/Metallurgical/Manufacturing Engineering) students to score high grades in the *Welding/Joining Technology* course. Besides UG students, graduate students and doctoral researchers may also benefit from this book because it addresses both fundamentals and advances in welding technologies. Additionally, welding and manufacturing engineers can reduce the manufacturing cost by minimizing rejections because the book teaches the prevention of welding defects and testing/QA as well as *case studies*.

This book explains traditional as well as advanced welding processes, including welding metallurgy, safety, welding design, testing, quality assurance (QA), and automation/robotic welding. Theory of welding is presented with reference to the physics of welding and its metallurgy, and the residual stresses and distortions with detailed mathematical analyses and calculations. Welding practices are discussed, covering fusion welding processes and solid-state welding processes. Advanced welding processes and composites welding are also discussed, including magnetic pulse welding, laser pulse welding, electron beam welding, and the like.

The book is divided into five parts. Part I (Chaps. 1–3) serves as the foundation of welding and joining engineering. Chapter 1 first emphasizes the technological and societal importance of welding, and then classifies and introduces all widely used welding and joining processes; here safety and its codes and robotic welding are also explained. Chapter 2 focuses on the physics of welding and joining. This chapter covers arc physics, metallography and microstructures, alloys in welding and joining, phase diagrams and their applications to joining, solidification of metals in welding, grain growth and its control, and heat transfer in welding. Chapter 3 explains the metallurgy of welding; it covers the microstructures and mechanical behaviors of various weld zones (fusion zone, partially melted zone, heat affected zone, etc.). This chapter also discusses dilution and weldabilities of steels and aluminum alloys; here, the microstructural control in welding of austenitic stainless steel is also elaborated.

Part II (Chaps. 4–6) discusses various types of fusion welding processes, including their working principles, equipment, advantages, and limitations. Chapter 4 focuses on arc welding processes (SMAW, GMAW, GTAW, SAW, FCAW, PAW, and ESW). The *SMAW*, shielded metal arc welding, is explained with reference to process parameters, heat input, polarity, and electrode selection. The *GMAW*, gas metal arc welding, is described covering the various droplet transfer modes, process parameters, heat transfer, and energy balance. The *GTAW*, gas tungsten arc welding, is illustrated by covering its process parameters, electrode selection, and heat input. The *SAW*, submerged arc welding, is also qualitatively and quantitatively explained. Chapter 5 discusses resistance welding processes (RSW, RSeW, RPW). The *RSW*, resistance spot welding, is described by covering its working principle and equipment, the steps of operation, applications, advantages, and limitations, and its engineering analysis; here, recent advances in *RSW* are also taken into consideration. The *RSeW*, resistance seam welding, is also discussed, covering its working principle, operating data, and engineering analysis. Chapter 6 teaches oxy-fuel welding (OFW) and thermit welding processes. The OFW is explained with reference to its working principle, thermochemistry, equipment, operation (flame adjustment), safety precautions, engineering analysis, and forehand and backhand welding. Thermit welding is described, covering its working principle, applications, and the steps of operation.

Part III (Chaps. 7–8) explains *solid-state welding processes*. Chapter 7 elaborates *cold-pressure welding processes* covering the cold-pressure welding of metal sheets, ultrasonic welding (USW), explosion welding (EXW), and roll welding. Chapter 8 teaches hot solid-state welding processes, which include friction welding (FrW), friction stir welding (FSW), diffusion bonding/welding, and inertia (friction) welding. Part IV (Chaps. 9–10) describes brazing, soldering, and advanced welding processes. Chapter 9 teaches joining by brazing and soldering; here, the role of capillarity and surface tension in brazing/soldering is first explained. Then brazing of metals and ceramics is discussed, covering its working principle, advantages and applications, brazing practice and process steps, and the steps of operation in brazing. The engineering analysis of brazing is also presented. Braze joint design considers the braze joint design for flat components, the design for tubular braze joints, and the relevant calculations. Soldering is illustrated with reference to its working principle and applications, solder alloys and the steps of operation, and its engineering analysis (effects of contact angle on wetting, wetting radius of a solder droplet, etc.). Chapter 10 focuses on advanced welding processes (MPW, EBW, and LBW). The *MPW*, magnetic pulse welding, is discussed covering its working principle and applications, equipment set up, and mathematical modeling. The EBW, electron beam welding, is described covering its working principle and equipment and engineering analysis. The *LBW*, laser beam welding, is explained taking into account its working principle, advantages and applications, and engineering analysis.

Part V (Chaps. 11–14) covers the most important aspects of welding—*welding design and quality assurance*; this part is, therefore, the heart of the book. Chapter 11 develops the skill of welding design by illustrating weld joint and design, welding symbols, welding codes, weld sizing, the designs of fillet welds and butt welds; here, the factor of safety (FoS) is also considered in solving the *design problems*. Chapter 12 explains residual stresses, distortion, and their prevention; here, thermal stresses, expansion, and residual stresses are first illustrated, followed by residual welding distortions in beams. The techniques to reduce residual stresses and distortion are also presented. This chapter also covers recent advances in reducing residual stresses in welded pipes and pressure vessels. Chapter 13 develops the skills of various testing techniques (both destructive and NDT) of weldments. Non-destructive testing (NDT) of welds is explained by covering visual examination (VE), dye-penetrant testing (DPT), radiographic testing (RT), and ultrasonic testing (UT). Metallographic examination of welds is also discussed, covering both macro- and microscopic examinations; here, the *grained microstructure of a weldment/HAZ* is also illustrated. The steps in metallographic specimen preparation are listed, and quantitative metallography for *microscope resolution and grain size measurement* are presented. Destructive mechanical testing of welds is illustrated covering weld parts, guided bend test, free bend test, back bend test, nick bend test, and tensile testing of weldments. The last chapter (Chap. 14) assures quality of a weldment by discussing various welding defects and the measures to prevent them. The discussion on external welding defects includes cracks, undercut, overlap, and porosity, along with their engineering analysis with reference to AWS codes. Internal welding defects (slag inclusion and incomplete penetration) are also explained with sketches. The role of heat input in a weld quality is also emphasized along with relevant calculations. Finally, *case studies on weld joint failures* are presented; these case studies include the failure of the Union Oil Co. amine-absorber tower/pressure vessel, the failure of weld joint of a centrifugal pump, and the weld failure in *Liberty Ships*.

Based on my 40-year association with academics/industry, I anticipate that this volume will prove a useful literature for both engineering students and practicing engineers. In particular, the readers can significantly benefit from 200 worked numerical examples on welding and joining. In order to best benefit from the numerical examples, the student/reader is advised to first prepare a list of relevant formulae and then attempt to solve the problem (numerical example) while hiding its solution; she or he must try to solve the problem on their own by analyzing the data. In case the reader is stuck during solving the problem, she or he may refer to its solution. *Question and Exercise Problems* are included at the end of each chapter. The answers to all 100 multiple-choice questions (MCQs) and selected problems are given at the end of the book.

Jeddah, Saudi Arabia Zainul Huda

Acknowledgments

All thanks to God who blessed me with the wisdom that enabled me to write this book's manuscript. I appreciate my wife's cooperation during the writing of this book at night. I would like to acknowledge Richard Jones, Vice President Sales and Marketing/Managing Director, *MTI Welding Technologies Limited,* UK, for granting permission to include the picture of his M40B/V friction welding machine in my book. I am thankful to Ghazi Alsoruji, PhD (Mechanical Engineering), Assistant Professor, Faculty of Engineering, King Abdulaziz University, Jeddah, Saudi Arabia, for valuable discussions with him regarding welding design (Chap. 11). I am also grateful to Tuan Zaharinie, PhD (Manufacturing Engineering), Senior Lecturer, Mechanical Engineering Department, University of Malaya, Malaysia, for reviewing Chap. 9 and for her valuable comments.

Competing Interests The author has no competing interests to declare that are relevant to the content of this manuscript.

Contents

Part III Solid-State Welding Processes

Part IV Brazing, Soldering, and Advanced Welding Processes

Part V Welding Design and Quality Assurance

About the Author

Zainul Huda is the author of over **100 publications, including 10 books** and **68** *Web of Science* articles published by reputed publishers from the USA, the UK, France, Switzerland, Germany, Poland, South Korea, Malaysia, Pakistan, and the UAE. He has worked as a **Professor** of Manufacturing/Materials/Mechanical Engineering at reputed/world-ranking universities (*e.g.,* King Abdulaziz University, University of Malaya, King Saud University) for **17 years** (2007–2024). He possesses over **40 academic/industrial experience**. He is the **World's Top 21st Scholar** in the field of *Materials and Manufacturing* (see www.scholar.google.com). His research work has been **cited over 1650 times** with **h-index of 18 and *i*-10 index of 31**. He (as a PI) has attracted nine **(9) research grants**, with a total worth of US$ 0.16 Million. He has supervised to completion several PhD and Master's research theses. Prof. em.Z. Huda has worked as a C. Plant Manager/Dev. Engineer/Metallurgist/Gr. Engineer at various manufacturing companies, including Pakistan Steel Mills Corporation Ltd., Karachi. He has also satisfactorily completed 17 industrial consultancy projects in the areas of manufacturing/failure analysis. He is the developer of Toyota Corolla cars' axle-hubs manufacturing cycle/heat treatment first ever implemented in Pakistan (through Indus Motor Company, Karachi). Prof. em. Z. Huda earned PhD (Materials Engineering/Tech) from Brunel University, London, UK in 1991. He is also a postgraduate in Manufacturing Engineering. He obtained BS (Metallurgical Engineering) from the University of Karachi, Pakistan. He has delivered guest lectures on metallurgy/mechanical engineering in the UK, South Africa, Malaysia, Pakistan, and Saudi Arabia. Prof. em. Huda is the recipient of UK-Singapore Partners in Science Collaboration Award, awarded to him by the British High Commission, Singapore. He is also the recipient of the University of Malaya's Vice Chancellor's Appreciation Certificate for an innovation in engineering design.

Part I
Fundamentals of Welding and Joining Engineering

Chapter 1
Introduction

Abstract This chapter first introduces welding and the benefits of its study to the reader, and then highlights the industrial and societal importance of welding with reference to historical engineering failures. A robust classification chart of welding/joining processes is presented by dividing all joining processes into five categories: (a) welding, (b) brazing, (c) soldering, (d) mechanical fastening, and (e) adhesive joining; here, an introduction to each category is also given. Welding processes have been divided into two groups: fusion welding and solid-state welding; the welding processes in each group are introduced. Since welding is a risky and hazardous activity, measures to ensure safety are presented in the form a comprehensive table to protect welders from hazards; here safety codes recommended by the ISO, the ASME, and the HSE are also included. The practical aspect of welding (welding positions) is also diagrammatically elaborated. Finally, the automation in welding is introduced by illustrating *robotic welding*. This chapter contains 9 diagrams, a numerical example, and 6 multiple-choice questions (MCQs) with their answers given at the end of the book.

Keywords Welding · Brazing · Soldering · Mechanical fastening · Adhesive bonding · Fusion welding · Solid-state welding · Weldment · Welding positions · Robotic welding

Abbreviations

AWS	American Welding Society
ASME	American Society of Mechanical Engineers
BFA	Brazing Filler Alloy
BPVC	Boiler and Pressure Vessel Code
ESW	Electro-slag welding
ExW	Explosion welding
FrW	Friction welding
GMAW	Gas metal arc welding

Z. Huda, *Welding and Joining Engineering*,
https://doi.org/10.1007/978-3-032-20462-2_1

GTAW	Gas tungsten arc welding
HAZ	Heat affected zone
MSDS	Material Safety Data Sheet
NIOSH	National Institute of Occupational Safety and Health
OFW	Oxy-fuel welding
OSHA	Occupational Safety and Health Administration
PAW	Plasma arc welding
PMZ	Partially-melted zone
RSeW	Resistance seam welding
RSW	Resistance spot welding
SAW	Submerged arc welding
SMAW	Shielded metal arc welding
TW	Thermite welding
USW	Ultra-sonic welding

1.1 What Is Welding and Joining? And Why Study It?

The term *welding* refers to a fabrication process whereby two or more parts are coalesced together by means of heat, pressure or both forming a joint as the parts cool. Although welding is generally used for joining of metallic materials, the term can also be used for joining thermoplastics. Welding produces a permanent joint; the completed welded joint is referred to as a *weldment* (see Fig. 1.1). It is evident in Fig. 1.1 that a weldment (resulting from fusion welding) comprises of four zones: (a) fusion zone, (b) partially-melted zone (PMZ), (c) heat-affected zone (HAZ), and (b) base metal zone; an in-depth explanation of these zones is given in Chap. 3. Like welding, brazing and soldering are also permanent joining processes. Welding involves direct transfer of heat (and/or pressure) to the work-pieces to be joined; on the other hand, brazing and soldering involves the application of heat to a filler metal resulting in its melting and spreading to the work to form a joint. The term *joining* is used in a broader context; it encompasses both permanent and temporary joining. Welding, brazing, soldering, and adhesive bonding processes are mainly used to provide permanent joints, whereas mechanical fastening can be used to provide either temporary or permanent joints.

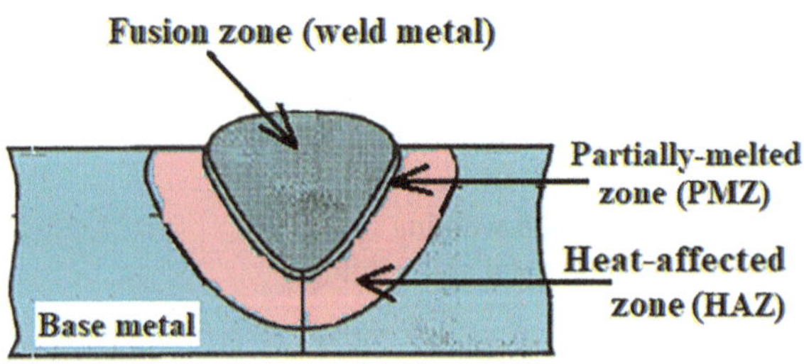

Fig. 1.1. A weldment showing its various zones

Welding and joining technologies form an integral part of modern manufacturing engineering. This is why all manufacturing/mechanical/metallurgical engineering students (both at undergraduate and graduate levels) are required to study "welding and joining" either as a core course or as a topic embedded in the manufacturing technology course. According to an American survey report, welding engineers earned an average base salary of $77,510 in 2021 (NEIT 2021). Hence, a strong knowledge of welding and joining is of great technological and industrial importance to a manufacturing/mechanical/metallurgical engineering student as well as to a practicing engineer.

1.2 Industrial and Societal Importance of Welding: *Historical Weld Failures*

Today's industrialized world heavily relies on precision manufactured products that involve a great deal of joining/assembling. Right from cell phones and computers through automobiles to aircraft and space-crafts, all machines are manufactured using welding/joining technologies. Welding, soldering, brazing, and adhesive bonding are all used in the production of household goods, vehicles, appliances, and electronic devices; welding and joining are also used in the construction of industrial plants and building structures. In fact, welding and joining technologies account for a substantial proportion of the entire manufacturing process of investment goods.

A faulty weld design or substandard welding practice would result in a poor quality product, posing a threat to society owing to its premature failure. For example, the failure of the Liberty ship that cracked in half during World War II is attributed to a brittle (fast) fracture that resulted due to poor weld quality of the fabricated ship structure. The Liberty ship was fabricated by welding steel plates to form a continuous body with no rivet joints. Cracks began at notches or arc-weld spots and spread all the way around the hull, eventually splitting the ship in half. A possible design improvement to avoid the fracture could be to provide crack arresters by incorporating some riveted joints into the ship structure (Huda 2022).

A number of historical disasters have resulted due to faulty design/practice in welding of pressure vessels. Hayes has reported six cases of historical failures in welded pressure vessels (Hayes 1996). These engineering failures include: Sizewell boiler (1963), John Thompson ammonia converter for ICI Immingham (1965), Cockenzie power station boiler drum (1966), Typpi Oy heat exchangers (1970), Robert Jenkins pressure vessel (1970) and the Union Oil Co. amine absorber tower (1984). The Union Oil Co. mine-absorber tower failure killed 17 people working at the refinery and heavy property damage (see Chap. 14). It is, therefore, utmost important that a welding engineer ensures the welded part or fabricated structure has high reliability and quality.

1.3 Welding and Joining Processes

1.3.1 Classification of Welding and Joining Processes

The classification charts of various joining/welding processes are presented in Figs. 1.2 and 1.3. It is evident in Fig. 1.2 that all joining processes can be broadly classified into five groups: (a) welding, (b) mechanical fastening, (c) brazing, (d) soldering, and (e) adhesive bonding. Mechanical fastening can be used to provide either temporary or permanent joints, while adhesive bonding, welding, brazing, and soldering processes mainly produce permanent joints (Campbell 2011). Fig. 1.3 illustrates that all welding processes can be classified into two groups: (a) fusion welding and (b) solid-state welding. The various types of joining and welding processes are explained in the following subsections.

1.3.2 Joining Processes

Metallic parts may be joined together by one of the following ways: welding, brazing, soldering, mechanical fastening, and adhesive bonding (see Figs. 1.2 and 1.3). *Welding* is, by far, the commonest and the most important joining process since it produces a reliable, strong joint. Welding is a permanent joining process that involves coalescence of two materials/parts at their contacting surfaces by a suitable

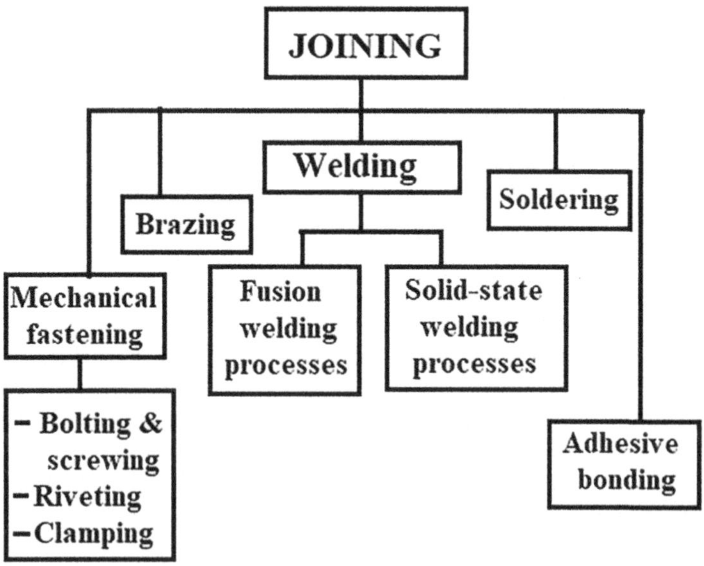

Fig. 1.2 Classification chart of joining processes

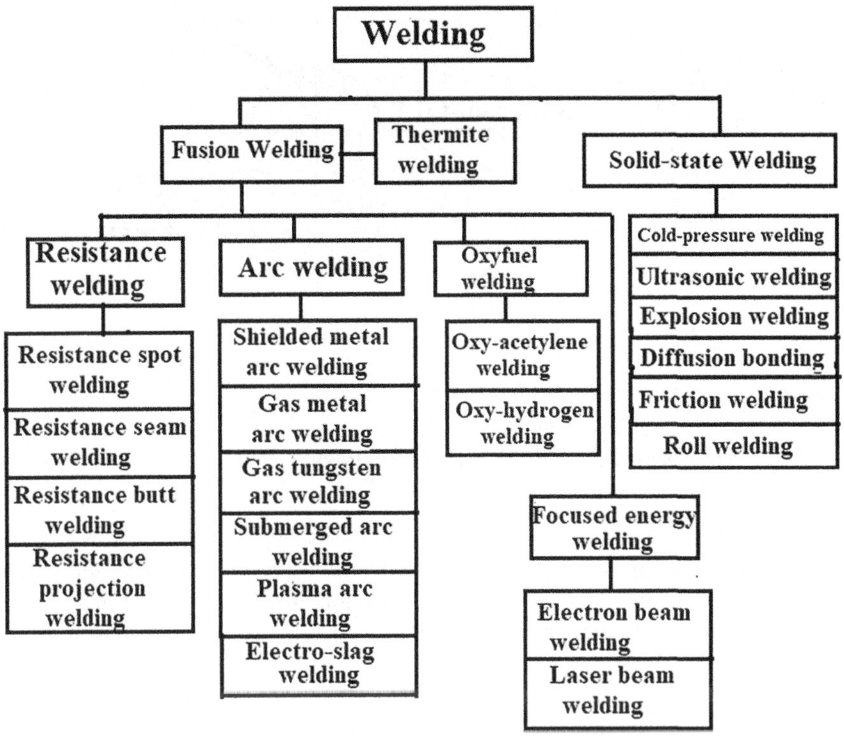

Fig. 1.3 Classification chart of welding processes

heat and/or pressure. In fusion welding, the parts are heated above their melting temperatures. *Brazing* is a joining process whereby a filler metal is melted and spread by a capillary action between the faying surfaces of the metallic parts (base metal). Brazing involves heating the brazing filler alloy (BFA) above 450 °C, which is above the melting temperature of BFA but below the melting temperature of the base metal. *Soldering* is a joining process in which a filler metal with a melting temperature below 450 °C is melted and spread between the faying surfaces of the metallic parts. Brazing and soldering processes are explained in detail in Chap. 9.

Mechanical fastening can be used to produce either a temporary or permanent joint. Mechanical fastening includes bolting and screwing, riveting, clamping, and the like. *Riveting* is a joining process in which two metallic parts are joined by the use of rivets with the application of force (see Fig. 1.4). Riveting is widely used in the automotive and aerospace industries, as well as in other applications that require permanent or semi-permanent bonding but do not allow for bolting or welding. *Adhesive bonding* involves the use of a glue, epoxy, or a polymeric agent, resulting in the joining of two surfaces together, usually with the creation of a smooth bond. A detailed account of mechanical fastening and adhesive bonding processes is given in Chap. 11.

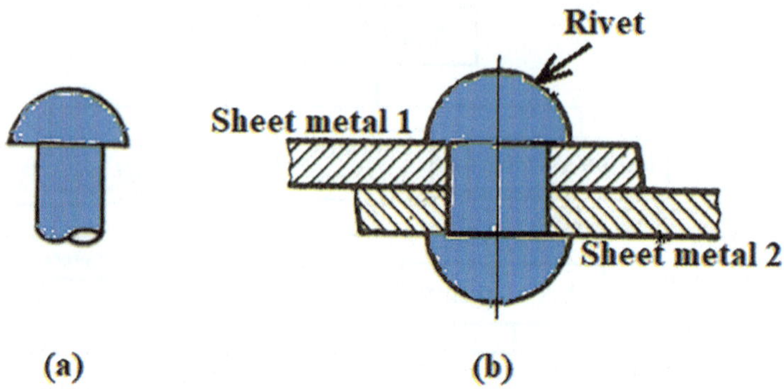

Fig. 1.4 Riveting; (**a**) a rivet, (**b**) riveted sheet metal part

1.3.3 Fundamentals of Welding Processes

All welding processes can be classified into two groups: (a) fusion welding processes and (b) solid–state welding processes (see Fig. 1.3). ***Fusion welding*** refers to a group of processes that involve fusion and joining of two or more materials by heating them to their melting points; these processes may or may not require the use of a filler material. With resistance welding as the exception, external pressure is not required for fusion welding processes. The examples of fusion welding processes include: oxy-fuel welding (OFW), thermite welding (TW), shielded metal arc welding (SMAW), gas metal arc welding (GMAW), gas tungsten arc welding (GTAW), submerged arc welding (SAW), plasma arc welding (PAW), electro-slag welding (ESW), resistance spot welding (RSW), resistance seam welding (RSeW), and the like.

Shielded metal arc welding (*SMAW*) is a commonly practiced welding process. The *SMAW* process involves the use of a flux-coated metal electrode, which is held in an electrode-holder that is connected to a power source; electricity passes through the electrode and instantly touches the base metal (Huda and Alostath 2021). Welding ground clamps (attached to the work-piece) allow current to create the arc required for welding (see Fig. 1.5). The heat of the arc melts the bottom of the electrode and base metal, thereby depositing the metal as a *weld* (see Fig. 1.6). The weld metal deposition rate can be calculated, as illustrated in Example 1.1. Fusion welding processes are discussed in detail in Chaps. 4–6.

Example 1.1 Calculating the Metal Deposition Rate in SMAW
A SMAW operation is performed, and 1.4 kg electrode is consumed in 45 min. Calculate the metal deposition rate (in kg/h). Assume no loss of electrode metal.

Solution

Mass of electrode consumed = 1.4 kg, welding time duration = 45 min.

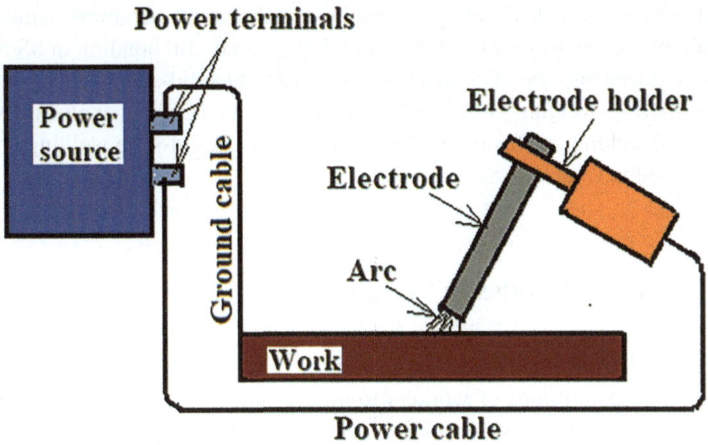

Fig. 1.5 Shielded metal arc welding (SMAW)—*basic components*

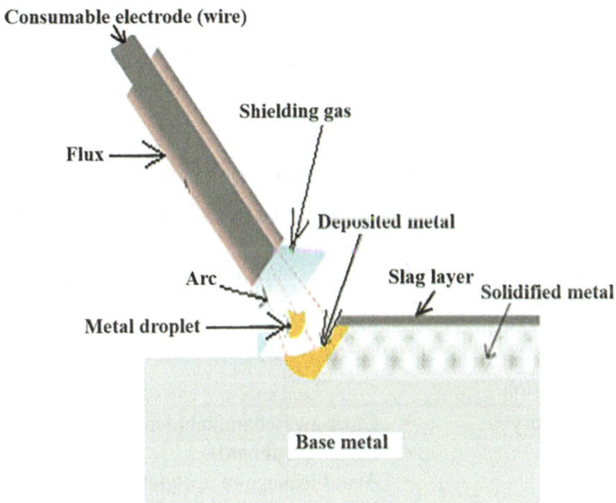

Fig. 1.6 The deposition of weld metal in SMAW

$$\text{Metal deposition rate} = \frac{Mass\ \ of\ \ electrode\ \ consumed}{Welding\ \ duration} = \frac{1.4}{45} = 0.031\ \text{kg/min}$$

Metal deposition rate = 0.031 kg/min = 0.031 × 60 = 1.86 kg/h.

Solid-state (SS) welding is a group of welding processes in which coalescence of the joining surface is achieved by either the application of pressure with no heat source (cold SS welding) or by the application of a combination of heat and pressure (hot SS welding), but the temperature is less than the melting temperature of base metal. Solid-state (SS) welding is advantageous over fusion welding, since neither filler metal is required in these processes, nor the undesirable heat-affected zone (HAZ) is formed. Bonding dissimilar metals can be achieved without problems of thermal expansions in SS welding. For a successful bonding in SS welding, the two surfaces must be very clean. The examples of solid-state welding processes include friction welding (FrW), ultra-sonic welding (USW), explosion welding (ExW), roll welding, diffusion bonding, and the like. Solid-state welding processes are discussed in Chaps. 7–8.

1.4 Safety and Codes in Welding

Preventive Measures in Welding. Welding is a risky and hazardous activity that globally exposes millions of workers to injuries and health issues each year. It is, therefore, important to ensure welding-safety measures that are designed to protect employees from welding hazards. Table 1.1 presents various safety risks and the techniques to prevent them.

Personal Protective Equipment (PPE). Although some safety risks and their remedies are summarized in Table 1.1, they are not exhaustive; the reader is advised to follow the safety manuals, published either by the Health and Safety Executive

Table 1.1 Safety Risks in Welding and their Prevention (HSE 2022)

#	Risk	Issues	Some prevention techniques
1.	Confined space working	• Lack of oxygen • Suffocation from smoke • Large amount of inert gases may cause death from asphyxiation	• Try to do the welding job in an open work environment • Avoid contact with argon, carbon dioxide, and helium
2.	Fire and explosion	Death or injury	• Clear away flammable material (e.g. wood, fabric, cardboard) • Avoid leakage of acetylene gas • Maintain a fire watch during and after the hot work. • Fire extinguishers and fire alarm
3.	Electrical Hazard	• Electric shock • Electric burn	For all arc welding, make sure that: • Welding equipment conforms to the appropriate ISO or regional standards • Proper installation of fixed welding equipment • The insulation on the welding and current return leads is undamaged and the conductor is thick enough to carry the current safely • All connectors are clean, undamaged and correctly rated • Insulation on the welding cables, plugs, clamps or torch/electrode holder on welding equipment is undamaged • Welders must use the appropriate personal protective equipment • Beware of stray welding currents

Fig. 1.7 The PPE used in welding

organization of the United Kingdom (HSE 2022), or by the ASME (see the following paragraph). The safety precautions to be followed in oxyfuel (oxy-acetylene) welding are presented in Chap. 6 (Sect. 6.1.4). In general, the following seven pieces of welding safety equipment are used in welding: (a) welding gloves, (b) helmet with eye protection, (c) heat-resistant welder's jacket, (d) welding pants, (e) welding respirator, (f) ear muff, and (g) welding work boots. The personal protective equipment (PPE) used in welding are shown in Fig. 1.7.

Safety Standards and Codes. Safety standards have been recommended and published as manuals by regional/international organizations; the companies carrying out welding operations should comply with the standards. These safety-standards organizations include (a) American Welding Society (AWS), (b) Occupational Safety and Health Administration (OSHA), (c) National Institute of Occupational Safety and Health (NIOSH), and the like. In particular, the OSHA recommends that employers must arrange to protect employees from potential workplace hazards through appropriate control methods. For example, OSHA's Hazard Communication Standard (HCS) requires that all employers with hazardous chemicals in their workplace must have labels and Material Safety Data Sheet (MSDS) for their exposed workers and train them to properly handle the materials (Hoffman et al. 2017).

In view of the health hazards discussed in the preceding paragraphs, a welding engineer must be committed to safety in welding by ensuring standard welding design and practices, as per welding codes recommended by licensed professional bodies.

The AWS, American Welding Society, USA, develops welding codes and standards. The Safety and Health section of the welding codes covers the codes from AWS Z-49.1 to AWS C4.2 (AWS 2025). The American Society of Mechanical

Engineers (ASME) has referred to an ISO welding standard in its pressure vessel code. The July 2015 edition of Section IX "Welding and Brazing Qualifications" of the ASME Boiler and Pressure Vessel Code (BPVC) refers to ISO 9606-1:2012 "Qualification testing of welders - Fusion welding - Part 1: Steels" in its Appendix ZZZ. It cites ISO 14732:2013 on the "Qualification testing of welding operators and weld setters for mechanized and automatic welding of metallic materials". In Germany, the latter standard is published as DIN EN ISO 14732 (Zernitz 2015)

1.5 Welding Positions

Welding is often done on structures in their existing positions, and in the position the fabricated part will be used. These structures or parts may be on the ceiling, in the corner, or on the floor. The American Welding Society has classified welding positions into four types: (1) flat welding position—1G, (2) horizontal welding position—2G, (3) vertical welding position—3G, and (4) overhead welding position—4G (see Fig. 1.8).

1.6 Robotic Welding: *Automation*

Robotic welding is an industrial automated technique that is performed by a welding robot, which incorporates an arm that is capable of welding metals by moving in three dimensions. Welding robots are programmable, multipurpose manipulators/ machines designed to repeatedly perform the same welding operation. Robotic welding was first introduced in the industrial sector during the 1960s. Although job-shop industries prefer to use manual welding owing to diversities in welding operations, mass production industries, especially the automotive sector prefer to automate welding operations by use of robots. Welding robots are generally used for resistance welding and arc welding in heavy industries, such as automobile manufacturers (see Fig. 1.9).

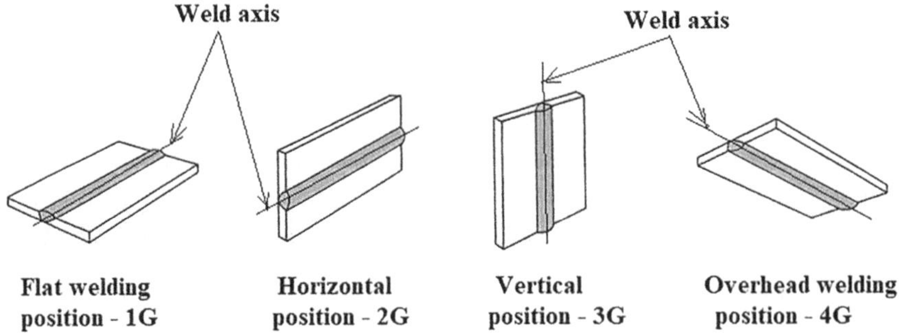

Fig. 1.8 The four basic welding positions

Fig. 1.9 Welding robots in a cars' manufacturing line using resistance spot welding (RSW)

Welding robots can work in multiple ways—a pre-programmed position, machine vision, or a combination of both. However, the most significant benefits of robotic welding technology include high accuracy, repeatability, and throughput.

Questions

1.1. Do you agree with the following statement: "welding is just an activity of repairing broken or damaged parts"? Justify your statement by giving real-world examples.

1.2. Briefly explain robotic welding technology and its benefits in today's industrial sector.

1.3. Distinguish between the following terms:

(a) Fusion welding and solid-state welding, and (b) Brazing and soldering.

1.4. Diagrammatically illustrate the classification chart for various joining processes.

1.5. (MCQs). Encircle the most appropriate answer for each of the following questions:

(1) Which one of the following welding positions is denoted by 3G?

(a) Horizontal, (b) vertical, (c) flat, (d) overhead.

(2) Which one of the following welding joints is most commonly used in welding?

(a) Butt joint, (b) lap joint, (c) T-joint, (d) edge joint, (e) corner joint.

(3) Which one of the following processes is a mechanical fastening process?

(a) Welding, (b) brazing, (c) soldering, (d) riveting.

(4) Which one of the following processes is a solid-state welding process?

(a) Arc welding, (b) oxyfuel welding, (c) friction welding, (d) RSW.

(5) Which one of the following processes is a fusion welding process?

(a) Ultrasonic welding, (b) oxyfuel welding, (c) friction welding, (d) roll welding.

(6) Which one of the following welding positions poses the greatest safety risk?

(a) Horizontal, (b) vertical, (c) flat, (d) overhead.

References

AWS (2025). AWS: Codes and Standards, Publisher: American Welding Society, USA Online: https://www.aws.org/standards-and-publications/codes-and-standards/#numberOfResults=5

Campbell, F.C. (ed.) (2011). Joining—Understanding the Basics, ASM International. Ohio, USA

Hayes, B. (1996). Six case histories of pressure vessel failures, *Engineering Failure Analysis,* Vol. 3, Issue 3, Pages 157-170

HSE (Health and Safety Executive) (2022). Safety Risks from Welding, Internet Source: https://www.hse.gov.uk/welding/other-welding-risks.htm#:~:text=Welders%20should%20wear%20clean%2C%20dry,it%20does%20provide%20some%20protection

Hoffman, D.J., Dahle, K.R., Fisher, D.J. (2017). Welding, 2nd edition, *Pearson India Education.*

Huda, Z, (2022), Mechanical Behavior of Materials, *Springer Nature*, Switzerland.

Huda, Z and Alostath, M.M. (2021), Effects of process parameters on SMAW weld quality using SK-260-M E6013 electrodes, *Journal of Technological Sciences and Engineering,* 2(1): 17–20. https://zenodo.org/records/6785432

NEIT, 2021. Welding Engineer – Salary, Career, and How to Become One. Internet Source: https://www.neit.edu/blog/welding-engineer Accesses on Dec. 17, 2025

Zernitz, H. (2015), ASME refers to ISO welding standard. Internet Source: https://www.din.de/en/din-and-our-partners/press/press-releases/asme-refers-to-iso-welding-standard-75636. Accessed on December, 17, 2025

Chapter 2
Physics of Welding and Joining

Abstract This chapter presents both qualitative and quantitative aspects of the physics of welding and joining. The chapter begins with a discussion on arc physics covering welding arc mechanism, free electrons, the types of emissions, voltage drop, and power in welding arc. The physical metallurgy of welding/joining is explained with reference to metallography and microstructures, alloys in welding and joining, and phase diagrams and their applications to joining. Solidification of metal in welding is also elaborated, including its mechanism and segregation. Grain growth, its control, and the kinetics are also discussed, including the issues of grain growth in the heat affected zone (HAZ). Finally, heat transfer in welding is explained. This chapter contains 13 diagrams, 19 worked numerical examples, 2 data tables, and 7 MCQs with their answers given at the end of the book.

Keywords Arc physics · Welding arc · Free electrons · Metallography · Microstructures · Alloys · Solidification · Segregation · Grain growth · Heat transfer in welding

Abbreviations

AISI	American Iron & Steel Institute
DCEN	Direct current electrode negative
BCC	Body centered cubic (crystal structure)
FCC	Face centered cubic
HAZ	Heat affected zone
OM	Optical microscope
PD	Power density
SEM	Scanning electron microscope
TEM	Transmission electron microscope

Z. Huda, *Welding and Joining Engineering*,
https://doi.org/10.1007/978-3-032-20462-2_2

Nomenclature

A_w	Area of the welded metal
$C_{Pearlite}$	Mass % carbon in pearlite
C_L	Composition of the liquid
C_o	Composition of the alloy under consideration
C_α	Composition of the α solid
C_α	Mass % carbon in ferrite
H_i	Heat input (J/mm)
M_{Fe3C}	Mass fraction of cementite (Fe_3C)
M_L	Mass fraction of the liquid in the alloy
M_s	Mass fraction of the solid phase
M_α	Mass fraction of α-phase (ferrite)
P_i	Power input of the heat source (W)
V_a	Voltage drop across the anode drop zone
V_c	Voltage drop across the cathode drop zone
V_p	Voltage drop across the plasma zone
v_w	Welding speed (or travel velocity of the heat source)
c	Specific heat of the solid metal (J/kg•K)
C_o	Mass % carbon in the steel
D	Average grain diameter
D_0	Pre-growth grain diameter
E_k	Kinetic energy (J)
H_f	Heat of fusion (J/kg),
I	arc current (A)
k	Equilibrium segregation coefficient
K	Temperature-dependent constant
n	Grain-growth exponent
P	Power of the arc (W),
Q	Volumetric specific heat
T_m	Melting temperature (K or °C)
V	Arc voltage (volt)

2.1 Arc Physics in Welding

2.1.1 *Welding Arc Mechanism*

A welding arc is an electric discharge that develops mainly due to the flow of charged particles (current) between cathode and anode. In the case of arc welding, the flow of current through the gap between electrode and work-piece needs column of charged particles for having reasonably high electrical conductivity. These charged particles are generated by various mechanisms, such as thermo-ionic

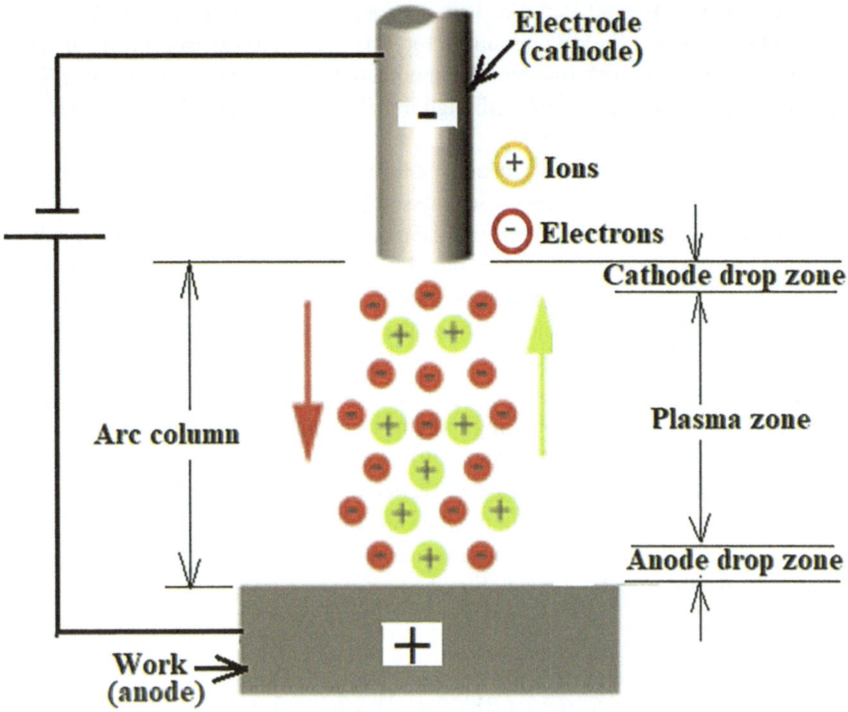

Fig. 2.1 The welding-arc with direct current electrode negative (DCEN) polarity

emission, field emission, etc. The density of charged particles in the arc gap governs the electrical conductivity of gaseous column.

In electric arc welding, electrons are released from the cathode due to electric field or thermo-ionic emission mechanism; these electrons are accelerated towards the anode because there exists a potential difference between the work piece and the electrode. The high velocity electrons moving from cathode toward anode collide with gaseous molecules and decompose them into charged particles, *i.e.* electrons and ions (see Fig. 2.1). These charged particles move towards electrode and work piece as per polarity and form a part of welding current, and eventually electrons merge into anode. The arc column between the electrode and the work-piece acts as pure resistance load. The heat generated in a welding arc depends on arc voltage and welding current.

2.1.2 Free Electrons and the Types of Emissions

The emission of electrons (free electrons) from the cathode metal strongly depends on the metal's work function, which is the energy (in eV or J) that is required to dislodge one electron from the surface of the metal (cathode). Another measure of ability of a metal to emit electrons is its ionization potential, which is defined as the energy/unit charge (in eV) required for removing an electron from an atom. Examples of metals with low ionization potentials (2.1–2.3 eV) include Ca, K, Na, etc. On the other hand, metals with high ionization potentials (4.0–4.5 eV) include *Al*, Fe, etc.

There are three common mechanisms through which free electrons are emitted during arc welding: (a) thermo-ionic emission, (b) field emission, and (c) secondary mechanism. *Thermo-ionic emission* is the mechanism of emission of electron due to heating of metal. This mechanism is based on the principle that increase in temperature of metal increases the kinetic energy (E_k) of free electrons; as the E_k exceeds a certain limit, electrons are ejected from the metal surface. This is why refractory materials (*e.g.* tungsten, carbon, etc.) with high melting points have a tendency of exhibiting thermo-ionic emission. In *field emission*, free electrons are pulled out of the metal surface by developing high-strength electromagnetic field. It is necessary to establish a high potential difference (10^7 V/cm) between the work piece and electrode for field emission. In the case of *secondary emission*, high-velocity electrons moving from cathode to anode in the arc gap collide with other gaseous molecules. This collision results in decomposition of gaseous molecules into atoms and charged particles (electrons and ions) (see Fig. 2.1).

2.1.3 Voltage Drop and Power in Welding Arc

The welding arc acts as impedance for flow of current; the impedance of arc is a function of temperature and inversely proportional to the density of charge particles and their mobility.

Three major zones can be noticed in the arc column that accounts for total voltage drop in the arc, *i.e. cathode-drop zone, plasma zone, and anode-drop zone*. The product of potential difference across the arc (V) and current (I) gives the power of the arc, which indicates the heat generated per unit time during arc welding. The arc power can be numerically expressed by:

$$\mathrm{P} = \mathrm{V} * \mathrm{I} = \mathrm{I} * \left(V_c + V_p + V_a\right) \tag{2.1}$$

where P is the power of the arc (in W), V is the arc voltage, I is the arc current (in A), V_c is the voltage drop across the cathode drop zone, V_p is the voltage drop across the plasma zone, and V_a is the voltage drop across the anode drop zone. It has been experimentally shown that $V_a \approx 4.8$ V for steel and $V_a = 4.2$ V for aluminum alloys, $V_c \approx 1$ V for thermionic cathodes (such as the GTAW electrode), and $V_c \approx 11$ V for pure Ar, and 13.8 V for Ar-5% CO_2 for non-thermionic cathodes (see Table 2.1). It has also been proved that helium shielded arcs melt more metal than argon shielded arcs (Eagar 1990). In most cases, 100% of the thermal power generated is absorbed

Table 2.1 Voltage drops across various zones and materials

Anode material	V_a	Cathode material/type	V_c
Steel	4.8 V	Thermo-ionic cathode	1 V
Aluminum alloys	4.2 V	Pure argon (Ar) (non-thermo-ionic cathodes)	11 V
–	–	Ar-5% CO_2 (non-thermo-ionic cathodes)	13.8 V

by the metal, and none of it is carried away by the plasma. The significance of Eq. 2.1 is illustrated in Example 2.1.

2.2 Metallography of Welding and Joining

2.2.1 Metallography and Microstructures

Welding, especially fusion welding, involves significant microstructural changes in the weld metal due to high-temperatures and/or pressure encountered during the welding process. Hence, it is important to be familiar with the microstructures of metals and alloys at specified compositions and temperatures. The field of physical metallurgy that deals with the study of the physical structure and components of metals by using microscopy is called *metallography*. In order to study a material's microstructure under a microscope, it is first important to prepare a metallographic specimen. The preparation of a metallographic specimen involves the following steps: (I) sectioning, (II) mounting, (III) grinding, (IV) polishing, (V) etching, and (VI) microscopy (Huda, 2020). There are generally three types of microscopes used for microstructural examination: (a) optical microscope (*OM*) or light microscope (*LM*), (b) scanning electron microscope (SEM), and (c) transmission electron microscope (TEM). The microstructure of an isomorphous (single-phase solid-solution) alloy, as revealed by an OM, is shown in Fig. 2.2. An alloy is formed by combining two or more metallic elements. An isomorphous alloy is formed when the atoms of a new element (solute) form a solid solution with the parent metal (solvent), but there still exists one phase.

Some important phases and microstructural constituents in steels include ferrite, cementite, pearlite, austenite, and the like (see Fig. 2.3). *Ferrite* is a solid solution of carbon in α-iron (BCC crystal structure). The maximum solid solubility of carbon in α-iron is 0.02 wt.%. Ferrite is the softest phase in steel. *Cementite* is an intermetallic compound of iron and carbon (Fe_3C) containing 6.7 wt.% carbon; it is the hardest phase in steel. *Pearlite* is a mechanical mixture of ferrite and cementite; it is a micro-constituent in steel. *Austenite* is a solid solution of carbon in γ-iron (FCC). The maximum solubility of carbon in FCC iron is 1.7 wt.%. Austenite, in carbon steels, is stable at temperatures above 727 °C. *Martensite*, a meta-stable phase, is hard and brittle; it is formed when steel is quenched from the austenitic range to below 727 °C.

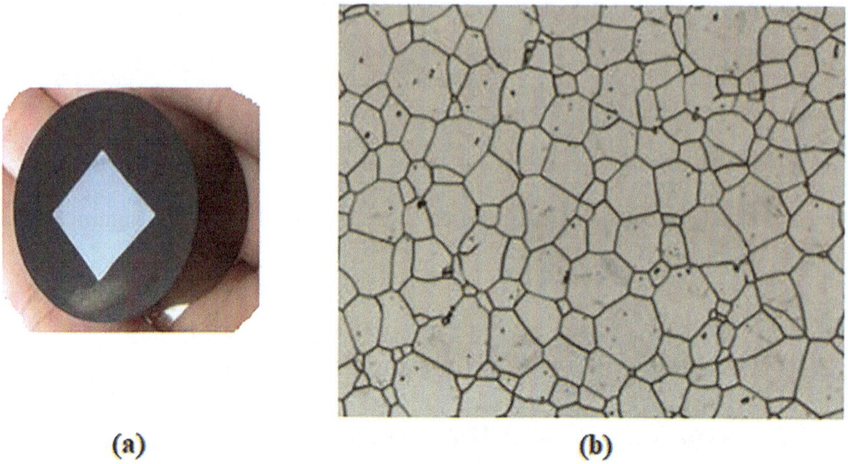

Fig 2.2 A polished metallographic specimen (**a**); the optical micrograph of the etched specimen showing the microstructure of a single-phase alloy (**b**)

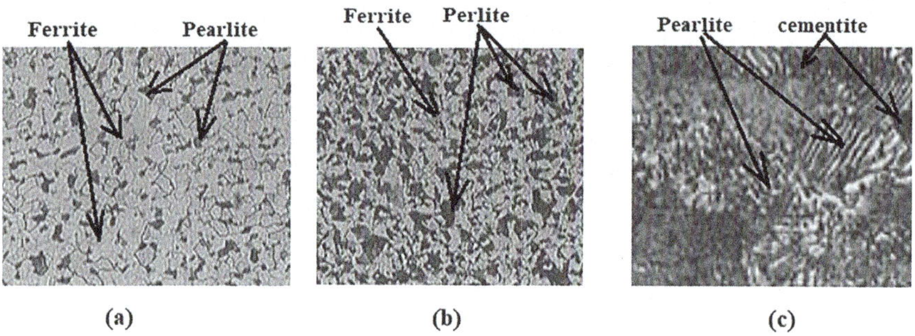

Fig. 2.3 Microstructures of carbon steels; (**a**) low carbon steel, (**b**) medium carbon steel, (**c**) high carbon steel

2.2.2 Alloys in Welding and Joining

All metallic materials may be classified into two main groups: (a) ferrous alloys, and (b) non-ferrous alloys. Ferrous alloys contain an appreciable amount of iron (usually > 90% Fe), whereas non-ferrous alloys do not. The examples of ferrous alloys include cast irons, steels, wrought iron, and the like. On the other hand, brasses, bronzes, aluminum alloys, nickel alloys are the examples of non-ferrous alloys. In industrial practice, the materials to be welded are generally ferrous alloys. It is, therefore, necessary for a welding engineer to be well acquainted with microstructures of ferrous alloy. All ferrous alloys can be broadly classified into three groups: (a) cast irons, (b) carbon steels, and (c) alloys steels (see Fig. 2.4). *Cast iron (CI)* is a ferrous alloy that contains higher than 2 wt% carbon and 0.5 wt% or more silicon. Cast irons find wide applications in automotive components (*e.g.* clutches,

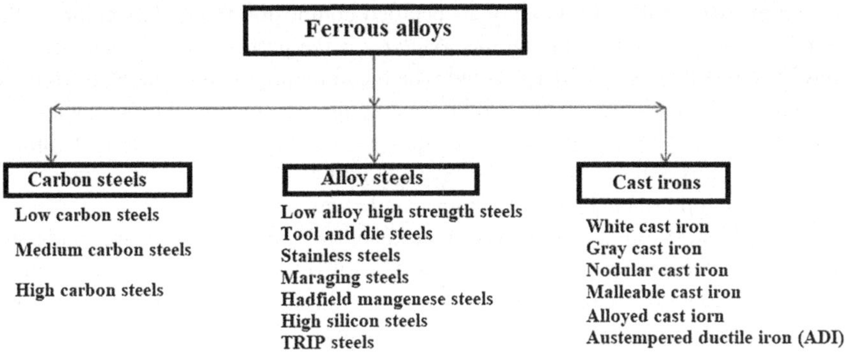

Fig. 2.4 Classification of Ferrous alloys

brake discs, cylinder blocks and heads, piston, liner, and the like) and other mechanical components (*e.g.* gears, pipes, water pipes, flywheels, etc.).

Carbon steels are iron-based alloys with carbon content in the range of 0.05–1.5% and up to 0.5% Si and 1.5% % Mn. *Low carbon steels* contain carbon in the range of 0.05–0.2 wt% (see Fig. 2.3a). The examples of low-carbon steels include AISI 1006 (0.06% C steel), AISI 1009, AISI 1020 steels, and the like. *Low-carbon steels* are the easiest to weld; another advantage is that no heat affected zones are (HAZ) formed in welding of these steels. *Medium carbon steels* contain carbon in the range of 0.2–0.8 wt.% (see Fig. 2.3b). The examples of medium-carbon steels include AISI 1030 (0.3% C), AISI 1055, AISI 1077, etc. *High carbon steels* contain carbon in the range of 0.8–1.5 wt.% (see Fig. 2.3c). High carbon steels are very difficult to weld because they readily form either cementite or martensite (both are hard and brittle phases) as the steel cools in fusion welding.

An alloy steel is a steel to which one or more elements (besides carbon) have been intentionally added for obtaining the desired mechanical properties. An *alloy steel* contains up to 50 wt% of alloying elements. The main alloying elements in alloy steels include silicon (Si), manganese (Mn), nickel (Ni), molybdenum (Mo), tungsten (W), chromium (Cr), vanadium (V), cobalt (Co), copper (Cu), aluminum (*Al*), titanium (Ti), boron (B), and niobium (Nb). The *AISI* system classifies alloy steels as follows: 2 for Ni steels, 3 for Ni-Cr steels, 4 for Mo steels, 5 for Cr steels, 6 for Cr-V steels, 7 for W-Cr steels, and 9 for Si-Mn steels. For example, alloy steel 4140 has the composition (mass %): 0.4C, 0.2Mo, 0.85Mn, 0.25 Si, 0.9Cr, balance: Fe (approx.).

2.2.3 Phase Diagrams and Their Applications to Joining

2.2.3.1 Phase Diagrams

A welding process (especially fusion welding and hot solid-state welding) involves significant changes in temperature causing phase transformations and microstructural changes. It is therefore important for the engineer to possess a good knowledge of the phase stabilities of an alloy at specified compositions and temperatures; the latter is achieved by phase diagrams. A *phase* is the part of a system with distinct physical and chemical properties; and which can be physically separated from the system. A *phase diagrams* (or equilibrium diagram) is a graphical representation of the stabilities of various phases (in equilibrium) at given compositions, temperatures, and/or pressure. The simplest phase diagram is the one for an isomorphous alloy (*e.g.* copper-nickel phase diagram), as shown in Fig. 2.5.

It is evident in Fig. 2.5 that a copper-nickel alloy forms a single-phase solid solutions (α) at all compositions at temperatures below the solidus curve. The two-phase field is represented by the liquid (L) and the solid α phases: (L + α). The compositions of the two phases can be determined by drawing a tie-line in the two-phase region (see Example 2.2). The *inverse lever rule* enables us to determine the mass fraction of each phase present in a two-phase field in the phase diagram. The *inverse lever rule* can be mathematically expressed as follows:

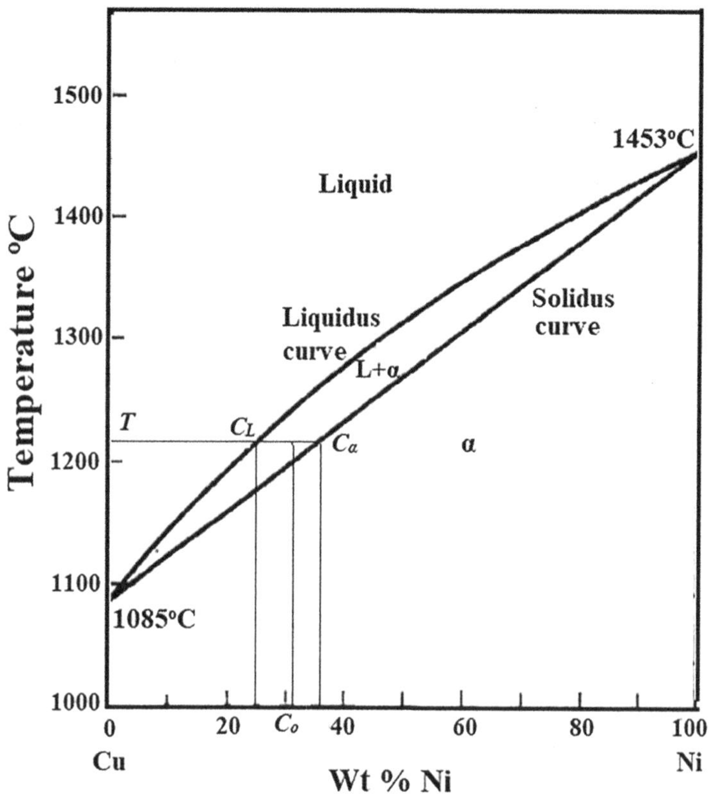

Fig. 2.5 Copper-nickel phase diagram (C_o is the composition of an ally under consideration, C_L is the composition of liquid, and C_α is the composition of solid)

$$M_L = \frac{C_\alpha - C_o}{C_\alpha - C_L} \tag{2.2}$$

where M_L is the mass fraction of the liquid in the alloy, C_o is the composition of the alloy under consideration; C_α is the composition of the α solid, and C_L is the composition of the liquid.

Similarly, the mass fraction of the solid phase (M_s) can be calculated as follows:

$$M_s = \frac{C_o - C_L}{C_\alpha - C_L} \tag{2.3}$$

The significance of Eqs. 2.1–2.3 is illustrated in Sub-Sect. 2.2.3.3.

2.2.3.2 Application of Phase Diagram to Soldering

Soldering process involves the use of a solder alloy with a low melting temperature. The selection of the solder alloy composition can be achieved by reference to lead-tin (Pb-Sn) phase diagram, which is shown in Fig. 2.6. It is evident in Fig. 2.6 that the maximum solid solubility of Sn in Pb is 18.3%., and that of Pb in Sn is 97.8%. The curve BC refers to the *solvus line,* which indicates that the solid solubility decreases with decreasing temperature (see Example 2.3). The selection of a suitable solder alloy composition is explained in the following paragraph.

The lead-tin phase diagram (Fig. 2.6) shows that Pb-Sn alloy has a eutectic temperature of 183 °C; and that the 61.9%Sn–39.1%Pb alloy (eutectic composition) will undergo eutectic reaction at the point *E* to form two phases (α + β). A lead-tin alloy containing less than 61.9% Sn is called a *hypoeutectic alloy*; the one containing greater than 61.9% Sn is called a *hypereutectic alloy*. The 62%Sn– 39%Pb alloy is attractive for use as a solder owing to the lowest melting temperature (eutectic temperature). However, lead is toxic thereby rendering the use of 62%Sn- 39%Pb as a solder hazardous and unsafe. This is why materials engineers have developed a safe lead-free eutectic solder alloy based on bismuth-tin (Bi-Sn) system: Sn–57% Bi (Huda 2020). Recently (2025), Choi and co-researchers have reported the development of a high-performance lead-free solder comprising of Sn-3.5 wt% Ag-0.8 wt% Cu (Choi, et al. 2025).

2.2.3.3 Application of Phase Diagram to Welding

Welding engineers generally deal with joining of steel parts; it is therefore important to study the application of welding with reference to the iron-cementite (Fe-Fe_3C) phase diagram, as shown in Fig. 2.7. Consider the cooling of carbon steel containing 0.76% C starting from the region of austenite (γ phase) stability. As the temperature reaches 727 °C, austenite phase transforms to two phases (ferrite and

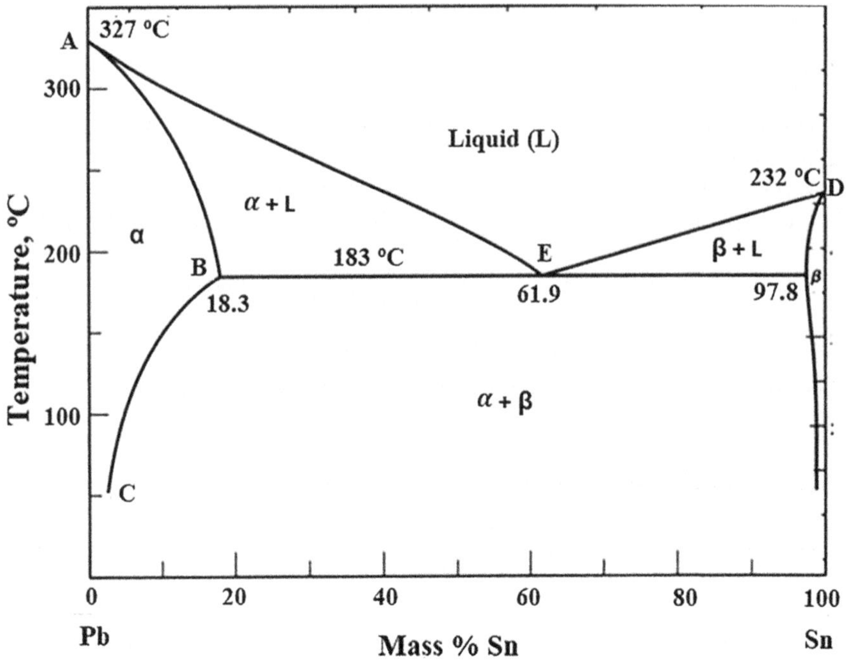

Fig. 2.6 Lead-tin phase diagram

cementite) (α + Fe_3C)—called *pearlite*. This phase transformation is called the eutectoid reaction. The temperature at which eutectoid reaction occurs (727 °C) is called the eutectoid temperature, and the specific composition of carbon steel (Fe–0.76% C) is called the eutectoid composition (see Fig. 2.7).

It is evident in Fig. 2.7 that a low-carbon steel (say Fe-0.18%C alloy) at a high temperature of 1500 °C exists as a two-phase mixture (Liq + δ); the δ phase refers to the solid solution of carbon in BCC iron. The maximum solid solubility of carbon in δ-iron is 0.11%.

The amounts (mass fraction) of each of the two phases can be calculated by using Eqs. 2.2–2.3 (see Example 2.4).

By reference to Fig. 2.3 and Fig. 2.7, the mass fractions of ferrite and pearlite in a low-carbon steel, at a temperature just below the eutectoid temperature, can be determined by modifying Eq. 2.2 (inverse lever rule), as follows:

$$M_{\alpha} = \frac{C_{\text{Pearlite}} - C_0}{C_{Pearlite} - C_{\alpha}} = \frac{0.76 - C_0}{0.76 - 0.02} \tag{2.4}$$

$$M_{Pearlite} = \frac{C_0 - C_{\alpha}}{C_{Pearlite} - C_{\alpha}} = \frac{C_0 - 0.02}{0.76 - 0.02} \tag{2.5}$$

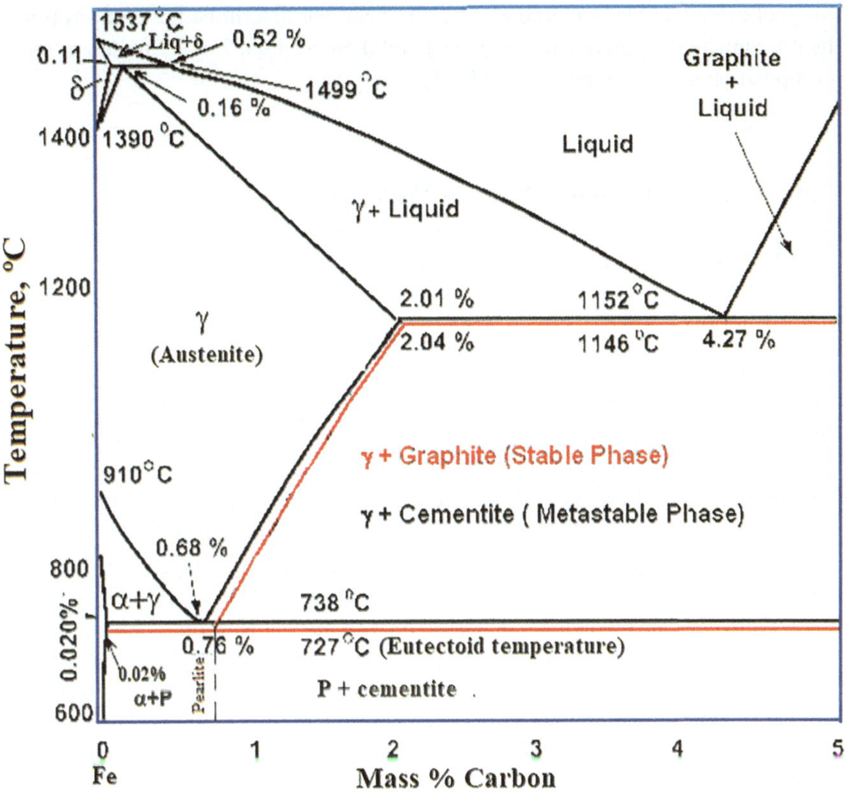

Fig. 2.7 Iron-iron carbide phase diagram (P = pearlite)

where M_α is the mass fraction of α-phase (ferrite), C_{Pearlite} is the mass % carbon in pearlite (=0.76), C_o is the mass % carbon in the steel, C_α is the mass % carbon in ferrite (=0.02), and $M_{Pearlite}$ is the mass fraction of pearlite in the steel. Similarly, the mass fractions of pearlite and cementite in a high-carbon steel, at a temperature just below the eutectoid temperature, can be computed by (see Fig. 2.7):

$$M_{Pearlite} = \frac{C_{Fe3C} - C_o^{'}}{C_{Fe3C} - C_{Pearlite}} = \frac{6.67 - C_o^{'}}{6.67 - 0.76} \tag{2.6}$$

$$M_{Fe3C} = \frac{C_o^{'} - C_{Pearlite}}{C_{Fe3C} - C_{Peralite}} = \frac{C_o^{'} - 0.76}{6.67 - 0.76} \tag{2.7}$$

where M_{Fe3C} is the mass fraction of cementite (Fe_3C), $C_o^{'}$ is the mass % C in the high-carbon steel, and C_{Fe3C} is mass % of carbon in Fe_3C (=6.67). The significance of Eqs. 2.4–2.7 is illustrated in Examples 2.5 and 2.6. The Fe-Fe_3C phase diagram (Fig. 2.7) and Eqs. 2.4–2.7 are useful to a welding engineer. In case the composition of steel to be welded is unknown, the engineer can determine the composition by heat treatment (by reference to Fig. 2.7) and then use Eqs. 2.4–2.7 to determine the composition (see Examples 2.7 and 2.8).

2.3 Solidification of Metal in Welding

2.3.1 Mechanism of Solidification

It is learnt in the preceding chapter that the parts, to be welded, are heated above their melting temperatures in fusion welding; this welding process involves both melting and solidification of metals. The solidification behavior of a metal, during fusion welding, is of great importance since it controls the grain size, which in turn influences the strength of the weld metal. In general, when a metal begins to solidify, multiple crystals begin to grow in the liquid resulting in the formation of many crystals thereby forming a polycrystalline solid (Huda and Bupett 2012). The solidification of a polycrystalline metal involves the following four stages: (a) *nucleation*: the formation of stable nuclei in the melt (Fig. 2.8a); (b) *crystal growth:* growth of nuclei into crystals involving a dendritic growth mechanism (Fig. 2.8b); (c) *formation of randomly oriented crystals* (Fig. 2.8c), and (d) grained microstructure with grain boundaries (Fig. 2.8d) (see also Fig. 2.2b). The dendritic crystal growth mechanism is further explained in Chap. 3 (see Fig. 3.3).

2.3.2 Segregation During Solidification

When a liquid metal of uniform composition solidifies, the resultant solidified metal is generally non-uniform in composition. The non-uniformity in the composition or the *segregation* is due to the re-distribution of the solute atoms within the melt. By reference to the phase diagram shown in Fig. 2.5, equilibrium between the solid and the liquid may be assumed during solidification. This means that at any temperature T, the composition of the solid at the solid-liquid interface (C_α) and the composition of the liquid at the interface (C_L) follow the solidus curve and the liquidus curve, respectively. At any temperature T, the ratio of the composition of the solid to that of the liquid is called the *equilibrium partition ratio*, or the *equilibrium segregation coefficient*, k, which is numerically expressed as:

$$\mathrm{k} = \frac{C_\alpha}{C_L} \tag{2.8}$$

The significance of Eq. 2.8 is illustrated in Example 2.9.

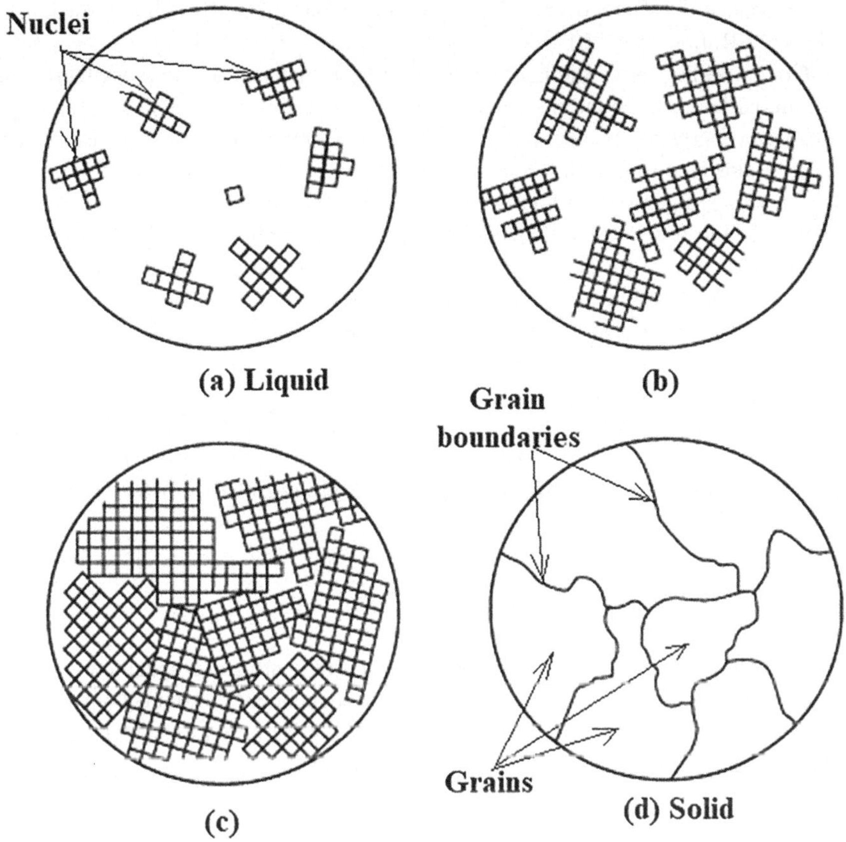

Fig. 2.8 Evolution of grained microstructure in a metal; (**a**) nucleation of crystals, (**b**) crystals grow by dendritic growth mechanism, (**c**) randomly oriented grains form, (**d**) formation of grained microstructure on completion of solidification

2.4 Grain Growth and Its Control

It is explained in the preceding section that solidification of metal involves the nucleation and growth of tiny-shaped crystals that result in the formation of a grained microstructure (see Fig. 2.8d). Figure 2.8 enables us to relate the grain size with the rate of solidification, as follow. If the rate of solidification is high (fast cooling), many nuclei are formed. It means that the rate of nucleation is greater than the rate of crystal growth; this mechanism results in fine grain size in the metal. On the other hand, slow cooling forms a few nuclei, which result in a coarse grain-sized material. It means that slow cooling of a weld, during fusion welding, would cause grain growth. Besides fusion welding, grain growth may also occur in hot solid-state welding processes. Heating, in hot solid-state welding, may cause three stages of mechanisms: (I) recovery, (II) recrystallization, and (III) grain growth (see Fig. 2.9) (Rollett et al. 2017).

Cold working raises the internal energy of the material due to work hardening (strain hardening); the internal energy is stored as strain energy. This stored energy creates an energetically unfavorable condition—a reduction in ductility due to elongated grains (see Fig. 2.9). Hence, heating at a temperature around $0.25{*}T_m$ (T_m is the melting temperature in K) results in the release of stored energy but there is no change in the microstructure. This heating stage is called *recovery*. Heating at a higher temperature (at about $0.4{*}T_m$) releases the strain energy, and restores the grain structure to fine equi-axed grains with restored ductility. This stage of heating is called *recrystallization*. Heating at a temperature significantly above the recrystallization temperature results in a pronounced increase in grain size; this phenomenon is called *grain growth* (see Fig. 2.9).

Since coarse grain size of a metal indicates a low strength, it is technologically important to restrict grain growth by mainly controlling the temperature and time (see Fig. 2.10). Metallic materials follow the grain-growth kinetic behavior as follows:

$$D^{\frac{1}{n}} - D_o^{\frac{1}{n}} = K\,t \tag{2.9}$$

where D is the average grain diameter at the end of heating at a specified temperature for a time duration of t, K is a temperature-dependent constant, D_0 is the pre-growth grain diameter, and n is the grain-growth exponent. For most metals, $n = 0.5$ (Huda et al., 2014); thus Eq. 2.9 may be re-written as:

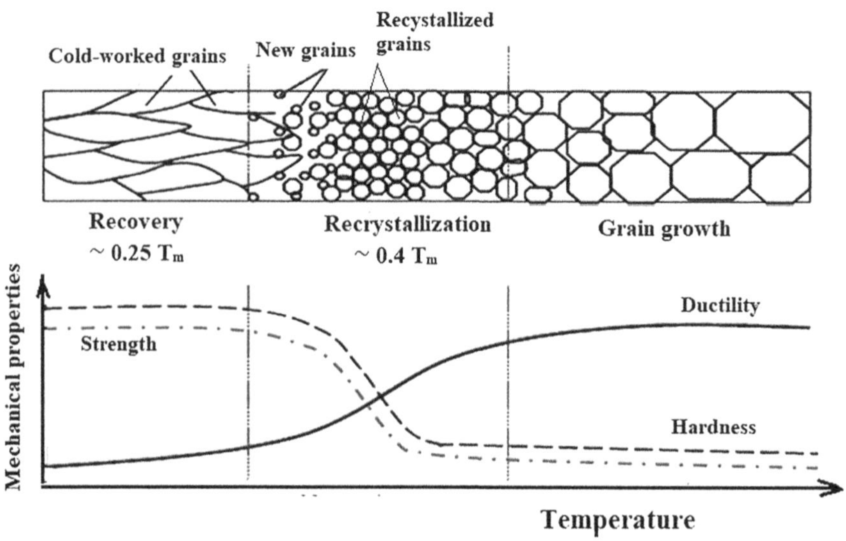

Fig. 2.9 The effects of temperature on the microstructure and properties

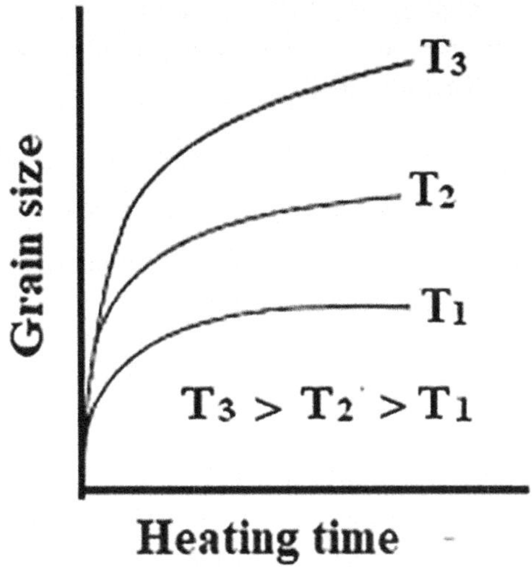

Fig. 2.10 Grain growth kinetics (T_1, T_2, and T_3 are three different temperatures)

$$D^2 - D_0^2 = Kt \tag{2.10}$$

The significance of Eq. 2.10 is illustrated in Example 2.10.

Welding process may cause significant changes in the grain size and mechanical properties of the metal. In particular a high-temperature heat-affected zone (HAZ) area, with a coarse grain size in a weldment, adversely affects the strength of the material. Hence, considerable attention needs to be paid to the issue of grain growth in the HAZ of welded joints (Moravec et al. 2019).

2.5 Heat Transfer in Welding

The term *welding* has been introduced to the reader in Chap. 1 (Sect. 1.1). In fusion welding (and hot solid-state welding), heat energy is generally transferred from a source of heat to the work-piece by way of convection and radiation; then the heat energy is transferred from the work-piece to the adjoining area by conduction. The energy input (from the heat source) can be related to the power input and the welding speed by the following formula:

$$H_i = \frac{P_i}{v_w} \tag{2.11}$$

where H_i is the heat input (J/mm), P_i is the power input of the heat source (W), and v_w is the welding speed or travel velocity of the heat source (mm/s) (see Example 2.11). The effective heat input takes into consideration "heat-transfer efficiency" or *thermal efficiency* (see Chap. 12),

Another useful term in welding is the *power density,* which is defined as the power transferred to the weld per unit surface area. The power density is expressed by (Groover 2019):

$$PD = \frac{P_i}{A} \tag{2.12}$$

where *PD* is the *power density* (W/mm^2), P_i is the power input to the surface of the metal part to which heat energy is concentrated (in W), and *A* is the surface area over which the energy is input (mm^2) (see Example 2.12). For good quality fusion-welding results, it is desirable to melt the metal with minimum energy and high *power density*. Table 2.2 presents the power densities of some fusion welding processes.

The quantity of metal to be melted, in fusion welding, depends on three factors: (a) the welding process, (b) the size and configuration of the weld joint, and (c) the number of passes. The completed weld joint (weldment) obtained after fusion welding is shown in Fig. 2.11. The area of fusion zone (due to the addition of filler metal during fusion welding) is represented by *A,* and the area of partially-melted zone (PMZ) (base metal that is welded) is *B*. The cross-sectional area of the welded metal (A_w) is the total weld area, which is calculated by:

$$A_w = A + B \tag{2.13}$$

In the case no filler metal is used in welding, Eq. 2.13 will simplify to $A_w = B$.

The amount of heat energy required to melt a unit volume of metal at ambient temperature, is called the *volumetric specific heat*, which can be calculated by:

$$Q = Q_1 + Q_2 = c * \rho * (T_m - T_o) + \rho * H_f \tag{2.14}$$

where Q is the volumetric specific heat (J/mm^3), Q_1 is the heat energy required to raise the temperature of unit volume of the solid metal at ambient temperature to its melting point (J), Q_2 is the heat energy required to convert the unit mass of the solid to liquid at the melting point (J), ***c*** is the specific heat of the solid metal (J/kg•K), H_f is the heat of fusion (J/kg), T_m is the melting temperature (*K*) and T_o is the starting temperature of the metal (*K*) (see Examples 2.13 and 2.14).

Equation 2.14 can be simplified to:

$$\text{Volumetric specific heat} = Q = \frac{T_m^2}{3 \times 10^5} \tag{2.15}$$

Table 2.2 Power densities of some welding processes

Welding process	OFW	Arc welding	Resistance welding	LBW	EBW
Power density, W/mm²	10	50	1000	9000	10,000

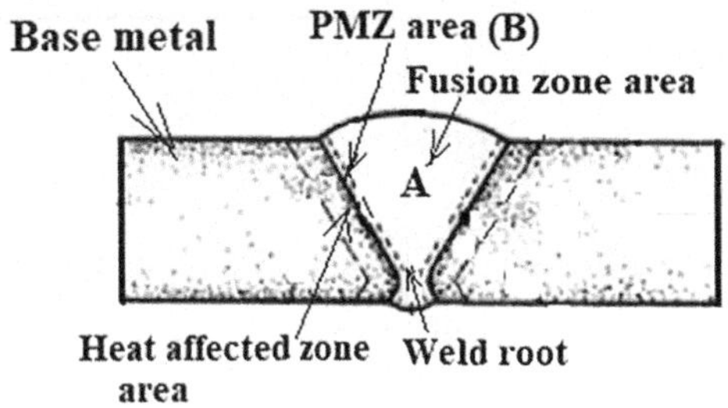

Fig. 2.11 A weldment obtained on completion of fusion welding. (PMZ = partially-melted zone)

The proof of Eq. 2.15 is illustrated in Example 2.15, and its significance is illustrated in Examples 2.15 and 2.16.

2.6 Worked Examples in Physics of Welding

Example 2.1 Determining the Power of a Welding Arc
Determine the power of a welding arc for a thermo-ionic electrode when the anode is made of steel. The arc welding is performed with an arc current of 100 A. Assume that 100% of the thermal power generated is absorbed by the metal, and none of it is carried away by the plasma.

Solution

With reference to Table 2.1,

$$I = 100\,\text{A}, V_a = 4.8\,\text{V}, V_c = 1\,\text{V}, V_p = 0, P = ?$$

By using Eq. 2.1,

$$P = I * (V_c + V_p + V_a) = 100 \times (1 + 0 + 4.8) = 100 \times 5.8 = 580\,\text{W}$$

Example 2.2 Identification of Phases in a Copper-Nickel Alloy
An alloy with a composition of 64%Ni-36%Cu is cooled from 1400 °C to room temperature. Determine the: (a) temperature of the liquid at which first solid forms, (b) temperature of the two-phase mixture at which the last liquid solidifies, (c) composition of the first solid to form, and (d) composition of the last liquid to solidify. Hint: Refer to Cu-Ni phase diagram (Fig. 2.5).

Solution

The modified Cu-Ni phase diagram for 64%Ni-36%Cu alloy is shown in Fig. 2.12. Point *X* represents 64%Ni-36%Cu alloy at 1400 °C. PQ and RS are the tie lines in the two-phase region.

(a) The temperature of the liquid at which first solid forms is around 1380 °C (see line PQ).

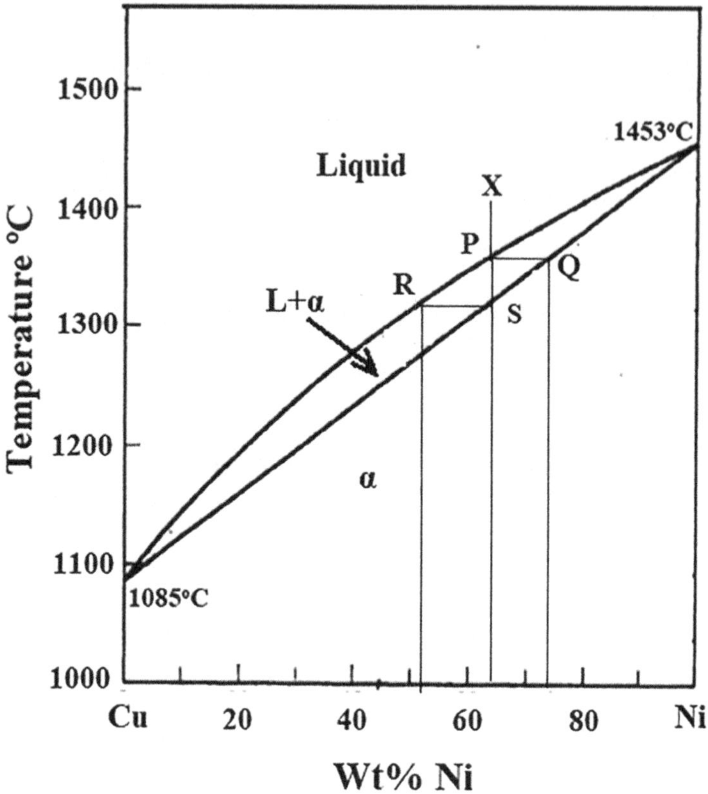

Fig. 2.12 The Cu-Ni phase diagram for Example 2.2

(b) The temperature at which the last liquid solidifies is around 1310 °C (see the tie line RS).
(c) Point Q gives the composition of the first solid to form; the composition is 73%Ni-27%Cu.
(d) Point *R* gives the composition of the last liquid to solidify; the composition: 52%Ni-48%Cu.

Example 2.3 Identifying Phases and Their Compositions for Lead-tin Phase Diagram
Consider the Pb–30 wt% Sn alloy. (a) What is the phase at a temperature of 320 °C? (b) What phases are stable when the alloy is cooled to 210 °C? (c) Identify the phase(s) when the alloy is cooled to 115 °C. (d) What phases are stable when the alloy is cooled to 50 °C? And what are their compositions? (e) Identify the composition of the alloy with the minimum melting temperature. Hint: Refer to Fig. 2.6.

Solution

(a) The stable phase in the Pb–30 wt% Sn alloy at 320 °C is liquid.

(b) The stable phases in the Pb–30 wt% Sn alloy at 210 °C are *α* (solid solution of Sn in Pb) and liquid. The solid *α* contains 18 wt% Sn and the liquid contains 52% Sn.
(c) The stable phases in the Pb–30 wt% Sn alloy at 115 °C are two solids: α + β. The solid *α* contains 8 wt% Sn and the solid β contains 98 wt% Sn.
(d) The stable phases in the Pb- 30 wt% Sn alloy at 50 °C are two solids: α + β; the solid *α* contains 10% Sn and the solid β contains 99% Sn.
(e) The alloy composition with the minimum melting temperature is 62%Sn-38%Pb.

Example 2.4 Determining the Composition of Equilibrium Phases for a Low Carbon Steel

Refer to Fe-Fe_3C phase diagram (Fig. 2.7). A specimen of low-carbon steel with 0.1 wt.% carbon is held at a temperature of 1510 °C. Assuming thermodynamic equilibrium, identify the equilibrium phases and their compositions.

Solution

The partial Fe-C phase diagram for 0.1%C steel at high temperatures is shown in Fig. 2.13.

With reference to Fig. 2.13, the equilibrium phases for 0.1 wt.% C steel at 1510 °C are liquid and δ solid solution (L+ δ). The composition of the liquid phase is 0.35%C. The composition of δ phase is 0.06%C.

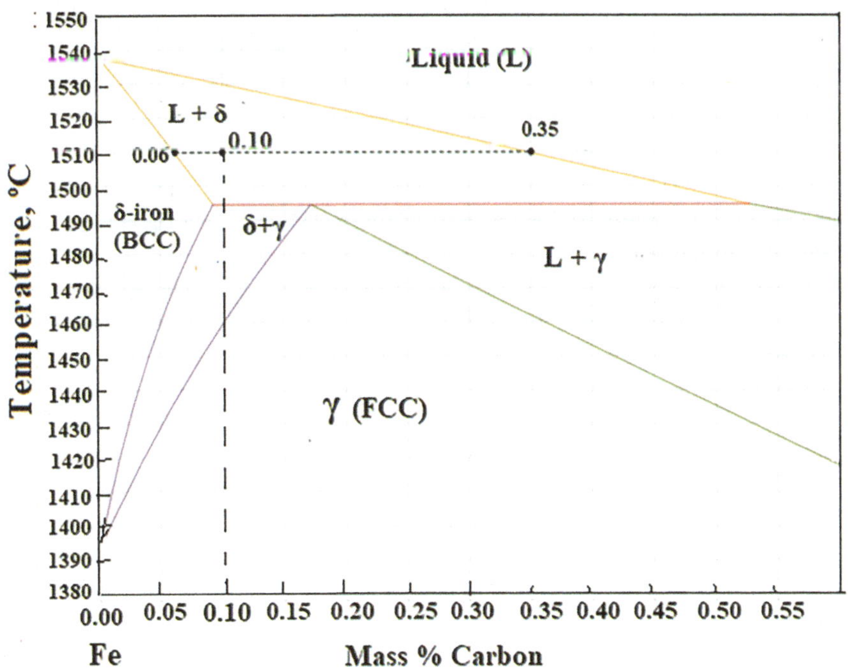

Fig. 2.13 Partial Fe-Fe_3C phase diagram for 0.1% C steel at 1510 °C for Example 2.4

Example 2.5 Calculating the Mass Fraction of Equilibrium Phases in a Low-carbon Steel
By reference to Fig. 2.13, calculate the mass fraction of each phase.

Solution

$$C_\alpha = C_\delta = 0.06,\ \mathrm{C_o} = 0.10,\ C_L = 0.35$$

By using Eq. 2.2,

Mass fraction of liquid phase = $M_L = \dfrac{C_\alpha - C_o}{C_\alpha - C_L}$ 0.138 (or 13.8%) = $\dfrac{0.06 - 0.10}{0.06 - 0.35}$

Mass fraction of δ phase = $M_\delta = 1 - M_L = 1 - 0.138 = 0.862$ (or 86.2%)

Example 2.6 Determining the Amounts of Phases/Micro-constituents in a Low-carbon Steel
Determine the mass fractions of phases/micro-constituents in a steel containing 0.08 wt.% C at a temperature just below the eutectoid temperature. Hint: Refer to Fig. 2.7.

Solution

Since the steel is a low-carbon steel, it comprises of α-phase and pearlite at just below 727 °C.

Here, $C_\alpha = 0.02$, $C_{\mathrm{Pearlite}} = 0.76$, and $\mathrm{C_0} = 0.08$

By using Eq. 2.4,

$$\text{Mass fraction of ferrite} = M_\alpha = \frac{C_{\mathrm{Pearlite}} - C_0}{C_{Pearlite} - C_\alpha} = \frac{0.76 - 0.08}{0.76 - 0.02} = \frac{0.68}{0.74} = 0.92\,(\text{or } 92\%)$$

$$\text{Mass fraction of pearlite} = 1 - 0.92 = 0.08\,(\text{or } 8\%)$$

Example 2.7 Calculating the Amounts of Phases/Micro-constituents in a High-carbon Steel
Calculate the mass fractions of phase/microconstituent in a carbon steel containing 1.3 mass % C at a temperature just below the eutectoid temperature. Hint: refer to Fig. 2.7.

Solution

Since the steel is a high-carbon steel, it comprises of Fe_3C and pearlite at just below 727 °C.

Here, $C_{Fe3C} = 6.67$, $C_{\mathrm{Pearlite}} = 0.76$, and $\mathrm{C_0} = 1.3$.

By using Eq. 2.6,

$$M_{Pearlite} = \frac{C_{Fe3C} - C_o'}{C_{Fe3C} - C_{Pearlite}} = \frac{6.67 - 1.3}{6.67 - 0.76} = 0.908\,(\text{or } 90.8\%)$$

$$M_{Fe3C} = 1 - 0.908 = 0.092 (\text{or } 9.2\%)$$

Example 2.8 Identifying an Steel to Assess Its Weldability

The composition of a carbon steel plate is unknown. An engineer heated a sample of the steel above 727 °C to obtain austenite, then held it for sufficiently long time. Finally, the steel sample was slowly cooled in a furnace to room temperature. The resulting microstructure was found to comprise of 75% ferrite and 25% pearlite. What is the composition and weldability of the steel?

Solution

We can determine the composition of the steel by using Eq. 2.4, as follows:

$$M_{\alpha} = \frac{0.76 - C_0}{0.76 - 0.02.}$$

$$75\% = \frac{0.76 - C_0}{0.77 - 0.02}$$

$$C_0 = 0.197$$

The carbon steel contains 0.197 mass% C. Since the material is a medium carbon steel, its weldability is good.

Example 2.9 Determining the Equilibrium Segregation Coefficient

Determine the equilibrium segregation coefficient for the Cu-31%Ni alloy at 1220 °C.

Solution

By reference to Fig. 2.5, $C_{\alpha} = 37\%\text{Ni} = 0.37$, and $C_L = 24\%\text{Ni} = 0.24$

By using Eq. 2.8,

Equilibrium segregation coefficient = $\text{k} = \frac{C_{\alpha}}{C_L} = \frac{0.37}{0.24} = 1.54$

Example 2.10 Grain Growth—Calculating the Time Required to Increase the Grain Size

The heating of brass at a temperature of 600 °C for 3000 min results in an increase of grain size from 0.03 to 0.3 mm. The grain-growth exponent of brass is approximately equal to 0.5. Determine the time required to increase the average grain diameter of brass from 0.3 to 0.6 mm at 600 °C.

Solution

By using Eq. 2.10 for $D_0 = 0.03$ mm and $D = 0.3$ mm, $t = 3000$ min,

$$Kt = \text{D}^2 - \text{D}_0^{\,2}$$

$$\text{K}(3000) = (0.3)^2 - (0.03)^2$$

$$\text{K} = 2.97 \times 10^{-5}$$

Since the temperature is the same, the time t can be computed for $D_0 = 0.3$ mm, $D = 0.6$ mm; and $K = 2.97 \times 10^{-5}$ by using Eq. 2.10, as follows:

$$\left(2.97 \times 10^{-5}\right) t = \left(0.6\right)^2 - \left(0.3\right)^2$$

$$\text{Time} = t = 9090\,\text{min} = 151.5\,\text{h}$$

Example 2.11 Calculating the Heat Input in Welding
A heat source is capable of transferring 2800 W to the surface of a metal part. The heat source travels a distance of 15 cm over the surface of the metal for welding in 7 s. What is the heat input, if 100% thermal efficiency is assumed?

Solution

$$\text{Power input} = P_i = 2800\text{W}, \text{welding speed} = v_w = \frac{150mm}{7s} = 21.4\text{mm/s}$$

By using Eq. 2.11,

$$\text{Heat input} = \text{H}_\text{i} = \frac{P_i}{v_w} = \frac{2800}{21.4} = 130.84\,\text{J/mm}$$

Example 2.12 Determining the Power Density in the Inner Circle of a Metallic Ring
A heat source has a power input of 2700 W. The heat strikes the surface of the metal (for welding) in a circular area with varying intensities of the power within the circle. About 75% of the power is transferred within a circle of 6-mm-diameter, while 90% of the power is transferred within a concentric circle of 13-mm-diameter. Determine the power density in the 6-mm-diameter inner circle.

Solution

$P_i = 0.75 \times 2700 = 2025$ W. The area of 6-mm-diameter circle = $D^2 = \frac{\pi}{4} \times 6^2 =$ 28.28 mm^2

By using Eq. 2.12,

$$\text{Power density in the 6-mm-diameter inner circle}$$
$$= \text{PD} = \frac{P_i}{A} = \frac{2025}{28.28} = 71.6\,\text{W/mm}^2$$

Example 2.13 Determining the Power Density in the Ring of Given Diameter
By using the data in Example 2.12, determine power density in the 13-mm-diameter ring lying around the inner circle.

Solution

$$\text{P}_\text{i} = 0.90 \times 2700 = 2430\,\text{W}.$$

The area of the 13-mm-diameter ring lying around the inner circle can be calculated by:

$$\text{Area of the ring} = A = \frac{\pi}{4}\left(D_o^2 - D_i^2\right) = \frac{\pi}{4}\left(13^2 - 6^2\right)$$
$$= 0.785 \times (169 - 36) = 104.4\,\text{mm}^2$$

$$\text{Power density in the ring} = \text{PD} = \frac{P_i}{A} = \frac{2430}{104.4} = 23.27\,\text{W}/\text{mm}^2$$

Example 2.14 Calculating the Volumetric Specific Heat of a Metal by Using Density Data

The specific heat and the latent heat of fusion of solid aluminum are 921 J/kg•K and 376 kJ/kg, respectively. The density of aluminum is 2.7 g/cm^3. The melting temperature of aluminum is 660 °C. Calculate the volumetric specific heat of aluminum.

Solution

$$c = 921\,\text{J}/\text{kg}\bullet\text{K}, \rho = 2.7\,\text{g}/\text{cm}^3 = 2.7\times10^{-6}\,\text{kg}/\text{mm}^3,$$
$$\text{H}_\text{f} = 376\,\text{kJ}/\text{kg} = 376{,}000\,\text{J}/\text{kg}$$

By using Eq. 2.14,

$$\text{Q}_1 = c\rho\left(\text{T}_\text{m} - \text{T}_\text{o}\right) = (921\,\text{J}/\text{kg K}) \times \left(2.7\times10^{-6}\,\text{kg}/\text{mm}^3\right)$$
$$\times\left(933\text{K} - 298\text{K}\right) = 1.58\,\text{J}/\text{mm}^3$$
$$\text{Q}_2 = \rho \times \text{H}_\text{f} = \left(2.7\times10^{-6}\,\text{kg}/\text{mm}^3\right)$$
$$\times\left(376{,}000\,\text{J}/\text{kg}\right) = 1.015\,\text{J}/\text{mm}^3$$

$$\text{The volumetric specific heat of aluminum} = \text{Q} = \text{Q}_1 + \text{Q}_2$$
$$= 1.58 + 1.015 \cong 2.6\,\text{J}/\text{mm}^3$$

Example 2.15 Proof of the Relationship: Volumetric Specific Heat = $\frac{T_m^2}{3\times10^5}$

Prove Eq. 2.15.

Solution

By reference to the worked Example 2.14, the volumetric specific heat of aluminum = 2.6 J/mm^3

Now, we calculate the volumetric specific heat of aluminum by using Eq. 2.15, as follows.

$$\text{The volumetric specific heat of aluminum} = \text{Q} = \frac{T_m^2}{3\times10^5} = \frac{933^2}{3\times10^5} = 2.9\,\text{J}/\text{mm}^3$$

Since the two answers are in close agreement, Eq. 2.15 stands proved.

Example 2.16 Calculating the Volumetric Specific Heat of Mild Steel

Calculate the volumetric specific heat of mild steel containing 0.4 mass% carbon.

Solution

By reference to Fe-Fe_3C phase diagram (Fig. 2.7), the melting temperature of Fe-0.4%C alloy is about 1508 °C. It means that T = 1508 + 273 = 1781 *K*.

By using Eq. 2.15,

$$\text{The volumetric specific heat of Fe} - 0.4\%\text{C alloy}$$
$$= Q = \frac{T_m^2}{3\times10^5} = \frac{1781^2}{3\times10^5} = 10.57\,\text{J}/\text{mm}^3$$

Example 2.17 Calculating Heat Energy for Melting the Given Volume of Metal in Welding

A welder wants to perform fusion welding by melting 90,000 mm^3 volume of the base metal made of carbon steel (Fe-0.4 mass %C) without using a filler metal. Assuming 100% efficiency, calculate the heat energy required for melting the given volume of the metal,

Solution

By reference to Example 2.16, the volumetric specific heat or the amount of heat energy required to melt a unit volume of the steel at ambient temperature is 10.57 J/mm^3.

Heat energy required for melting the given volume of the metal = 90,000 × 10.57 = 951300 J

Example 2.18 Determining the Power Dissipated in Melting a Metal

By using the data in Example 2.17, determine the power dissipated in melting the metal in 40 s.

Solution

$$\text{Power dissipated} = \text{energy consumed per unit time} = \frac{951300\ J}{40\ s} = 23{,}782\,\text{W}$$

Example 2.19 Determining the Cost of Energy Consumed in Melting of a Metal

By using the data in Examples 2.17 and 2.18, determine the cost of energy consumed, if the heat source is electrically powered and the price per unit kwh is $ 0.18.

Solution

$$\text{The cost of energy consumed} = (\text{energy consumed in } kwh)\times(\text{cost of energy } per\ kwh)$$
$$= \left(951300\times2.78\times10^{-7}\right)kwh\times(\$\ 0.18/kwh) = \$\ 0.0476$$

Questions and Problems

2.1. (a) Illustrate the welding-arc mechanism with the aid of a labeled sketch.

(b) Explain the three mechanisms through which free electrons are emitted during arc welding

2.2. (a) Why is metallographic examination of welds important?

(b) List the steps in the preparation of a metallographic specimen.

2.3. (a) Differentiate between low-carbon-, medium-carbon-, and high-carbon steels.

(b) Which type of carbon steel has poor weldability? And why?

2.4. Define the following terms:

(a) α-ferrite, (b) cementite, (c) pearlite, (d) austenite, (e) δ-ferrite.

2.5. Define the following terms in the Fe-Fe_3C phase diagram:

(a) Eutectoid steel, (b) eutectoid temperature, (c) eutectic composition, (d) eutectic temperature.

2.6. (a) Explain how the lead-tin phase diagram is useful in selecting an alloy for soldering.

(b) What is the issue with a lead-tin solder? And how has it been resolved?

2.7. Explain the effects of rate of cooling (during solidification) on the grain size of a metal, with reference to the mechanism of solidification.

2.8. (a) What is the effect of cold working on the microstructure and mechanical properties?

(b) What is meant by recrystallization? And why is it done?

(c) What is the meant by grain growth? Is the mechanism desirable? Explain.

(d) Draw a sketch showing the effects of temperature and time on the grain size.

2.9. (MCQs). Select the most appropriate answers for the following statements:

(1) Which of the following types of processing does not result in microstructural change?

(a) Cold working, (b) recovery, (c) recrystallization, (d) grain growth.

(2) Which of the following phases/microconstituents in steel has the highest hardness?

(a) Cementite, (b) ferrite, (c) pearlite, (d) austenite.

(3) Which of the following types of steel has the poorest weldability?

(a) Mild steel, (b) low-carbon steel, (c) medium-carbon steel, (d) high-carbon steel.

(4) Which of the following elements is the main alloying element in AISI-4140 steel?

(a) Nickel, (b) manganese, (c) molybdenum, (d) vanadium.

(5) Which of the following alloys is the safest and the most appropriate for use as a solder?

(a) Sn–40%Pb, (b) Sb–40%Pb, (c) Sn-57%Bi, (d) Sn-40%Sb.

(6) Which of the following fusion welding processes has the highest power density?

(a) Electron beam welding, (b) laser beam welding, (c) oxyfuel welding, (d) arc welding.

(7) Which of the following fusion welding processes has the lowest power density?

(a) Electron beam welding, (b) laser beam welding, (c) oxyfuel welding, (d) arc welding.

P-2.10. An specimen of carbon steel with 0.2% mass of carbon is held at 1515 °C. Assuming thermodynamic equilibrium, (a) identify the equilibrium phases and their compositions, and (b) calculate the mass fraction of each phase.

P-2.11. A heat source can transfer 2500 W to the surface of a metal part. The heat source travels 25 cm over the surface of the metal for welding in 10 s. Determine the heat input.

P-2.12. A heat source has a power input of 3000 W. The heat strikes the surface of the metal (for welding) in a circular area within a diameter of 8 mm. About 80% of the power is transferred within the circular area. Determine the power density.

P-2.13. Calculate the volumetric specific heat of mild steel containing 0.6 mass% carbon.

P-2.14. A welder wants to perform fusion welding by melting 95,000 mm^3 volume of the base metal made of mild steel (Fe-0.3 mass %C alloy) without using a filler metal. Assuming 100% efficiency, calculate the: (a) volumetric specific heat of the mild steel, (b) heat energy required for melting the given volume of the metal, (c) power dissipated in melting the metal in 50 s, and (d) cost of energy consumed, if the heat source is electrically powered and the price per unit kwh is $ 0.20.

References

Choi, S., Lim, S., Hanifa, M.M.M., et al. (2025). An Introductory Overview of Various Typical Lead-Free Solders for TSV Technology", MDPI - Inorganic Materials

Eagar, T.W. (1990), The Physics of Arc Welding Processes; *In*: Advanced Joining Technologies, (T.H. North, ed.), *Chapman and Hall*, London, U.K., 61, 1990.

Groover, M. (2019). Fundamentals of Modern Manufacturing, *Wiley*, New York

Huda, Z. (2020). Metallurgy for Physicists and Engineers, CRC Press, Boca Raton, FL

Huda, Z. and Bupett, R. (2012), Materials Science and Design for Engineers, Trans Tech Publications, Switzerland

Huda, Z., Zaharinie, T., Metselaar, I.H.S.C, et al. (2014). Kinetics of Grain Growth in 718 Ni-base Superalloy; ARCHIVES OF METALLURGY AND MATERIALS, Vol. 59, Issue 3, pp. 847–852; Polish Academy of Sciences, Poland

Moravec, J., Novakova, I., Sobotka, J., and Neumann, H. (2019), Determination of grain growth kinetics and assessment of welding effect on properties of S700MC steel in the HAZ of welded joints, *Metals,* Vol. 9, 707–727.

Rollett, A., Rohrer, G.S., Humphreys, J. (2017). Recrystallization and Related Annealing Phenomena. New York: *Elsevier Publishers*

Chapter 3
Metallurgy of Welding

Abstract A working knowledge of metallurgy is essential to understand welded structures, especially the mechanisms that control weldment performance. The success of welding not only depends on the weldability of the metals being joined, but also on the microstructural evolution occurring during welding. This chapter presents metallurgical aspects of welding by highlighting the effects of temperature on the microstructure of the metal; it also explains the weldability of various metals and alloys. The different zones in welding and the resulting phase transformations have been discussed with particular reference to the fusion zone, partially-melted zone (PMZ), and heat-affected zone (HAZ). Additionally, weldabilities of carbon steels, stainless steels, and non-ferrous alloys have been explained. This chapter includes 15 diagrams, 20 worked examples, 6 data tables, and 7 MCQs with answers.

Keywords Weldment · Weldability · Fusion zone · Partially-melted zone · Heat-affected zone · Columnar structure

Abbreviations

CET	Columnar-to-equiaxed transition
CEV	Carbon equivalent value
CGHAZ	Coarse-grained HAZ
EBW	Electron beam welding
FGHAZ	Fine-grained HAZ
FZ	Fusion zone
GTAW	Gas tungsten arc welding
HAZ	Heat affected zone
HISC	Hydrogen-induced stress cracking
ICHAZ	Inter-critical HAZ
PDAS	Primary dendrite arm spacing
PMZ	Partially-melted zone
SCHAZ	Sub-critical HAZ

Z. Huda, *Welding and Joining Engineering*,
https://doi.org/10.1007/978-3-032-20462-2_3

SDAS	Secondary dendrite arm spacing
SMAW	Shielded metal arc welding

Nomenclature

C_o	Alloy composition
G	Temperature gradient (K/mm)
K	Diffusivity coefficient ($m^2\ s^{-1}$)
R	Growth rate (or welding speed) (mm/s)
T_L	Liquidus temperature (K)
T_o	Temperature at the center of the welding pool (K)
x	Distance between the center and the corresponding point at the fusion boundary (mm)
ε	Cooling rate (K/s)
ΔT	Undercooling

3.1 Weld Zones in Fusion Welding

A working knowledge of metallurgy is essential to understanding welded structures, especially the mechanisms that control weldment performance. Fundamental to joining metallurgy are the microstructures of a weld joint, which determine the mechanical properties, and welding variables such as weld thermal cycle, chemical reactions in the molten pool, alloying, flux composition, and contaminants, which significantly affect the weld and the heat affected zone.

It has been learnt in the previous chapters that fusion welding involves heating of the base metal and the filler metal (if used) above their melting temperatures. In general, a weldment, resulting from fusion welding, consists of the following four zones: (a) fusion zone, (b) partially-melted zone (PMZ), (c) heat affected zone (HAZ), and (d) base metal, as illustrated in Fig. 3.1. Fusion zone (FZ) is the part of a weldment which is liquid during the welding process; it comprises the metal from the original work-piece (base metal) and may contain filler metal which is melted

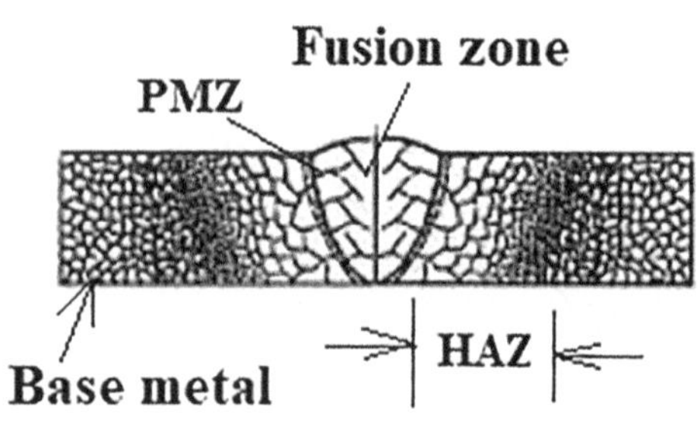

Fig. 3.1 The four zones in a fusion weldment (PMZ = partially melted zone, HAZ = heat affected zone)

during the process. In the weldment, there is a high-temperature area where no melting of the metal occurs, but there are significant microstructural changes due to the heat produced during the welding process; this region is called the *heat affected zone* (HAZ). The region between the FZ and the HAZ is called the *partially-melted zone* (PMZ) (see Fig. 3.1). The fourth zone, next to the HAZ, is the base metal, which is unaffected by any heating operation during the welding process (David et al. 2003).

3.2 The Fusion Zone

3.2.1 *Microstructural Evolution and the Mechanical Behavior*

The solidification behavior of metal in the fusion zone controls the microstructural evolution, such as size and shape of grains, the extent of segregation, the distribution of inclusions, and the defects (*e.g.* porosity, hot cracks, etc.). Figure 3.2 integrates the various fusion weldment zones with the temperature and the microstructures resulting from welding of mild-steel work-piece(s). It is evident in Fig. 3.2 that the fusion zone experiences the peak temperature (T_o) resulting in melting of the material and a columnar-grained structure on solidification; T_L is the liquidus temperature. The length x_f is the distance between the location of the center and the corresponding point at the fusion boundary, which is half the width of the weld metal.

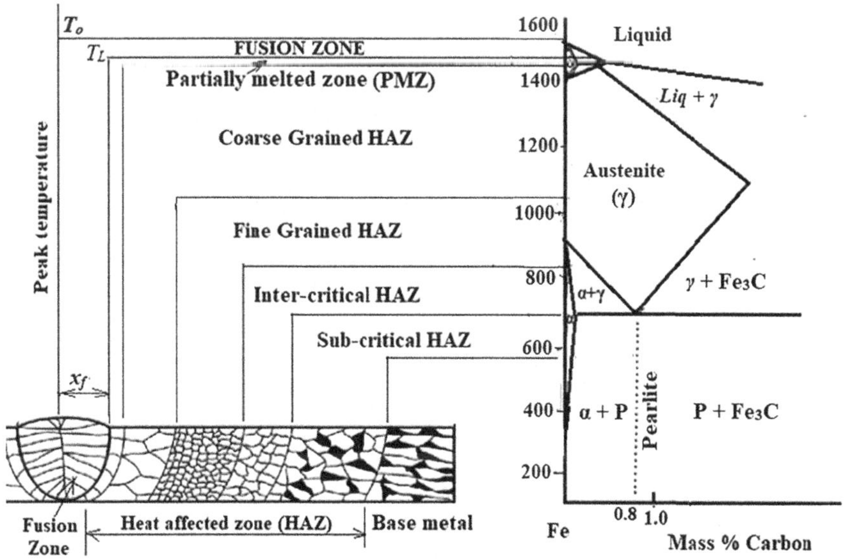

Fig. 3.2 The microstructure of a weldment from fusion welding of mild steel and the corresponding temperatures in the partial Fe-Fe_3C phase diagram

The metallurgical aspects of the fusion zone, the PMZ, and the HAZ are explained in detail in Sects. 3.2–3.4. The weld pool shape strongly influences the fusion-zone grain structure. For example, in a tear-drop-shaped weld pool, there is almost an invariant direction of maximum thermal gradient at all points on the pool edge from the fusion boundary to the weld centerline. This thermal behavior will result in a coarse-grained columnar structure (see Fig. 3.2), which is susceptible to hot cracking. On the other hand, in an elliptically- or circularly-shaped weld pool, the grains grow towards the centerline, resulting in a finer equi-axed grain structure in the fusion zone; such observations are typical in high-energy electron beam welding (EBW), or arc welding with low welding speeds (David and Vitek 1993). The columnar-to-equiaxed transition (CET) is commonly observed in laser-beam welds. The CET is able to prevent the growth of large columnar grains and consequently improves the mechanical properties of welded joints (Guo et al. 2022).

3.2.2 *Analysis:* Dendritic Crystal Growth and Calculations

The partial Fe-Fe_3C phase diagram in Fig. 3.2 enables us to predict the temperatures, phases and their amounts at specified points within the fusion zone (see Examples 3.1 and 3.2). The temperature at a point with coordinates (x,y) within the fusion zone can be precisely determined by taking a suitable point (0,0) as an origin within the fusion zone and by using the following formula:

$$T = 1494 + a\ x^{m} + \mathrm{b}\ y^{n} \tag{3.1}$$

where T is the temperature (°C), *a*, *b*, *m*, and *n* are constants, and *x* and *y* are the coordinates of the point under consideration in the fusion zone in a weldment (see Fig. 3.3). The significance of Eq. 3.1 is illustrated in Examples 3.3 and 3.4.

Solidification behavior of metal in the fusion zone is largely dependent on the following factors: the temperature gradient (*G*), the growth rate (R), undercooling (ΔT), and alloy composition (C_o). The temperature gradient can be determined by (Guo et al. 2022):

$$\mathrm{G} = \frac{T_o - T_L}{x} \tag{3.2}$$

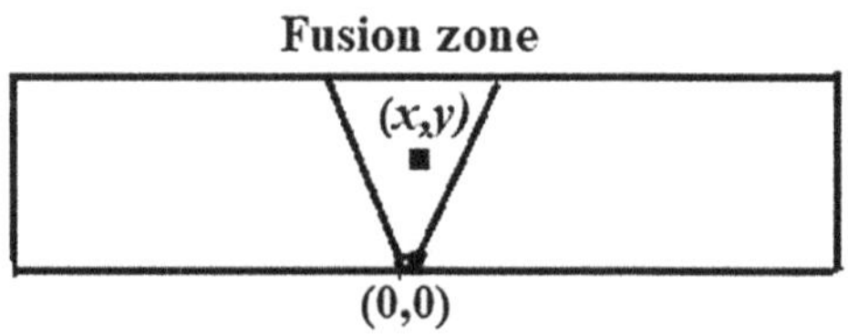

Fig. 3.3 A weldment showing the origin and the point (x,y) within the fusion zone

where G is the temperature gradient (K/mm), T_o is the temperature at the center of the welding pool, T_L is the liquidus temperature (K), and x is the distance between the location of the center and the corresponding point at the fusion boundary (mm). The product of the temperature gradient and the growth rate is called the cooling rate, *i.e.*

$$\varepsilon = G\ R \tag{3.3}$$

where ε is the cooling rate (K/s) and R is the growth rate at the center line (or the welding speed) (mm/s) (see Example 3.5).

It is mentioned in the preceding chapter that a pure metal solidifies by a dendritic growth mechanism (see Fig. 2.8). A dendritic crystal is a crystal that develops into a typical multi-branching form, as shown in Fig. 3.4. Three types of dendritic arm spacing can be identified in dendritic growth of a crystal: primary dendrite arm spacing (λ_1), secondary arm spacing (λ_2), and tertiary arm spacing (λ_3) (see Fig. 3.4). It has been shown in aluminum alloys that the finer the dendrite arm spacing, the higher the ductility and yield strength of the weld metal (Kou 2003).

The primary dendrite arm spacing (PDAS) strongly influences the mechanical properties; it can be related to the temperature gradient and the growth rate by (Bunaziv et al. 2019):

$$\lambda_1 = A G^{-0.5} R^{-0.25} \tag{3.4}$$

where λ_1 is the PDAS, A is a composition-dependent material (alloy) parameter. Because the values of G and R are obtainable and since λ_1 can be measured from a micrograph, the parameter A can be estimated based on Eq. 3.4 (see Example 3.6). The secondary dendrite arm spacing (SDAS) also influences the mechanical properties; its size can be determined by:

$$\lambda_2 = B \varepsilon^{-0.33} \tag{3.5}$$

where λ_2 is the SDAS (in mm), B is a material-dependent constant, and ε is the cooling rate (in K/s). Once the SDAS (from the micrograph) and the constant B are known, the cooling rate can be determined by using Eq. 3.5 (see Example 3.8). The peak temperature (T_o) for fusion welding (see Fig. 3.1) can be calculated by modifying Eq. 3.2, as follows:

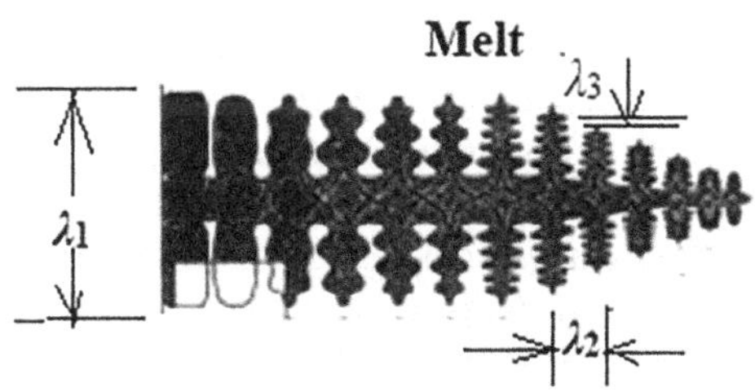

Fig. 3.4 Dendritic growth of crystal during solidification of a metal in fusion zone

$$T_o = T_L + G\ x \tag{3.6}$$

The significance of Eq. 3.6 is illustrated in Example 3.9.

3.3 Partially Melted Zone (PMZ)

The partially melted zone (PMZ) is the region between the fusion zone and the high-temperature HAZ. The size of the PMZ can be determined by engineering analysis (see Sect. 3.4.2). Liquation can occur in PMZ during welding due to heating above the eutectic temperature. In particular, intermetallic phases in aluminum alloys induce liquation that can occur partially along the grain boundary and partially in the grain interiors. The grain boundary liquation can make PMZ weak and may lead to inter-granular cracking under the tensile strains developed during welding (Rao et al. 2005). The heat treatable aluminum alloys and nickel-base superalloys are especially known to be susceptible to cracking in the PMZ of the weld.

Liquation cracking, in the PMZ, can be reduced by controlling the following factors: (a) filler metal, (b) heat source, (c) degree of restraint, and (d) base metal. For example, liquation cracking occurs in the 6061 aluminum-alloy welds produced with *Al–Mg* fillers at high dilution ratios, but it does not occur in welds made with *Al–Si* fillers at any dilution ratios (Kou 2003).

3.4 Heat-Affected Zone (HAZ)

3.4.1 Phase Transformations and Microstructural Evolution in the HAZ

The heat affected zone (HAZ) is the area of the weldment where the metal has not been melted and has undergone microstructural changes as a result of being exposed to relatively high temperatures during welding. The HAZ is located between the PMZ and the unaffected base metal (see Fig. 3.5). An in-depth look at Fig. 3.5 reveals the following areas in the HAZ: (a) CGHAZ, (b) FGHAZ, (c) ICHAZ, and (d) SCHAZ; these subzones are explained below.

The coarse-grained HAZ (*CGHAZ*) is the coarse-grained microstructure resulting from heating in the temperature range from 1100 °C to just below the melting temperature of the steel. The fine-grained HAZ (*FGHAZ*) is the fine-grained microstructure that results when the steel has been heated in the temperature range from A_3 (upper critical temperature) to 1100 °C. The inter-critical HAZ (*ICHAZ*) is the area resulting from heating in the temperature range of the lower critical temperature and the upper critical temperature (A_1–A_3). The sub-critical HAZ (*SCHAZ*) is the zone resulting from heating in the temperature range from 600 °C to just below the A_1 (lower critical temperature).

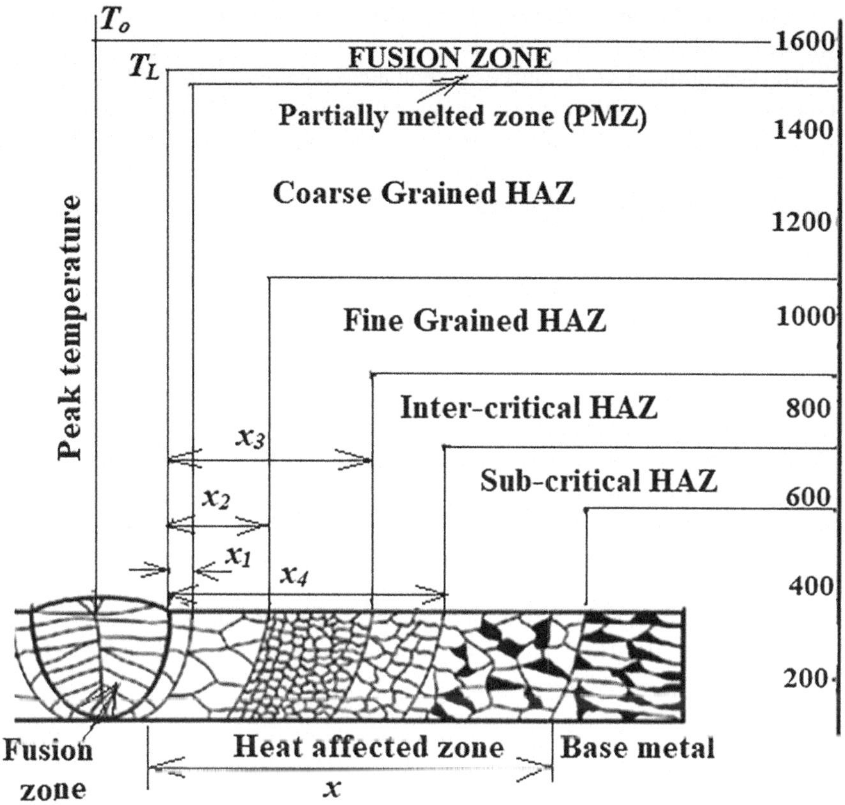

Fig. 3.5 The microstructures and the lengths of various zones and HAZ sub-zones and their corresponding temperatures in a fusion weldment of mild steel

The hypo-eutectoid steels have better weldability than hyper-eutectoid steels. For hypo-eutectoid steels, heating at temperatures above A_1 results in the transformation of ferrite (α) to austenite (γ), whereas heating above A_3 results in the homogenization of austenite (see Fig. 3.2).

The understanding of the microstructural changes in the HAZ involves a great deal of recrystallization and grain growth, which has been discussed in the preceding chapter (see Sect. 2.5, Fig. 2.9). It is evident in Fig. 3.5 that the coarse-grained HAZ (CGHAZ) indicates extensive grain growth resulting from high temperature during welding. Since the CGHAZ is adjacent to very high temperature zones (fusion zone and PMZ), the rate of cooling is slow, resulting in a coarse grain size, which causes loss of strength and fracture toughness.

The fine-grained HAZ (FGHAZ) is the result of normalizing of steel. Since this zone is closer to base metal (as compared to CGHAZ), the rate of cooling is a bit faster, thereby producing a fine-grained microstructure indicating fairly good strength. The microstructure of sub-critical HAZ (SCHAZ) indicates recrystallization of steel; here the cold-worked structure of the base metal changes to fairly fine, equi-axed grains due to recrystallization (see Fig. 2.9). As a result, ductility is restored without sacrificing strength.

The size of the HAZ and its constituent areas are affected by the following factors: material welded, the type of welding (OFW, SMAW, GMAW, GTAW, etc.), the welding speed, the maximum temperature, welding conditions, and the like. The length of the HAZ and its constituent zones can be determined with reference to Fig. 3.5 and the engineering analysis presented in Sect. 3.4.2.

3.4.2 Engineering Analysis of Sub-zones in the HAZ

It is useful to determine the size (length) of the HAZ and its constituents, since mechanical properties are dependent on the microstructure in each sub-zone of the HAZ. For this purpose, certain assumptions are made. Firstly, it is assumed that the temperature distribution in the half-space (on the surface $x = 0$ in Fig. 3.5) is even when heated in time t with constant maximum temperature ($T_0 = T_{max.}$ = constant). Secondly, the temperature T inside the material only depends on x and t, and it does not depend on y and z.

It can be numerically shown that for the peak temperature $T_o = 1550$ °C, the total length of the HAZ can be calculated by the formula (Śloderbach and Pająk 2015):

$$x \approx 1.22 \sqrt{K \cdot t} \tag{3.7}$$

where x is the total length of the HAZ (m), K is the diffusivity coefficient of steel ($m^2\ s^{-1}$), and t is the time of exposure (s) of heat surface at the peak temperature T_o (see Example 3.10).

The size of the PMZ, at a temperature of 1480 °C, can be determined by the following formula:

$$x_1 \approx 0.1 \sqrt{K \cdot t} \tag{3.8}$$

where x_1 is the length (in m) of the PMZ (see Fig. 3.5, see Example 3.11).

The length x_2 in Fig. 3.5 (at a temperature of 1100 °C) can be determined by the formula (Śloderbach and Pająk 2015; Carslaw and Jaeger 1980):

$$x_2 \approx 0.51 \sqrt{K \cdot t} \tag{3.9}$$

By reference to Fig. 3.5, the length of the CGHAZ can be calculated by combining Eqs. 3.8 and 3.9; thus we obtain:

$$\text{Length of CGHAZ} = x_2 - x_1 \approx 0.41 \sqrt{K \cdot t} \tag{3.10}$$

The size of the CGHAZ increases with the increase in the heat input (in J/mm); it means that the loss of strength in the HAZ increases with the heat input of the weld (Kou 2003).

Similarly, the length x_3 in Fig. 3.5 (at a temperature of 850 °C) can be determined by:

$$x_3 \approx 0.834 \sqrt{K \cdot t} \tag{3.11}$$

By reference to Fig. 3.5, the length of the fine-grained HAZ (FGHAZ) can be calculated by combining Eqs. 3.9 and 3.11 to obtain:

$$\text{Length of FGHAZ} = x_3 - x_2 \approx 0.324 \sqrt{K \cdot t} \tag{3.12}$$

The significance of Eqs. 3.10 and 3.12 is illustrated in Example 3.12.

Similarly, the length x_4 in Fig. 3.5 (at a temperature of 710 °C) can be determined by:

$$x_4 \approx 1.1.3 \sqrt{K \cdot t} \tag{3.13}$$

$$\text{Length of ICHAZ} = x_4 - x_3 \approx 0.296 \sqrt{K \cdot t} \tag{3.14}$$

The length of the SCHAZ can be determined by:

$$\text{Length of the SCHAZ} = x - x_4 \simeq 0.1 \sqrt{K \cdot t} \tag{3.15}$$

The significance of Eqs. 3.14 and 3.15 is illustrated in Example 3.13.

3.5 Weldability of Metals

3.5.1 What Is Weldability?

Weldability is the ease with which a metal can be properly welded. A metal's weldability is used to select the welding process and to compare the final weld quality with other metals and alloys. According to the ISO-standard 581-1980, weldability is defined as follows: "metallic material is considered to be susceptible to welding to an established extent through given processes and for given purposes when welding provides metal integrity by a corresponding technological process for welded parts to meet technical requirements as to their own qualities as well as to their influence on a structure they form" (Lippoid 2014). Metals with good weldability include low-carbon steels, medium-carbon steels, stainless steels, aluminum and its alloys, nickel and its alloys, copper and its alloys, titanium and its alloys, and cast irons (except white cast iron).

3.5.2 Dilution and Weldability

The term *dilution* refers to the proportion of base materials in the resultant weld metal. The *dilution ratio*, for a single bead deposit, is the ratio of the cross-sectional area of melted base material (and the backing) to the total cross-sectional area of the fusion zone (DuPont and Marder 1996). Consider a fusion welding process that involves the use of a filler metal (F) to join a base metal X to another base metal Y with a backing Z (see Fig. 3.6). It is evident in Fig. 3.6 that the total cross-sectional area of the melted base metals and backing is $(x + y + z)$ and the total cross-sectional area of the fusion zone is $(x + y + z + f)$. Numerically, the dilution ratio can be expressed as:

$$\text{Dilution ratio} = \frac{x+y+z}{x+y+z+f} \tag{3.16}$$

Accordingly, the percentage of dilution can be expressed by:

$$\%\text{dilution} = \frac{x+y+z}{x+y+z+f} \times 100 \tag{3.17}$$

The significance of Eqs. 3.16 and 3.17 is illustrated in Example 3.14.

Dilution can affect weldability. A high dilution can be detrimental to the mechanical properties of the weld metal. The dilution, during welding, strongly depends on the heat input. For example, SMAW and GTAW processes (involving stick electrodes) with lower current and voltage have lower heat input, thereby resulting in a lower dilution in the range of 20–40%. On the other hand, a high input-energy welding process may result in a high dilution of up to 80%.

3.5.3 Weldability of Carbon/Alloy Steels

Carbon steels have been introduced to the reader in the preceding chapter (Sect. 2.3.1). Low-carbon steels have the best weldability as they are the easiest to weld; another advantage of these steels is that no heat affected zone (HAZ) is formed after

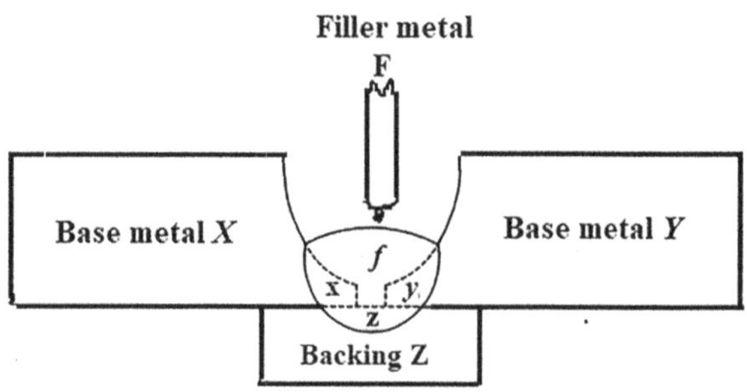

Fig. 3.6 Welding of dissimilar metal involving filler metal dilution

welding. A high carbon content in steel is undesirable because it promotes the formation of brittle martensite phase, thereby resulting in the post-solidification cracking of the weld metal. However, this problem can be overcome by "buttering" the groove faces of the base metal with austenitic stainless steel electrodes before welding. Medium carbon steels have intermediate weldability.

During welding of steels, carbon equivalent value (CEV) is used to determine how various alloying elements affect the hardness of the steel being welded. Since hydrogen-induced stress cracking (HISC) is the most frequent weld flaw in steel, this is directly tied to it, which is why it is the most popular method for assessing weldability. Higher carbon content and other alloying elements, including manganese (Mn), chromium (Cr), silicon (Si), molybdenum (Mo), vanadium (V), copper (Cu), and nickel (Ni) have the tendency to increase hardness while lowering weldability (Modulus Metals 2025). However, as each of these elements has a different tendency to have an impact on the steel's hardness and weldability, it is required to compare the hardness of two steels based on their CEV. According to the AWS, there is a risk of cracking in the HAZ on flame cut edges and welds for CEV >0.40%. The CEV can be calculated by:

$$\mathrm{CEV} = \mathrm{C} + \frac{Mn}{6} + \frac{Cr + Mo + V}{5} + \frac{Ni + Cu}{15} \tag{3.18}$$

where the symbols denote the wt% of the elements (see Example 3.15).

3.6 Weldability of Stainless Steel

3.6.1 Stainless Steels: An Introduction

Stainless steels are high nickel-chromium steels that are noted for their excellent corrosion resistance and strength. Alloying elements, in stainless steels, may be divided into two groups: (a) austenite formers (Ni, C, and Mn) and ferrite formers (Cr, Mo, Si, and Nb). The AISI system for steels has been introduced to the reader in the preceding chapter (see Sect. 2.3.1). An example of stainless steel is the 304 stainless steel with composition (mass %): 18Cr-8Ni-0.05C-2Mn. The chromium equivalent and nickel equivalent, in an alloy/stainless steel, can be calculated by the following formulas (Kotecki and Siewert 1992):

$$\mathrm{Cr\,Equiv.} = \%\mathrm{Cr} + \%\mathrm{Mo} + 0.7\ (\%\mathrm{Nb}) \tag{3.19}$$

$$\mathrm{Ni\,Equiv.} = \%\mathrm{Ni} + 35\ (\%\mathrm{C}) + 20\ (\%\mathrm{N}) + 0.25(\%\mathrm{Cu}) \tag{3.20}$$

The significance of Eqs. 3.19 and 3.20 is illustrated in Example 3.16.

3.6.2 *Microstructural Control in Welding of Austenitic Stainless Steel*

The weld-metal post-solidification microstructure of high-nickel stainless steels (*e.g.* 310 stainless steel) (25%Cr-20%Ni-55%Fe) consists of austenite (FCC) as the primary solidification phase with inter-dendritic δ-ferrite (BCC) between the primary and secondary dendrite arms. On the other hand, in high-chromium stainless steels, δ-ferrite is the primary solidification phase in the microstructure. It is essential to have a proper amount (around in the range of 5–10 vol.%) of δ-ferrite in the weld metal of an austenitic stainless steel; too much δ-ferrite (>10 vol.%) reduces the ductility, toughness, and corrosion resistance, whereas too little δ-ferrite (<5 vol.%) may result in solidification cracking (Kou 2003). The amount of δ-ferrite (or the ferrite number), in an austenitic stainless steel, can be predicted by using *WRC-1992 constitution diagram* (Kotecki and Siewert 1992) (see Example 3.17).

Welding of stainless steel requires great care and a strong understanding of metallurgy. In particular, austenitic stainless steels may develop *sensitization* problems, which is explained as follows. When an austenitic stainless steel cools slowly in the temperature range from 870 to 425 °C during welding, chromium carbide ($Cr_{23}C_6$) precipitates at the grain boundaries. As a result, the austenite at the grain boundaries may contain less than 12% Cr (the minimum required to produce a passive oxide layer); the steel is said to have *sensitized* (see Fig. 3.7). Since the grain boundary regions are small and highly anodic, rapid corrosion of the austenite at the grain boundaries occurs, resulting in inter-granular (*IG*) corrosion.

The inter-granular (IG) corrosion problem due to sensitization in austenitic stainless steels can be overcome by adopting one of the following techniques: (a) keep the chromium (Cr) content reasonably high in steel, (b) keep the carbon content less than 0.03% to prevent the formation of $Cr_{23}C_6$, (c) *stabilize* the steel by addition of titanium (Ti) or niobium (Nb), and (d) avoid the sensitization temperature range (425–870 °C) during welding.

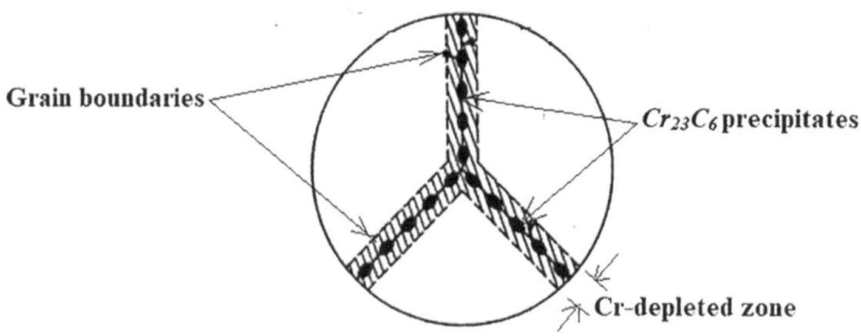

Fig. 3.7 Schematic of the microstructure of sensitized austenitic stainless steel

3.6.3 The Dilution Ratio and Electrode Factors in Welding of Stainless Steels

Besides other factors, weldability of stainless steels strongly depends on *dilution.* A high dilution can have a detrimental effect on mechanical properties and corrosion resistance. Dilution can be reduced, during welding, by the following ways: (a) increasing the gap between the electrode and work-piece during arc welding, (b) using a lower welding current, and (c) preferring austenitic stainless-steel electrodes as compared to ferritic electrodes (see Examples 3.18–3.20).

3.7 Weldability of Aluminum Alloys

Commercially pure aluminum is relatively soft; this is why small amounts of alloying elements are added to produce alloys with a range of mechanical properties. Welding of aluminum alloys is more common than any other types of nonferrous alloys because of their fairly good weldability and wide applications. The series 1xxx (*e.g.* 1100) aluminum alloys have no major alloying elements; they are not heat-treatable and have high electrical and thermal conductivities.

The characteristics of various series of aluminum alloys are presented in Table 3.1.

Wrought (non-heat-treatable, NHT) aluminum alloys are usually strengthened by solid-solution strengthening and work hardening. Welding of NHT work-hardened alloys may result in the loss of strength in the HAZ of the weldment; however, the series 5xxx *Al* alloys have excellent weldability by using OFW, GTAW, and GMAW processes. Welding of heat treatable *Al* alloys also results in the loss of strength in the HAZ due to the redistribution of the hardening constituents. In particular, heat treatable *Al* alloys may develop solidification cracking during fusion welding; these alloys are, therefore, best welded by friction stir welding (TWI 2022).

Table 3.1 Aluminum alloys' series and their characteristics

Series	Example	Principal alloying element	Characteristics
1xxx	1100	Nil	Not heat-treatable
2xxx	2024	Copper	Heat treatable, good strength
3xxx	3003	Manganese	Not heat-treatable, good formability
4xxx	4043	Silicon	Not heat-treatable, used as filler wires
5xxx	5052	Magnesium	Not heat treatable and have good strength after welding
6xxx	6061	Magnesium, silicon	Heat treatable, good strength and extrudability
7xxx	7075	Zinc	Heat treatable, good strength

3.8 Worked Examples on Metallurgy of Welding

Example 3.1 Identifying the Temperature and Composition of Phases in Fusion Zone

A low-carbon high-purity steel plate with 0.3 mass% C is being welded. Considering the thermodynamics equilibrium, identify the temperature, the phases, and their compositions for point *A*, as shown in Fig. 3.8.

Solution

By reference to Fig. 3.2, the temperature at point *A* is around 1500 °C, and the existing phases are liquid and δ-ferrite (BCC). By drawing a tie-line for the 1500 °C temperature in the partial phase diagram (Fig. 3.9), we obtain:

The composition of the liquid is Fe-0.46%C and that of δ-ferrite is Fe-0.08%C.

Fig. 3.8 A weldment showing a specified point within the fusion zone

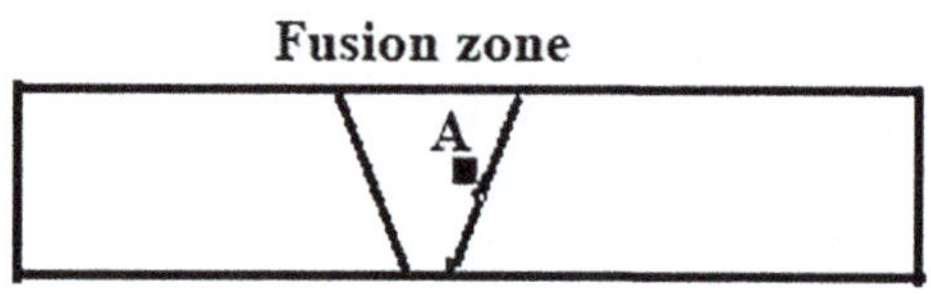

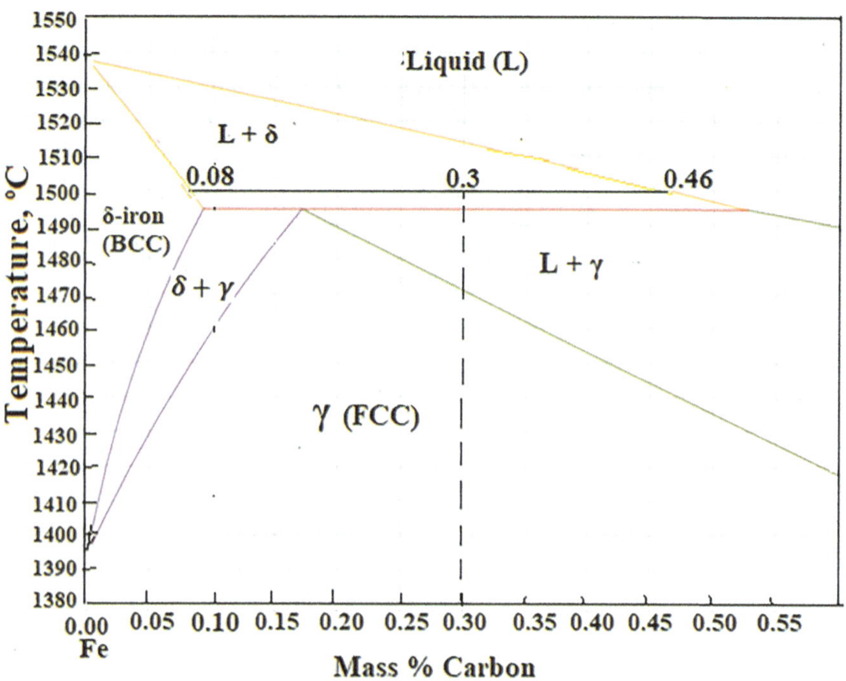

Fig. 3.9 Partial Fe-Fe_3C phase diagram for point A in the fusion zone

Example 3.2 Determining the Mass Fraction of Phases at a Point in Fusion Zone
By using the data in Example 3.1, determine the mass fraction of the phases at point A.

Solution

By reference to Figure 3.9, $C_L = 0.46$, $C_\delta = 0.08$, $C_o = 0.3$.

By using the *inverse lever rule* (Eq. 2.2), we can write:

$$\text{Mass fraction of liquid} = M_L = \frac{C_\delta - C_o}{C_\delta - C_L} = \frac{0.08 - 0.30}{0.08 - 0.46} = 0.579(\text{or } 57.9\%)$$

$$\text{Mass fraction of } \delta \text{ ferrite} = M_\delta = 1 - 0.579 = 0.421(\text{or } 42.1\%)$$

Example 3.3 Identifying Temperature and Composition of Phases with Known Coordinates
A low-carbon steel (Fe-0.18%C) plate is being welded. Assume that the solidification is at thermodynamics equilibrium and a temperature field of $T = 1494 + y^2$ exists for the fusion zone. (a) Determine the temperature at point (1,4) (see Fig. 3.10). (b) What phases exist at (1,4)?

(c) What are the compositions of the phases? (d) Calculate the mass fractions of the phases?

Solution

Given Equation: $T = 1494 + y^2$ Eq. 3.1: $T = 1494 + a\,x^m + b\,y^n$.

By comparing the given equation with Eq. 3.1, we obtain: $a = 0$, $x = 1$, $b = 1$, $y = 4$, and $n = 2$.

The temperature field can be numerically determined by using Eq. 3.1, as follows:

$$T = 1494 + 0 + (1) * 4^2 = 1510\ ^\circ\text{C}$$

Hence, the temperature at point (1.4) in the fusion zone is 1510 °C.

(a) By reference to the partial Fe-Fe_3C phase diagram (Fig. 3.9), the existing phases for 0.18%C steel at 1510 °C are liquid (Liq) and δ-ferrite (BCC).
(b) By reference to Fig. 3.9, the composition of δ-ferrite (BCC) is Fe-0.06%C, and the composition of the liquid phase is Fe-0.35%C.
(c) By reference to Fig. 3.9, $C_L = 0.35$, $C_\delta = 0.06$, $C_o = 0.18$.
    By using the *inverse lever rule* (Eq. 2.2), we can write:

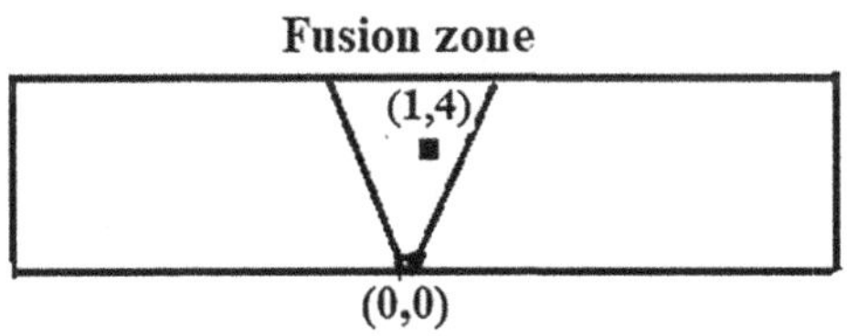

Fig. 3.10 A weldment showing the point (1,4) within the fusion zone

$$\text{Mass fraction of liquid} = M_L = \frac{C_\delta - C_o}{C_\delta - C_L} = \frac{0.06 - 0.18}{0.06 - 0.35} = 0.41(\text{or } 41\%)$$

$$\text{Mass fraction of } \delta \text{ ferrite} = M_\delta = 1 - 0.41 = 0.59(\text{or } 59\%)$$

Example 3.4 Identifying Temperature and Composition of Phases with Another Coordinates

A low-carbon steel plate with 0.14% carbon is being welded. Assume that the solidification is at equilibrium and a temperature field of $T = 1494 - x^2 + y^{2.2}$ exists for the fusion zone.

(a) Calculate the temperature at point (3,5) (see Fig. 3.11). (b) What phases exist at (3,5)? (c) What are the compositions of the phases? (d) Calculate the mass fractions of the phases.

Solution

By comparing the given data with Eq. 3.1, a = −1, $x = 3$, m = 2, b = 1, y = 5, and $n = 2.2$.

(a) $$T = 1494 - x^2 + y^{2.2} = 1494 - 3^2 + (5)^{2.2} = 1519.5°C$$

Hence, the temperature at point (3,5) in the fusion zone is 1519.5 °C.

(b) By reference to Fig. 3.9, the existing phases for the 0.14% carbon steel at 1519.5 °C are Liquid and δ-ferrite (BCC).
(c) By reference to Fig. 3.9, the composition of δ-ferrite (BCC) is Fe-0.04%C, and the composition of the liquid phase is Fe-0.24%C.
(d) By reference to Fig. 3.9, $C_L = 0.24$, $C_\delta = 0.04$, $C_o = 0.14$.

By using the *inverse lever rule* (Eq. 2.2), we can write:

$$\text{Mass fraction of liquid} = M_L = \frac{C_\delta - C_o}{C_\delta - C_L} = \frac{0.04 - 0.14}{0.04 - 0.24} = 0.5(\text{or } 50\%)$$

$$\text{Mass fraction of } \delta \text{ ferrite} = M_\delta = 1 - 0.5 = 0.5(\text{or } 50\%)$$

Example 3.5 Calculating the Cooling Rate During Solidification

A deep penetration laser beam welding experiment of an *Al–Mg* alloy is being performed using the maximum power of 4000 W. The temperature at the center of the welding pool is 1373 K and the liquidus temperature is 773 K. The distance between the location of the center and the corresponding point at the fusion boundary is

Fig. 3.11 A weldment showing point (3,5) in the fusion zone

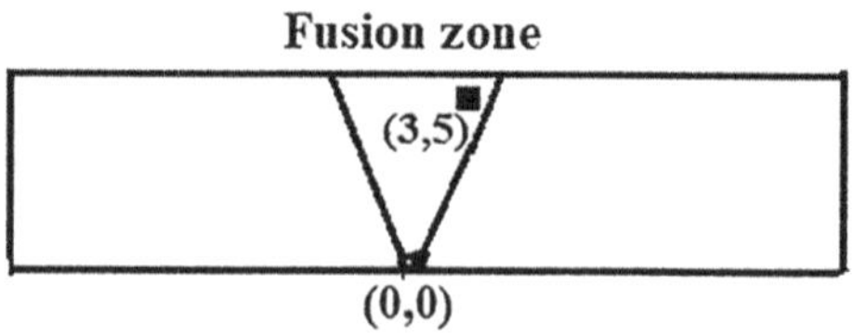

2 mm. The growth rate is 3 mm/s. Determine the: (a) crystal growth rate, and (b) cooling rate of the metal in the fusion zone.

Solution

$T_o = 1373$ K, $T_L = 773$ K, $x = 2$ mm, R = 3 mm/s.

By using Eqs. 3.2 and 3.3,

$$\text{(a)}\ \ G = \frac{T_o - T_L}{x} = \frac{1373 - 773}{2} = 300\,\text{K}/\text{mm}$$

(b) Cooling rate = ε = G R = 300 × 3 = 900 K/s.

Example 3.6 Determining the Material-Dependent Constant When the PDAS Is Known

A measurement from a micrograph for solidification of an alloy showed the primary dendrite arm spacing (PDAS) to be 13 μm. By using the solidification data given in Example 3.5, calculate the parameter *A* in Eq. 3.4.

Solution

PDAS = λ_1 = 13 μm, G = 300 K/mm, R = 3 mm/s, A =?

By using the modified form of Eq. 3.4,

$$A = \frac{\lambda_1}{G^{-0.5} R^{-0.25}} = \frac{13\times10^{-6}\,m}{300^{-0.5}\times3^{-0.25}} = \frac{13\times10^{-3}\,mm}{0.0577\times0.76} = 0.296\left(\frac{mm}{s}\right)^{0.25}\sqrt{K\cdot mm}$$

Example 3.7 Calculating the Primary Dendrite Arm Spacing for Solidification Process

Refer to the solidification data for the same alloy mentioned in Examples 3.5 and 3.6. In another experimental condition, the temperature gradient was 280 K/mm and the growth rate was 2.7 mm/s. Calculate the primary dendrite arm spacing.

Solution

$$G = 280\,\text{K}/\text{mm}, \text{R} = 2.7\,\text{mm}/\text{s}, \text{A} = 0.295\left(\frac{mm}{s}\right)^{0.25}\sqrt{K\cdot mm}.$$

By using Eq. 3.4,

$$\text{PDAS} = \lambda_1 = \text{A}\ G^{-0.5}\ R^{-0.25}$$
$$= 0.295\left(\frac{mm}{s}\right)^{0.25}\sqrt{K\cdot mm}\times280^{-0.5}\ \frac{K^{-0.5}}{mm^{-0.5}}\times2.7^{-0.25}\ \frac{mm^{-0.25}}{s^{-0.25}}$$
$$\lambda_1 = 0.295\times0.0598\times0.78 = 0.0138\,\text{mm} = 13.8\,\mu m$$

Example 3.8 Determining the Material-Dependent Constant When the SDAS Is Known

Refer to the weld solidification data in Example 3.5. A measurement from a micrograph showed the SDAS to be 4 μm. Determine the constant B in Eq. 3.5.

Solution

ε = 900 K/s, λ_2 = 4 μm = 0.004 mm, B =?

By using the modified form of Eq. 3.5,

$$B = \frac{\lambda_2}{\varepsilon^{-0.33}} = \frac{0.004}{900^{-0.33}} = 0.0377\,\text{mm}\left(\frac{K}{s}\right)^{0.33}$$

Example 3.9 Determining the Peak Temperature in Fusion Welding

The weld metal cooling rate, in a fusion welding process, is 500 K/s and the welding speed is 2.8 mm/s. The liquidus temperature with reference to point 1.8 mm from the center line (in the fusion zone) is 1500 °C. Determine the peak temperature at the center line of the fusion zone.

Solution

ε = 500 K/s, R = 2.8 mm/s. x = 1.8 mm, T_L = 1500 °C = 1773 K.

By using the modified form of Eq. 3.3,

$$G = \frac{\varepsilon}{R} = \frac{500}{2.8} = 178.6\,\text{K}/\text{mm}$$

By using Eq. 3.6,

$$\text{Peak temperature} = T_o = T_L + Gx = 1773 + (178.6 \times 1.8) = 2094\text{K} = 1821°\text{C}$$

Example 3.10 Calculating the Length of the HAZ

A cast steel work-piece is being fusion welded by evenly heating the surface with x = 0 at the peak temperature of 1550 °C for 30 s. Calculate the length of the HAZ. The diffusivity coefficient of cast steel is $0.12 \times 10^{-4}\, m^2 \bullet s^{-1}$.

Solution

$K = 0.12 \times 10^{-4}\, m^2 \bullet s^{-1}$, t = 30 s. By using Eq. 3.7,

$$\text{Length of the HAZ} = x \approx 1.22\sqrt{K \cdot t} = 1.22\sqrt{0.12 \times 10^{-4} \times 30} = 0.0231\,\text{m} = 2.31\,\text{cm}$$

Example 3.11 Calculating the Size of the PMZ

By using the data in Example 3.10, calculate the length of the PMZ.

Solution

$K = 0.12 \times 10^{-4}\, m^2 \bullet s^{-1}$, t = 30 s. By using Eq. 3.8,

$$\text{Length of PMZ} = x_1 = 0.1\sqrt{K \cdot t} = 0.1 \times \sqrt{0.12 \times 10^{-4} \times 30}$$
$$= 189.7 \times 10^{-5}\,\text{m} = 1.9\,\text{mm}$$

Example 3.12 Calculating the Sizes of the CGHAZ and FGHAZ
By using the data in Example 3.10, calculate the lengths of the CGHAZ and FGHAZ.

Solution

$K = 0.12 \times 10^{-4}\, m^2 \bullet s^{-1}$, $t = 30$ s.

By using Eqs. 3.10 and 3.12,

$$\text{Length of CGHAZ} = x_2 - x_1 = 0.41\sqrt{K \cdot t}$$
$$= 0.41 \times \sqrt{0.12 \times 10^{-4} \times 30} = 0.0078\,\text{m} = 7.8\,\text{mm}$$
$$\text{Length of FGHAZ} = x_3 - x_2 = 0.32\sqrt{K \cdot t}$$
$$= 0.32 \times \sqrt{0.12 \times 10^{-4} \times 30} = 0.0061\,\text{m} = 6.1\,\text{mm}$$

Example 3.13 Calculating the Sizes of ICHAZ and SCHAZ
By using the data in Example 3.10, calculate the lengths of the ICHAZ and the SCHAZ.

Solution

$K = 0.12 \times 10^{-4}\, m^2 \bullet s^{-1}$, $t = 30$ s. By using Eqs. 3.14 and 3.15,

$$\text{Length of ICHAZ} = x_4 - x_3 \approx 0.296\sqrt{K \cdot t} = 0.296 \times \sqrt{0.12 \times 10^{-4} \times 30} = 5.62\,\text{mm}$$
$$\text{Length of SCHAZ} = x - x_4 \cong 0.01\sqrt{K \cdot t} = 0.1 \times \sqrt{0.12 \times 10^{-4} \times 30} = 1.897\,\text{mm}$$

Example 3.14 Calculating the Dilution Ratio and Percent Dilution
A fusion weldment was cross sectioned, and the various cross-sectional areas are shown in Fig. 3.12. Calculate the (a) dilution ratio, and (b) percent dilution.

Solution

x = 12, y = 14, z = 8, f = 20.

By using Eqs. 3.16 and 3.17,

$$\text{(a) Dilution ratio} = \frac{x+y+z}{x+y+z+f} = \frac{12+14+8}{12+14+8+20} = 0.63$$

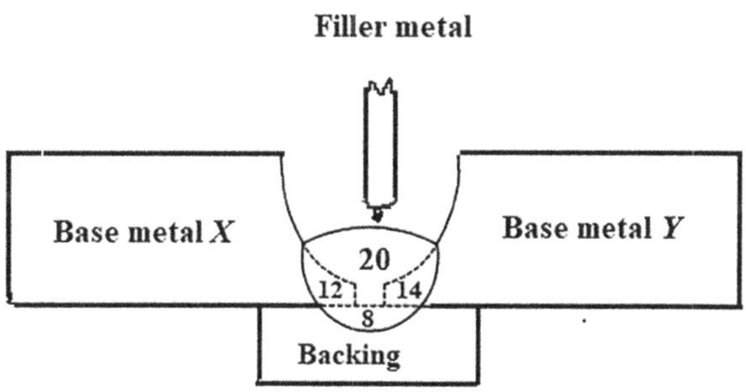

Fig. 3.12 The areas (in mm²) in the fusion weldment

(b) $\%\text{dilution} = \dfrac{x+y+z}{x+y+z+f} \times 100 = 0.63 \times 100 = 63$

Example 3.15 Evaluating the Weldability of an Steel by Calculating the CEV
Evaluate the weldability of the steel with the following chemical composition (%wt.): 0.17C-0.12Si-0.7Mn-0.06P-0.03S-0.05Cu-0.04Cr-0.04Sn0.03Ni-0.01Mo.

Solution

By using Eq. 3.18

$$\begin{aligned} \text{CEV} &= \text{C} + \frac{Mn}{6} + \frac{Cr+Mo+V}{5} + \frac{Ni+Cu}{15} \\ &= 0.17 + \frac{0.7}{6} + \frac{0.04+0.01+0}{5} + \frac{0.03+0.05}{15} = 0.302 \end{aligned}$$

The weldability is good since CEV is less than 0.40.

Example 3.16 Calculating the Chromium Equiv. and Nickel Equiv. in a Stainless Steel
Calculate the chromium equivalent and nickel equivalent in the type 304 stainless steel with a composition (mass %): 18Cr-8Ni-0.05C-2Mn.

Solution

%Cr = 18, % Ni = 8, %C = 0.05, %Mn = 2.0. By using Eqs. 3.18 and 3.19,

$$\text{Cr Equiv.} = \%\text{Cr} + \%\text{Mo} + 0.7(\%\text{Nb}) = 18 + 0 + 0 = 18$$

$$\text{Ni Equiv.} = \%\text{Ni} + 35(\%\text{C}) + 20(\%\text{N}) + 0.25(\%\text{Cu}) = 8 + (35 \times 0.05) + 0 + 0 = 9.75$$

Example 3.17 Predicting the Ferrite Number of a Weld Metal for Austenitic Stainless Steel
By using the data in Example 3.16, predict the ferrite number in the weld metal of the stainless steel, and hence determine the weldability of the steel.

Solution

Cr Equiv. = 18, Ni Equiv. = 9.75. By reference to the *WRC-1992 diagram*, the ferrite number for the weld metal is 12. Since the ferrite number >10, the meld metal would have reduced ductility, toughness, and corrosion resistance. Hence, the weldability of the material is poor.

Example 3.18 Calculating Weld Metal Composition by Using 310 Stainless-Steel Electrode
It is required to weld a 304 stainless steel to an AISI-1015 steel by arc welding. The dilution ratio is 26%, half from the 304 steel and half from the 1015 steel. Calculate the weld metal composition by using a type 310 stainless-steel electrode that contains 0.12% carbon by mass. The material composition data is given in Table 3.2.

Solution

By using the dilution data, welding of the 304 steel with 1015 steel is shown in Fig. 3.13.

Table 3.2 Compositions of carbon and stainless steels (mass %)

Material	% Cr	% Ni	% C	% Mn	% S
304 steel	18	8	0.05	2.0	0
1015 steel	0	0	0.15	0.4	0.2
310 electrode	26	21	0.12	1.75	0.4

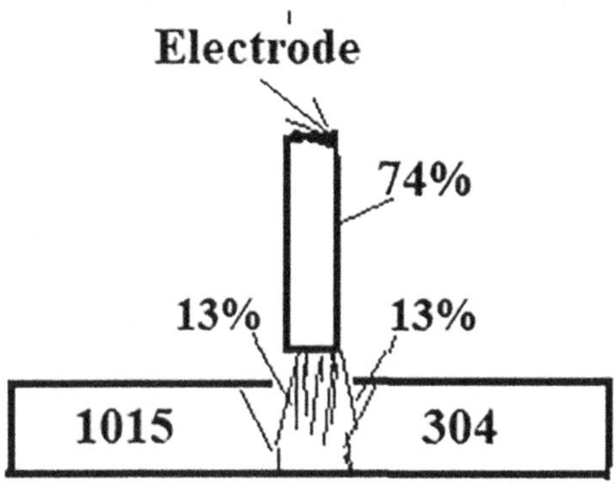

Fig. 3.13 Dilution data in welding an 1015 steel to a 304 stainless steel

Table 3.3 Determination of the weld metal composition (mass %)

Element	1015 steel	×13%	304 steel	×13%	310 electrode	×74%	Weld metal
Cr	0	0	18	2.34	26	19.24	21.58
Ni	0	0	8	1.04	21	15.54	16.58
C	0.15	0.0195	0.05	0.0065	0.12	0.0888	0.1148
Mn	0.4	0.052	2.0	0.26	1.75	1.295	1.607
S	0.2	0.026	0	0	0.4	0.296	0.322

By using the data in Table 3.2 and Fig. 3.13, the weld metal composition is calculated as shown in Table 3.3.

Thus the weld metal composition (mass %) is as follows: 21.6Cr-16.6Ni-0.115C-1.61Mn-0.32S.

Example 3.19 Assessing the Suitability of Type 310 SS Welding Electrode
Refer to the welding data in Example 3.18.

(a) Calculate the chromium equivalent and nickel equivalent for the weld metal,
(b) predict the ferrite number and solidification cracking by use of WRC-1992 diagram, and
(c) assess the suitability of the type 310 stainless steel electrode for the welding operation.

Solution

(a) By using the data for the weld metal composition in Table 3.3 and by using Eqs. 3.16 and 3.17, the Cr Equiv. and Ni Equiv. are calculated as follows:

Table 3.4 Compositions of carbon and stainless steels (mass %)

Material	% Cr	% Ni	% C	% Mn	% S
304 steel	18	8	0.05	2.0	0
1015 steel	0	0	0.15	0.4	0.2
309 electrode	23.8	14	0.04	1.9	0.6

Table 3.5 Determination of the weld metal composition (mass %)

Element	1015 steel	×13%	304 steel	×13%	309 electrode	×74%	Weld metal
Cr	0	0	18	2.34	23.8	17.61	19.95
Ni	0	0	8	1.04	14	10.36	11.4
C	0.15	0.0195	0.05	0.0065	0.04	0.03	0.056
Mn	0.4	0.052	2.0	0.26	1.9	1.41	1.72
S	0.2	0.026	0	0	0.6	0.44	0.47

$$\begin{aligned}\text{Cr Equiv.} &= \%\text{Cr} + \%\text{Mo} + 0.7(\%\text{Nb}) \\ &= 21.58 + 0 + 0 = 21.58\end{aligned}$$

$$\begin{aligned}\text{Ni Equiv.} &= \%\text{Ni} + 35(\%\text{C}) + 20(\%\text{N}) + 0.25(\%\text{Cu}) \\ &= 16.58 + (35 \times 0.1148) = 20.598\end{aligned}$$

(b) By using the WRC-1992 diagram, the ferrite number < 0 *i.e.* the weld metal is fully austenitic. Hence, the weld metal has high tendency of solidification cracking.

(c) Since the weld metal has high tendency of solidification cracking, the weldability by using the type 310 stainless steel electrode is poor.

Example 3.20 Assessing the Suitability of 309 Stainless Steel Welding Electrode

It is required to weld a 304 stainless steel to an AISI-1015 steel by arc welding. The dilution ratio is 26%, half from the 304 steel and half from the 1015 steel. Calculate the weld metal composition by using a type 309 extra-low-carbon stainless-steel electrode that contains 0.04% carbon by mass. The material composition data is given in Table 3.4. Predict the ferrite number and hence assess the suitability of the type 309 SS electrode for the welding operation.

Solution

By using the data in Table 3.4 and Fig. 3.13, the weld metal composition is calculated as shown in Table 3.5.

Thus the weld metal composition (mass %) is: 19.95Cr-11.4Ni-0.056C-1.72Mn-0.47S.

By using the data for the weld metal composition, and by using Eqs. 3.18 and 3.19,

$$\text{Cr Equiv.} = \%\text{Cr} + \%\text{Mo} + 0.7(\%\text{Nb}) = 19.95 + 0 + 0 = 19.95$$

$$\text{Ni Equiv.} = \%\text{Ni} + 35(\%\text{C}) + 20(\%\text{N}) + 0.25(\%\text{Cu}) = 11.4 + (35 \times 0.056) = 13.36$$

By using the *WRC-1992 diagram*, the ferrite number = 6, which is well within the 5–10 range. Hence, the weld metal has good ability to resist solidification cracking. It can be assessed that the weldability by using the type 309 SS electrode is good.

Questions and Problems

3.1. Draw a sketch showing various microstructural zones in a fusion weldment of mild steel, and the corresponding temperatures in the Fe-Fe_3C phase diagram

3.2. Explain the welding conditions for the following microstructural evolutions in the fusion zone: (a) coarse-grained columnar structure, (b) fine grain microstructure, and (c) CET.

3.3. Draw a sketch showing the three types of arm spacing in dendritic growth during solidification in fusion welding. What is the effect of dendritic arm spacing on the mechanical properties?

3.4. (a) What is meant by PMZ? (b) What is the effect of grain-boundary liquation in PMZ? (c) Which alloys generally develop liquation cracking in the PMZ? (d) What are the techniques to reduce liquation cracking in the PMZ?

3.5. (a) Compare and discuss the microstructures, by relating them to the strength, of various sub-zones in the HAZ. (b) Which is the weakest sub-zone in the HAZ? Explain with reference to the microstructure and its evolution. (c) What are the factors that control the size of HAZ and its constituent areas?

3.6. (a) What is meant by "weldability"?
(b) List the metals with good weldability in descending order.

3.7. (a) Compare the weldabilities of low-, medium, and high-carbon steels. (b) What problems do occur in fusion welding of high-carbon steels? And how are they overcome?

3.8. Why may inter-granular (IG) corrosion occur in welded austenitic stainless steels? Explain with the aid of a sketch.

3.9. What is meant by dilution in welding? What is the effect of high dilution? What factors tend to reduce dilution in welding?

3.10(MCQs). Select the most appropriate answers for the following statements.

(1) Which of the following term is used for the region between fusion zone and the HAZ?

(a) CGHAZ, (b) PMZ, (c) FGHAZ, (d) SCHAZ.

(2) Which of the following zones/sub-zones indicates recrystallization in welded carbon steel?

(a) CGHAZ, (b) PMZ, (c) FGHAZ, (d) SCHAZ, (e) fusion zone.

(3) Which of the following zones/subzones is the weakest sub-zone in welded carbon steels?

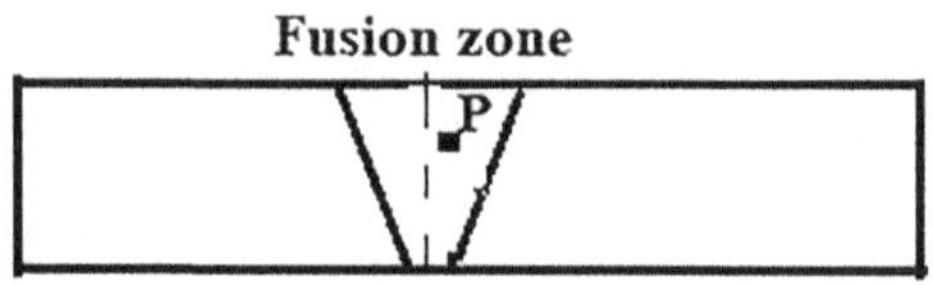

Fig. 3.14 Point P in the fusion zone of a weldment in P3.11

(a) CGHAZ, (b) PMZ, (c) FGHAZ, (d) SCHAZ.

(4) Which microstructure is evolved by a tear-drop-shaped weld pool in fusion welding?

(a) Columnar grains, (b) fine grains, (c) CET microstructure.

(5) Which of the following alloys has the best weldability?

(a) Stainless steel, (b) high-carbon steel, (c) aluminum alloy, (d) low-carbon steel.

(6) Which of the following aluminum alloys has the best weldability by using OFW, GTAW, and GMAW processes?

(a) 2xxx series *Al* alloy, (b) 5xxx series alloy, (c) 6xxx alloy, and (d) 7xxx series *Al* alloy.

(7) Which of the following processes is the best for welding heat-treatable aluminum alloys?

(a) Oxyfuel welding, (b) GTAW, (c) GMAW, (d) friction stir welding.

P-3.11. A high-purity carbon steel plate with 0.4 mass% C is being welded. Point P, in Fig. 3.14, is in the thermodynamics equilibrium. (a) Identify the temperature, the phases, and their compositions, and (b) calculate the mass fraction of the phases.

Hint: refer to Figs. 3.2 and 3.9.

P-3.12. A low-carbon steel plate with 0.18% carbon is being welded. Assume that the solidification is in thermodynamics equilibrium and a temperature field of $T = 1494 + y^2$ exists for the fusion zone. (a) Calculate the temperature at the point (2,3) in the fusion zone. (b) What phases exist at (2,3)? (c) What are the compositions of the phases? (d) What are the mass fractions of the phases? Hint: refer to Fig. 3.15.

P3.13. A deep-penetration laser beam welding experiment of an alloy is being performed using the maximum power of 3800 W. The temperature at the center of the welding pool is 1300 K, and the liquidus temperature is 700 K. The distance between the location of the center and the corresponding point at the fusion boundary is 1.7 mm. The growth rate is 2.8 mm/s. Calculate the cooling rate of the metal in the fusion zone.

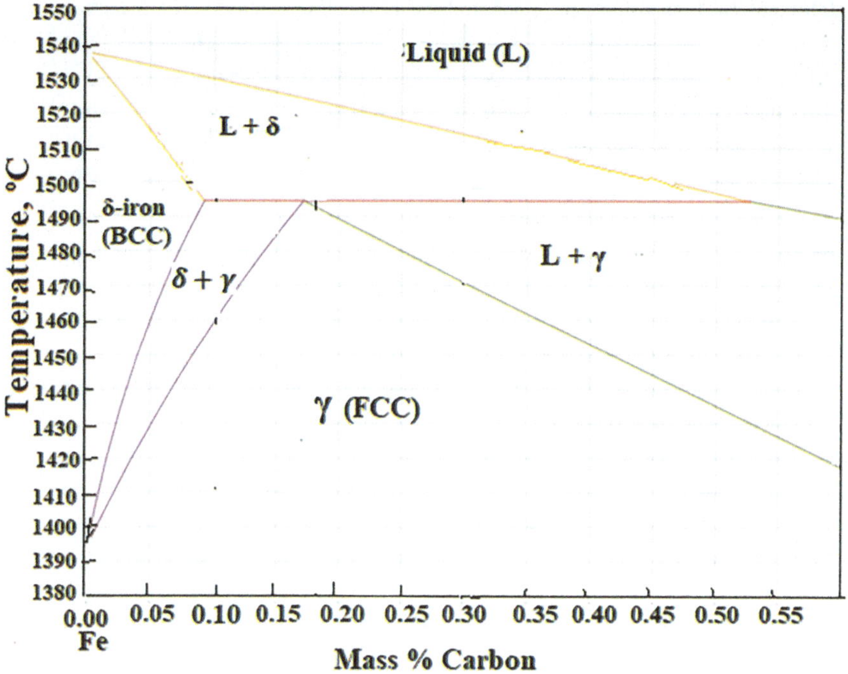

Fig. 3.15 Partial Fe-Fe_3C phase diagram

P3.14. In a solidification experiment, the temperature gradient was 265 K/mm and the growth rate was 2.5 mm/s. The constant A = 0.295 $\left(\frac{mm}{s}\right)^{0.25}\sqrt{K \bullet mm}$. Calculate the primary dendrite arm spacing (PDAS).

P3.15. Refer to the weld solidification data in P3.13. A measurement from a micrograph showed the SDAS to be 4.2 μm. Calculate the constant B in Eq. 3.5.

P3.16. The weld metal cooling rate, in a fusion welding process, is 470 K/s and the welding speed is 3.3 mm/s. The liquidus temperature with reference to point 1.8 mm from the center line is 1500 °C. Calculate the peak temperature at the center line of the fusion zone.

P3.17. A cast steel work-piece is being fusion welded by evenly heating the surface with x = 0 at the peak temperature of 1550 °C for 50 s. The diffusivity coefficient of cast steel is $0.12 \times 10^{-4}\, m^2 \bullet s^{-1}$. Calculate the: (a) length of the HAZ, (b) size of the PMZ, (c) size of the CGHAZ, (d) size of the FGHAZ, (e) size of the ICHAZ, and (f) size of the SCHAZ.

P3.18. A type 316L stainless steel has the following composition (mass %): 17Cr-12Ni-0.1N-2.2Mo. Calculate the (a) chromium equivalent, and (b) nickel equivalent.

Table 3.6 The Compositions of carbon steel and the stainless steel for P-3.19 (mass %)

Material	% Cr	% Ni	% C	% Mn	% S
304 steel	18	8	0.05	2.0	0
1017 steel	0	0	0.17	0.4	0.2
310 electrode	26	21	0.12	1.75	0.4
309 electrode	23.8	14	0.04	1.9	0.6

P3.19. It is required to weld a 304 stainless steel plate to 1017 steel plate by arc welding. The dilution ratio is 30%, half from the 304 steel and half from the 1017 steel. There are two choices of welding electrodes: the type 310 stainless-steel electrode and the type 309 electrode. Which electrode will result in better weld quality in terms of the resistance to solidification cracking? The composition data is given in Table 3.6.

References

Carslaw, H.S., Jaeger, J.C. (1980), Conduction of Heat in Solids, Oxford University Press, Oxford, UK.

David, S.A. and Vitek, J. (1993), Principles of Weld Solidification and Microstructures, *In*: International Trends in Welding Science and Technology, ASM, 1993, 147-157

David, S.A., Babu, S.S. and Vitek, J.M. (2003). Welding: Solidification and microstructure. A hyper-text-enhanced article. *JOM Journal*, 55(6), 14.

DuPont, J.N., Marder, A.R. (1996), Dilution in single pass arc welds, *Metall. Mater. Trans. B*, 27, pp. 481-489

Guo, L., Han, C., Ren, L., Yang, W., Yin, A. (2022), Effect of transient thermal conditions on columnar-to-equiaxed transition during laser welding: A Phase-Field Study, *Metals, 12*(4), 571

Bunaziv, I., Olden, V., Akselsen, O.M. (2019). Metallurgical aspects in the welding of clad pipelines—A global outlook, *Appl. Sci.* 2019, *9*(15), 3118

Kou, S. (2003), Welding Metallurgy, John Wiley & Sons, Inc., New York.

Kotecki, D.J. and Siewert, T.A. (1992), WRC-1992 Constitution Diagram for stainless steel weld metals: A modification of the WRC-1988 Diagram. *Welding Journal*, Vol. 71, 171s–178s.

Lippoid, J.C. (2014), Welding Metallurgy and Weldability, John Wiley & Sons, Inc., New York.

Modulus Metals (2025). What is Caron Equivalent? (accessed on Nov. 12, 2025) Online: https://www.modulusmetal.com/2022/12/24/what-is-carbon-equivalent/

Rao, K.S., Reddy, G.M., Rao, K.P. (2005), Studies on partially melted zone in aluminum–copper alloy welds—effect of techniques and prior thermal temper, *Materials Scie. & Engineering A*, 403(1–2), 69-76

Śloderbach, Z., Pająk, J. (2015). Determination of ranges of components of heat affected zone including changes of structure, *Archives of Metallurgy and Materials*, 60(4), 2607–2612.

TWI (2022). Aluminum Alloys, Internet Source: https://www.twi-global.com/technical-knowledge/job-knowledge/weldability-of-materials-aluminium-alloys-021 accessed: Dec 7, 2025

Part II
Fusion Welding Processes

Chapter 4
Arc Welding Processes

Abstract This chapter first reviews the physics of arc welding, and then presents both qualitative and quantitative aspects (engineering analyses) of various arc welding processes (SMAW, GMAW, GTAW, SAW, FCAW, PAW, and ESW), with particular references to their working principles and equipment, advantages and limitations, process parameters, and the like. The discussion on the SMAW process includes current, voltage, heat input, polarity selection, and the electrode selection. The GMAW process is explained taken into consideration the droplet transfer modes, process parameters, heat transfer and energy balance. The description of GTAW process involves its working principle and equipment, process parameters and electrode selection, heat input, and advantages and limitations. Similar aspects of SAW, FCAW, PAW, and ESW are also explained. This chapter contains 19 worked numerical examples, 13 diagrams, 4 tables, and 10 MCQs with their answers given at the end of this book.

Keywords Shielded metal arc welding · Gas metal arc welding · Gas tungsten arc welding · Submerged arc welding · Fluxed core arc welding · Plasma arc welding · Electro-slag welding · Droplet transfer

Abbreviations

AWS	American Welding Society
CF	Coating factor
DCEN	Direct current electrode negative
DCEP	Direct current electrode positive
ESW	Electro-slag welding
FCAW	Fluxed core arc welding
GMAW	Gas metal arc welding
GTAW	Gas tungsten arc welding
MAG	Metal active gas

Z. Huda, *Welding and Joining Engineering*,
https://doi.org/10.1007/978-3-032-20462-2_4

MIG	Metal inert gas
PAW	Plasma arc welding
SAW	Submerged arc welding
SMAW	Shielded metal arc welding

Nomenclature

A_e	Cross-sectional area of the electrode
A_f	Cross-sectional areas of the flux
A_m	Cross-sectional area of metal in the electrode
D_e	Diameter of the electrode
D_f	Diameters of the flux
f_1	Heat transfer factor
H_i	Heat input (kJ/mm)
$H_i^{pc-GTAW}$	Heat input for pulsed current GTAW (kJ/mm)
H_{net}	Net heat received by the weld joint
H_w	Actual heat available for welding to melt the joint (J)
I_m	Mean current (A)
R_H	Rate of heat energy input (or power) generated by the welding source
R_{Hw}	Rate of heat energy delivered for melting for welding operation
U_m	Unit melting heat energy of the metal (J/mm^3)
v_w	Welding speed (mm/min)
a	Anode or cathode constant of proportionality for heating
b	Constant of proportionality for electrical resistance heating and electrode resistivity
C	Circumference of the tube (mm)
C	Fraction of time for which the power is on
d	Electrode core diameter
D	Overall diameter of the electrode
I	current (A)
L	Electrode extension (stick out) (mm)
MR	Melting rate (kg/(h·mm))
N	Rotational speed of the tube (rev/min)
Q	Power input (W)
U	Arc voltage (V)
V	Volume of weld metal

4.1 Arc Welding Fundamentals

Arc welding (or electric arc welding) is a commonly practiced welding process; it involves the creation of an arc by use of an electrode. The reader has been familiarized with the physics of arc welding in Chap. 2 (Sect. 2.1), which includes the discussion on voltage drops in various zones and the power in arc welding. It has been learnt that in a direct current electrode negative (DCEN) arc welding process, electrons are released from the electrode (cathode) due to an electric field or thermo-ionic emission mechanism; these electrons are accelerated towards the anode (work-piece) because there exists a potential difference between the work and the electrode. These high velocity electrons moving from cathode toward anode collide with gaseous molecules and decompose them into charged particles, *i.e.* electrons and ions (see Chap. 2, Fig. 2.1).

There is a variety of arc welding processes, which include shielded metal arc welding (SMAW), gas metal arc welding (GMAW), gas tungsten arc welding (GTAW), submerged arc welding (SAW), fluxed core arc welding (FCAW), plasma arc welding (PAW), and electro-slag welding (ESW). These arc welding processes are discussed in the following sections.

4.2 Shielded Metal Arc Welding

4.2.1 Working Principle and Equipment

Shielded metal arc welding (*SMAW*) process involves the generation of heat by the use of a flux-coated metal electrode, which is held in an electrode-holder that is connected to a power source (see Fig. 4.1). The work cable is connected to the work-piece by use of a work clamp. On the completion of circuit, electric current passes through the electrode for welding. Thus, the arc is struck between the metal rod (flux-coated electrode) and the work-piece; both the rod and work piece surface melt to form a weld pool. In the meantime, melting of the flux coating on the rod forms gas and slag, which protects the weld pool from the surrounding atmosphere (Fig. 4.1).

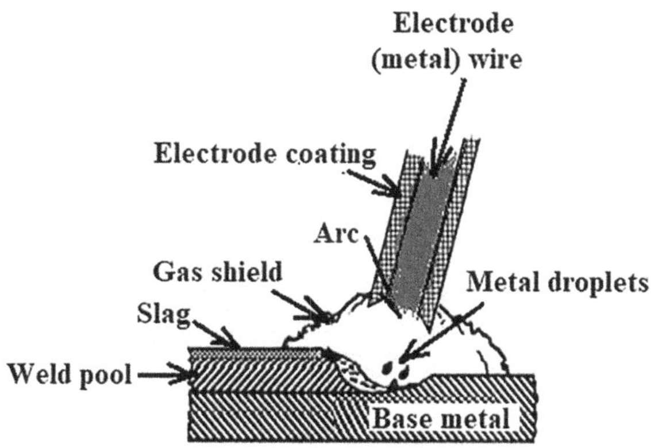

Fig. 4.1 The formation of weld pool by shielded metal arc welding

The SMAW equipment include (a) power source, (b) flux-coated electrodes, (c) electrode cable, (d) electrode holder, (e) ground cable, (f) work clamp, and (g) work-piece. Firstly, a power source and an electrode are selected; the flux-coated metal electrode is held in an electrode-holder that is connected to a power source (see Fig. 4.2). The work cable is connected to the work-piece by use of a work clamp. On the completion of circuit, electric current passes through the electrode to generate the arc for welding.

A welding power source must have sufficient amperage and duty cycle to perform the work requirement. The SMAW process requires a constant current supply from a welding power source, which may be alternating current (AC), direct current (DC), or AC/DC. Although simple AC output welding power sources are less expensive, the operating characteristics are less favorable due to amperage fluctuations. In order to ensure smooth current supply, DC welding power source is preferred. A DC power source could be either a DC generator or a rectifier; the latter converts AC output from a transformer into DC for welding.

4.2.2 Advantages and Limitations of SMAW

The SMAW equipment is simpler to set-up and use as compared to gas-shielded welding processes (*e.g.* GMAW, GTAW, etc.). The electrical circuitry and components are cheaper and less complex. The portability of the SMAW equipment enables us to perform onsite welding for erection of structural members, piping systems, and other general fabrication jobs. Another advantage is that SMAW can be done in confined places because there is no external gas supply. The SMAW process is also capable to weld both similar metals and dissimilar metals.

The major limitation of SMAW is its inability to weld aluminum and its alloys, copper & its alloys, and stainless steels; this is because aluminum and chromium are highly oxidizing elements, and the gas shield from the flux coating is ineffective to

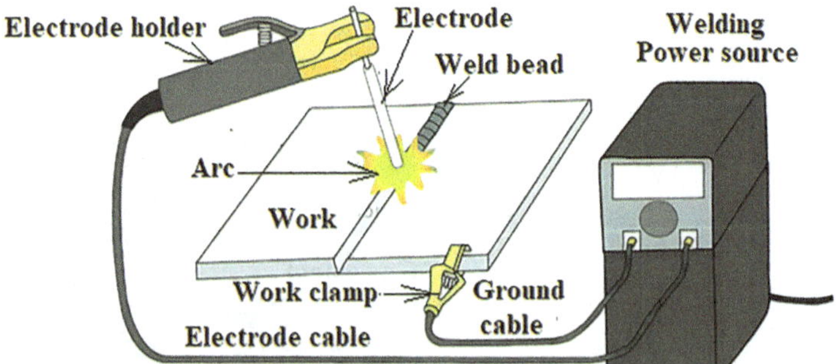

Fig. 4.2 SMAW equipment set-up

protect the loss of these elements by oxidation. Another limitation of SMAW is the weld quality issues, since this process in manually controlled.

4.2.3 SMAW Process Parameters

The welding process parameters strongly influence the weld deposition rate and its quality. The following are the principal process variables in welding: (a) welding current and speed, (b) arc voltage, (c) polarity selection, (d) electrode selection, (e) electrode diameter and feed rate, and (f) joint geometry. These process parameters are discussed in the following sections.

4.2.4 Current, Voltage, and Heat Input in SMAW

The SMAW current (amperage) varies from 70 A (for light work) to 500 A (for heavy work). The amperage can be increased or decreased by adjusting the amperage control on the front of the power source. In the case of shorting (short arc length), the amperage can be increased by using the arc control setting, which refers to the percent increase in the actual power source amperage setting. For example, if the power source is set to operate at 150 A and the arc control setting is 40, it means that 40 additional amperage is supplied at the instance of shorting. The use of arc control setting for all applications is limited to electrodes in the F-3 filler metal group (see Sect. 4.2.6). The correct welding current setting mainly depends on the electrode core diameter (see Fig. 4.3a), according to the formula:

$$I = 40d \tag{4.1}$$

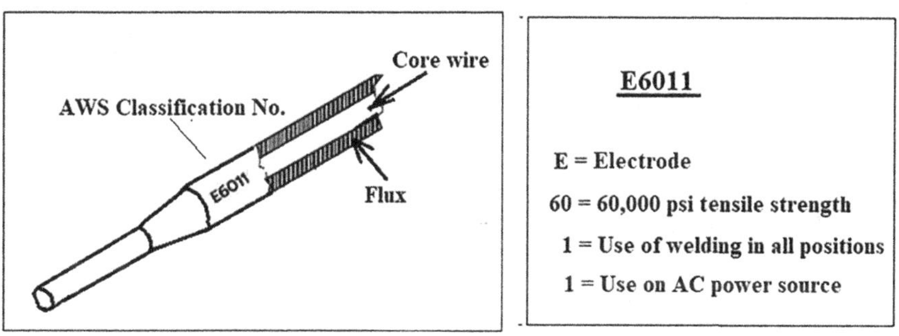

(a) (b)

Fig. 4.3 (**a**) SMAW electrode with AWS classification number; (**b**) nomenclature

where I is current (A) and d is the core diameter of mild-steel electrode (mm) (see Example 4.1).

Heat input is the amount of heat energy applied to the base metal per unit distance travelled by the electrode during arc welding. The heat input, in SMAW, depends on three factors: (a) current, (b) arc voltage, and (c) welding/travel speed. The arc voltage is normally in the range of 20–30 V. An increase in the heat input (H_i) results in an increase of the melting rate of the electrode, thereby producing a wider and deeper weld. However, excessive heat input will cause welding defects, such as sputtering. The heat input (H_i) can be numerically expressed, by modifying Eq. 2.11, as follows:

$$H_i = \frac{Power}{travel\ speed} = \frac{0.06U * I}{v_w} \tag{4.2}$$

where H_i is the heat input (kJ/mm), U is the arc voltage, I is the current (A), and v_w is the welding speed (mm/min) (see Examples 4.2 and 4.3). Equation 4.2 implies that the heat input varies directly as the arc voltage and current, and varies inversely as the welding speed.

4.2.5 Polarity Selection in SMAW

Polarity refers to the selection of electrical cable setting for welding current. There are three polarity choices for arc welding: (a) direct current electrode positive (DCEP), (b) direct current electrode negative (DCEN), and (c) alternating current (AC). In the DCEP, the electrode cable is connected to the positive terminal on the power source. With DCEP setting, the heat input on the electrode is higher resulting in deeper weld penetration which is desirable in thicker base metals (Hoffman et al. 2017). In the DCEN polarity, the electrode cable is connected to the negative terminal on the power source (see Chap. 2, Fig. 2.1); welding with DCEN results in poorer penetration; however, the globular droplet transfer (the *weld metal transfer across the arc in large droplets)* is better. The DCEN is recommended for welding thin sheet metals (see Table 4.1).

Table 4.1 The difference between DCEP and DCEN

#	DCEP	DCEN
1	The electrode cable is connected to the positive terminal on the power source	The electrode cable is connected to the negative terminal on the power source
2	Higher heat input on the electrode	Lower heat input on the electrode
3	Deeper weld penetration	Poorer weld penetration
4	Globular droplet transfer is poorer	Globular droplet transfer is better
5	Recommended for thicker work	Recommended for thin sheet metals

4.2.6 Electrode Selection in SMAW

The electrode, for SMAW, is composed of an alloy core covered by a flux coating. The core wire carries the current for welding; the current supply should be sufficiently high to melt the base metal, and coalesce with the molten base metal to form a weld bead. The SMAW electrodes are manufactured in accordance with the AWS filler metal specifications (see Fig. 4.3). In general, the low-hydrogen mild-steel electrodes with AWS classification number E7018-H4R are used; the E7018-H4R code is illustrated in Fig. 4.4. In the case of welding dissimilar metals, special non-ferrous metal electrodes are used. For example, Inconel ENiCrMo3 electrodes are used in welding copper and stainless steel (Baghel 2022).

The flux coatings on electrodes serve several purposes, which include (a) provision of shielding gas, (b) controlling the cooling rate of the weld pool, (c) influencing weld bead shape, (d) initiation and stabilizing the arc, and (e) addition of alloying elements to improve weld mechanical properties. The AWS has established the AWS-A5.1 filler metal grouping for low-carbon steel electrodes for SMAW, which is presented in Table 4.2. The flux covering thickness depends on the electrode classification number; it is usually expressed as *coating factor* by:

$$\mathrm{CF} = \frac{D}{d} \tag{4.3}$$

where *CF* is the coating factor; *D* is the overall diameter of the electrode, and *d* is the electrode core diameter (see Example 4.4).

It is mentioned in Sect. 4.2 that both the metal rod (in the electrode) and the work-piece surface melt to form a weld pool during SMAW. The electrode melting rate can be determined by (Khan 2007):

$$\mathrm{MR} = a * I + b * L * I^2 \tag{4.4}$$

where *MR* is the melting rate, kg/(h•mm); *a* is the anode or cathode constant of proportionality for heating, which depends on the polarity and composition in case of DCEP and on the emissivity of cathode in case of DCEN; *b* is the constant of

Fig. 4.4 Nomenclature of AWS E7018-H4R electrode

E7018-H4R

E = Electrode

70 = 70,000 psi tensile strength

1 = All welding positions

8 = Low-hydrogen steel, iron powder, DC or AC

H4 = Hydrogen < 4 ml/100 g of deposited weld metal

R = Meets requirements of absorbed moisture test

Table 4.2 AWS-A5.1 Filler metal grouping for SMAW (AWS 2004)

Grouping	Electrodes	Coating type	Characteristics
F-1	E7024, E7027, E7028, E6020, E6022, E6027	High iron-oxide powder	High deposition rate, shallow-to-medium penetration
F-2	E6012, E6013, E6019, E7014	High titania	Shallow penetration
F-3	E6010, E6011	Cellulose	Deep penetration
F-4	E7015, E7016, E7018, E7048	Low hydrogen	Low hydrogen, medium penetration

Table 4.3 The proportionality constants' values for 1.6-mm-dimaeter wire electrodes

Wire metal	*a*, kg/(h•A)	*b*, kg/(h•A^2•mm)
Aluminum (DCEP)	5.4×10^{-3}	4.4×10^{-6}
Mild steel (DCEP)	8.6×10^{-3}	2.5×10^{-5}
Mild steel (DCEN)	1.8×10^{-2}	2.5×10^{-5}

proportionality for electrical resistance heating and electrode resistivity; and L is the electrode extension (stick out), mm (see Table 4.3). The significance of Eq. 4.4 is illustrated in Example 4.5.

4.3 Gas Metal Arc Welding (GMAW)

4.3.1 Working Principle and Equipment

Gas metal arc welding (GMAW) is a semi-automatic arc welding process that uses a wire (consumable) electrode fed through a welding gun with the admission of inert/shielding gas. In GMAW, the heat for melting the metal is generated by an electric arc between the wire electrode and the work-piece (see Fig. 4.5). There are two variations of GMAW: (a) metal inert gas (MIG) welding and (b) metal active gas (MAG) welding; the former type involve the use of inert gas (*e.g.* Ar, He), whereas the latter variation uses active gases (*e.g.* CO_2, Ar + 2–5% O_2, Ar + 5–25% CO_2 and Ar + CO_2 + O_2) (TWI 2025). In order to perform GMAW operation, the following five basic equipment components are required: (1) a welding gun, (2) a wire feed unit, (3) a welding power supply, (4) a wire electrode wire, and (5) an inert/shielding gas supply.

4.3.2 Droplet Transfer Modes in GMAW

There are four metal-droplet transfer modes in GMAW: (a) short circuiting (or dip) transfer, (b) globular transfer, (c) spray transfer, and (d) pulsed transfer (see Fig. 4.6). *Short circuiting GMAW* (or GMAW-S) is a low-energy process, which generates its heat from an electric arc but the weld metal is transferred only partially across an arc

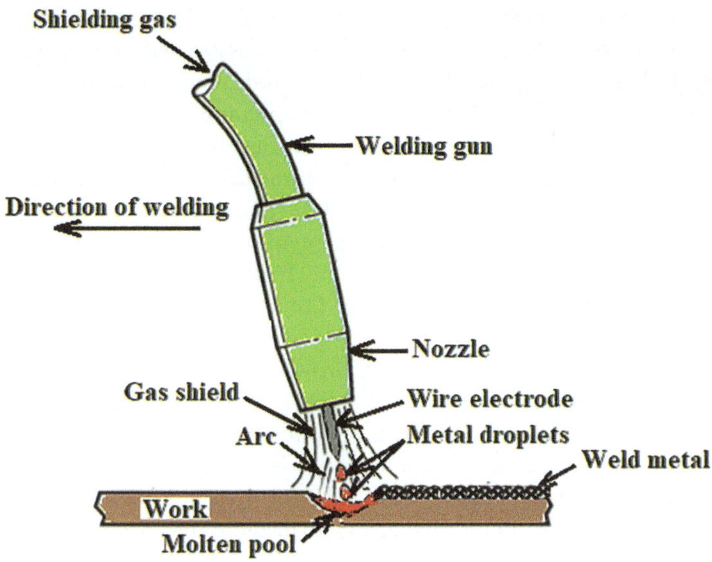

Fig. 4.5 Working principle of GMAW

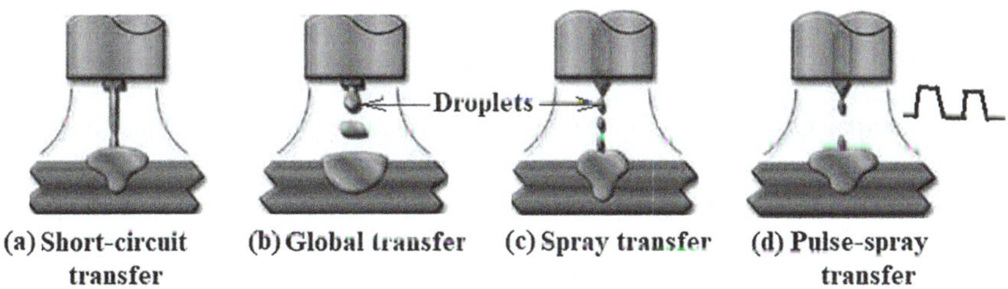

Fig. 4.6 The four droplet transfer modes in GMAW

and partially when the filler metal touches the base metal and the arc is short circuited (see Fig 4.6a). The short circuiting occurs 20–200 times per second which results in a small molten weld pool. When the arc is short circuited, the molten weld pool is able to freeze more quickly than when welding with a continuous arc. This quick-freezing gives GMAW-S a unique ability to weld out-of-position, to weld thin base metals, and to weld open butt-root passes without backing and without "blowing through" (Houle 1985). However, welder's skill is required to avoid lack of fusion in GMAW-S.

In GMAW with short-circuiting transfer (GMAW-S), the power input (or rate of heat energy input) can be determined by:

$$Q = C * U * I \tag{4.5}$$

where Q is the power input (W), C is the fraction of time for which the power is on, *U* is the arc voltage, and I is the current (A) (see Example 4.6).

Globular transfer GMAW occurs as the current and voltage increase beyond those for GMAW-S. The globular transfer mode is associated with the use of either 100% CO_2 shielding or Ar/CO_2 blends. The electrode wire forms larger droplets that are transferred to the work-piece without short-circuiting (see Fig. 4.6b). Globular-transfer GMAW has very good penetration characteristics. However, the main drawbacks of globular transfer are its high spatter formation together with irregular bead shapes and formation of numerous ceramic/slag islands.

Spray-transfer GMAW is a high heat-energy process that involves a stable, continuous stream of small droplets (Fig. 4.6c). The droplets are propelled from the electrode to the work-piece in a spray-like fashion due to the high current and the use of a suitable shielding gas.

Pulsed-spray transfer GMAW (or *GMAW-P*) combines the benefits of spray transfer and short-circuiting transfer. It involves alternating between high and low currents (see Fig. 4.6d). During the high-current phase, small droplets are transferred in a spray-like manner, similar to spray transfer. The low-current phase allows for cooling and solidification of the weld pool, reducing the overall heat input energy. GMAW-P is useful for welding thin to medium-thickness materials, providing good control and reduced distortion.

4.3.3 Process Parameters in GMAW

The principal GMAW process parameters include electrode feed rate, electrode diameter & its type, the arc voltage, the current, and the welding speed. With semi-automatic or automatic GMAW processes, the arc voltage is selected on the power supply and it determines the arc length. GMAW arc voltages are typically in the range of 15–40 V, depending on the shape of the weld bead, and the type of metal transfer mode. The typical current values are in the range of 40–350 A. The electrode feed rate depends on the electrode diameter and current (see Table 4.4).

4.3.4 Heat Transfer and Energy Balance in GMAW

The quantity of heat energy required to melt a given volume of metal, in GMAW/fusion welding, mainly depends on: (a) the heat to raise the temperature of the solid metal to its melting point, and (b) the heat to transform the metal from solid to liquid

Table 4.4 The electrode feed rate for various electrode diameters and current in GMAW

#	Electrode diameter (mm)	Current (A)	Electrode feed rate (m/min)
1.	0.6	40–100	2–5
2.	0.8	60–100	3–6
3.	1.0	100–270	4–12
4.	1.2	120–350	4–18

phase at the melting point. There are two heat transfer mechanisms from the heat source to the work. The first mechanism involves how effectively heat is transferred from the electrode tip (heat source) to the work surface (joint); this mechanism is expressed by the heat transfer factor f_1, as follows:

$$f_1 = \frac{\textit{Net heat received by the joint}}{\textit{Total heat generated at the electrode tip}} = \frac{H_{net}}{H_i} \tag{4.6}$$

$$\text{or} \quad H_{net} = f_1^* H_i \tag{4.7}$$

where H_i is the heat input, and H_{net} is the net heat received by the joint (work surface).

The second heat-transfer mechanism considers the conduction of heat away from the weld joint to be dissipated throughout the work metal, so that only a portion of the heat transferred to the surface is available for melting. This melting factor f_2 is the proportion of heat received at the joint available for melting. Thus, the melting factor f_2 can be expressed as:

$$f_2 = \frac{\textit{Actual heat required to melt the joint}}{\textit{Net heat received by the joint}} = \frac{H_w}{H_{net}} \tag{4.8}$$

$$\text{or} \quad H_w = f_2^* H_{net} \tag{4.9}$$

By the combination of Eqs. 4.7 and 4.9, we obtain:

$$H_w = f_1 * f_2 * H_i \tag{4.10}$$

where H_w is the actual heat available for welding to melt the joint (J), f_1 is the heat transfer factor, f_2 is the melting factor, and H_i is the heat input *i.e.*, the total heat generated at the heat source (electrode tip). The values of factors f_1 and f_2 are in the range of 0–1. Equation 4.9 expresses the actual effective heat input for welding.

In order to strike a heat-energy balance between the actual effective heat input and the heat for welding, the actual heat available for welding to melt the joint (H_w) can be written as:

$$H_w = U_m * \mathrm{V} \tag{4.11}$$

where U_m is the unit melting heat energy of the metal (J/mm^3) and V is the volume of weld metal. In the welding of a continuous bead, the volume rate of metal welded is the product of weld area A_w and travel velocity v_w. Accordingly, the rate of heat energy delivered for melting for welding operation (R_{Hw}) can be expressed, by modifying Eq. 4.11, as follows:

$$R_{Hw} = U_m * A_w * v_w \tag{4.12}$$

By the combination of Eqs. 4.10 and 4.12, we obtain:

$$R_{Hw} = U_m * A_w * v_w = f_1 * f_2 * R_H \tag{4.13}$$

where R_H is the rate of heat energy input (or power) generated by the welding source (electrode tip). The significance of Eqs. 4.6–4.13 is illustrated in Examples 4.7–4.12.

4.3.5 Advantages and Limitations of GMAW

The following are the principal advantages of GMAW. It can be operated semi-automatically and fully automatically, including robotics. The metal transfer rate is higher, which allows for the fast production of high quality welds. Since no/little flux is used, there is no chance of slag being trapped in the weld metal. GMAW is a versatile process that can be used to join a variety of metals and alloys. Metal active gas (MAG) welding can be performed in all positions, making it one of the most widely-used welding processes.

There are some limitations of GMAW. Welding in vertical and overhead positions requires short-circuiting transfer (GMAW-S). In order to hold the molten weld-pool in position, a fast-freezing flux is required. Unlike SMAW, GMAW cannot be performed outdoors without enclosures as the welding gas needs to be protected from the wind. Another issue is the excessive spatter during high current welding (spray-transfer GMAW). Since limited deoxidants are available in this process, it is necessary to remove all rust from the work-piece before the commencement of welding.

4.4 Gas Tungsten Arc Welding (GTAW)

4.4.1 Working Principle and Equipment

Gas Tungsten Arc Welding (*GTAW*) (also called tungsten inert gas welding or *TIG* welding) involves the use of a non-consumable tungsten electrode to create the arc; an inert gas is supplied to shield the weld (molten) metal (see Fig. 4.7). There is better precise controls of heat input and weld pool, since the arc and filler wire are independent of each other. Unlike GMAW, there is no spatter in GTAW because no molten filler metal passes through the arc. Owing to the use of an inert shielding gas, GTAW is capable of producing the highest quality welds of all the arc welding processes for joining stainless steels, aluminum, and other non-ferrous metals.

The GTAW equipment set-up mainly consists of a constant-current power supply and a torch; the latter may sometimes uses water cooling (in the case of high current, I >200 A, applications). Some GTAW machines are equipped with a foot pedal

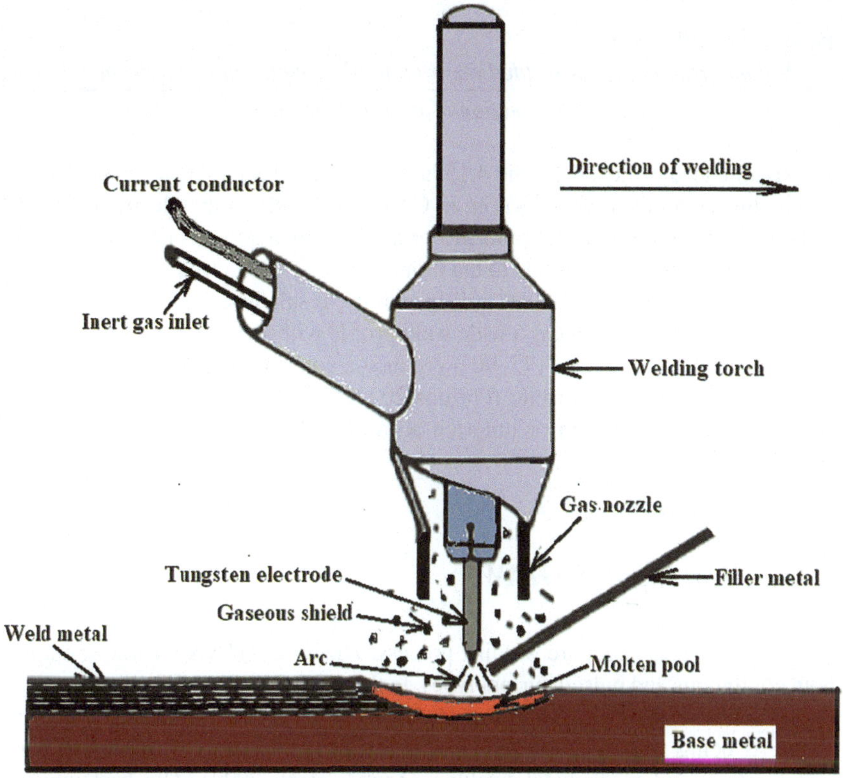

Fig. 4.7 GTAW—principle of operation

for a better welder control by providing very precise adjustments of current as the weld is being made.

4.4.2 Process Parameters and Electrode Selection

The GTAW process parameters are almost similar to GMAW, with exceptions of electrode selection and the use of filler wire (metal). The weld metal (bead), in GTAW, comprises of a mixture of filler metal and base metal The volume rate of metal welded, in GTAW/arc welding, can be calculated by:

$$\text{Volume rate of metal welded} = A_w * v_w \tag{4.14}$$

where A_w is the cross-sectional area of the weld metal (bead) (mm^2), and v_w is the welding speed (or travel speed) (mm/s). The feed rate of filler wire (mm/s), in GTAW, can be determined by:

$$\text{Feed rate of filler wire} = \frac{\textit{Volume rate of metal welded} \times \%\textit{volume filler metal in the weld metal}}{\textit{cross}-\textit{sectional area of filler wire}} \quad (4.15)$$

The significance of Eqs. 4.14 and 4.15 is illustrated in Examples 4.13 and 4.14.

The tungsten electrode selection, in GTAW, is based on the electrode tip angle and the composition. In order to form a properly shaped arc, it is necessary to grind the tip of the tungsten electrode to the proper angle. The electrodes with smaller tip angles (in the range of 15–30°) will, generally, result in a wider arc that is less dense, which in turn produces a wide weld profile with shallow penetration. On the other hand, blunt tip angles (75–90°) produce a concentrated arc with narrow and deeper penetrating weld profile (Phillips 2023). The GTAW electrodes consist of either 100% pure tungsten or tungsten alloyed with up to 5% alloying additions (AWS 2004).

4.4.3 Heat Input in GTAW

GTAW is low-heat-input arc welding process. The novel GTAW technique employs both continuous and pulsed current types. In the *pulsed-current GTAW*, the peak and background current fluctuate with a predetermined frequency. The *pulsing-current GTAW* is being used by industries to join different materials to provide superior metallurgical and mechanical qualities (Kumar et al. 2024). The heat input for pulsed current GTAW can be determined by (Srinivasan et al. 2024):

$$H_i^{pc-GTAW} = f_1 * \left(\frac{0.06 * U * I_m}{v_w} \right) \quad (4.16)$$

where $H_i^{pc-GTAW}$ is the heat input for pulsed current GTAW (kJ/mm), f_1 is the heat-transfer factor, U is the arc voltage, I_m is the mean current (A), and v_w is the welding speed (or travel speed) (mm/min). The mean current is calculated by:

$$I_m = \frac{(I_p * t_p) + (I_b * t_b)}{t_p + t_b} \quad (4.17)$$

where I_p is the pulsed welding current, I_b is the background current, t_p is the pulse-current time (%), and t_b is the background-current time (%). The significance of Eqs. 4.16 and 4.17 is illustrated in Example 4.15.

4.4.4 Advantages and Limitations of GTAW

The GTAW Process offers the following advantages. It produces the highest quality welds of all arc welding processes. The reasons of the high-quality welds include the precise heat and puddle control, the absence of spatter, and the inert gas shielding. For the same reasons, GTAW is an ideal arc welding process for welding stainless steel, aluminum, and other non-ferrous metals in all positions. Additionally, a minimal weld cleaning is required and there is no possibility of slag defect, because no flux is used in GTAW. This process is capable to weld complex geometries and thin sheets with excellent root passes.

The following are the limitations of GTAW. The weld metal deposition rate is lower as compared to GMAW. This process requires very high welder's skill. The operating cost is higher due to the additional arc-starting cost. There is also a possibility of tungsten inclusions in welds.

4.5 Submerged Arc Welding

4.5.1 Principle of Operation

Submerged arc welding (SAW) is an arc welding process that involves the use of a continuously-fed consumable electrode and a blanket of granular flux, which surrounds and covers the arc (see Fig. 4.8). The blanket of flux, when molten, provides electrical conduction between the metal to be joined and the electrode; it also generates a protective gas shield and a slag, all of which protects the weld zone. A 3/32-in.-dia. to a 5/32-in.-dia. electrode-wire will run in the 300–900 A range using direct-current electrode positive (DCEP) polarity.

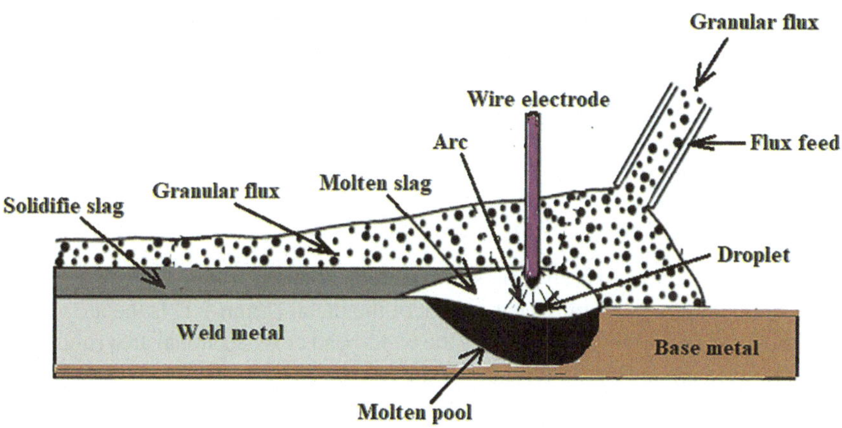

Fig. 4.8 Submerged arc welding—*principle of operation*

4.5.2 Advantages and Limitations of SAW

Since the arc is "submerged" beneath a blanket of flux in the SAW process, it is not usually visible during the welding operation itself; this fact is advantageous from health and safety viewpoint. The SAW process is typically suited to the longitudinal and circumferential butt welds required in the manufacture of pressure vessels, welded I-beams, and for joining plating and stiffeners in shipyards. Low-carbon, low-alloy, and stainless steels can be readily welded by SAW. A limitation of SAW is that the welding is positionally restricted as it is normally carried out in flat or horizontal positions; this restriction is owing to the highly fluid weld pool, the molten slag, and the need to maintain a flux covering over the arc. Another disadvantage of SAW is the difficulty in welding high-carbon steels, tool steels, and nonferrous metals.

4.5.3 Engineering Analysis for SAW

In the manufacture of pressure vessels using the SAW process, a continuous weld is generally made around the circumference of a round steel tube of a large diameter under automatic control.

The tube is slowly rotated under a stationary welding head. The rotational speed of the tube (N) can be related to the linear speed (v) by:

$$\mathrm{v} = \mathrm{R} * \omega = 2\pi * \mathrm{R} * \mathrm{N} = \pi * \mathrm{D} * \mathrm{N} = \mathrm{C} * \mathrm{N} \tag{4.18}$$

$$\text{or} \quad \mathrm{N} = \frac{v_w}{\pi * D} \tag{4.19}$$

where N is the rotational speed (rev/min), v_w is the travel speed or welding velocity (mm/min), D is the diameter (mm), and C is the circumference of the tube. The travel speed (v_w) can be determined by modifying Eq. 4.13, as follows:

$$U_m * A_w * v_w = f_1 * f_2 * R_H = f_1 * f_2 * \mathrm{U} * \mathrm{I} \tag{4.20}$$

$$v_w = \frac{f_1 * f_2 * \mathrm{U} * \mathrm{I}}{U_m * A_w} \tag{4.21}$$

where v_w is the travel speed (mm/s), f_1 is the heat transfer factor, f_2 is the melting factor, U_m is the unit melting heat energy of the metal (J/mm^3), U is the arc voltage, I is the welding current (A), and A_w is the weld-bead cross-sectional area (mm^2). The time to weld around circumference can be calculated by:

$$\text{Time to weld around circumference} = \frac{C}{v_w} = \frac{\pi * D}{v_w} \quad (4.22)$$

The significance of Eqs. 4.18–4.22 is illustrated in Examples 4.16 and 4.17.

4.6 Flux Cored Arc Welding

Flux-cored arc welding (FCAW) is an arc-welding process that involves the use of a tubular consumable electrode with a flux core inside acting as a shielding agent while the metal (from the electrode) is deposited in the weld zone (see Fig. 4.9). The flux-core electrode is continuously fed during the welding operation. There are two versions of FCAW: (a) FCAW-S (self-shielded), and (b) FCAW-G (gas/dual-shielded). The self-shielded FCAW solely relies on gaseous protection from the flux core and its slag deposits; it is mostly used for outdoor projects that have unpredictable conditions. The dual/shield FCAW uses flux-cored wires along with external shielding gas to provide additional protection for the weld area. This process allows higher deposition rates and offers better penetration against thicker metals. FCAW can be performed semi-automatically wherein a welding professional controls the welding torch, individually adjusting the parameters. It can also be performed fully automatically by implementing robots.

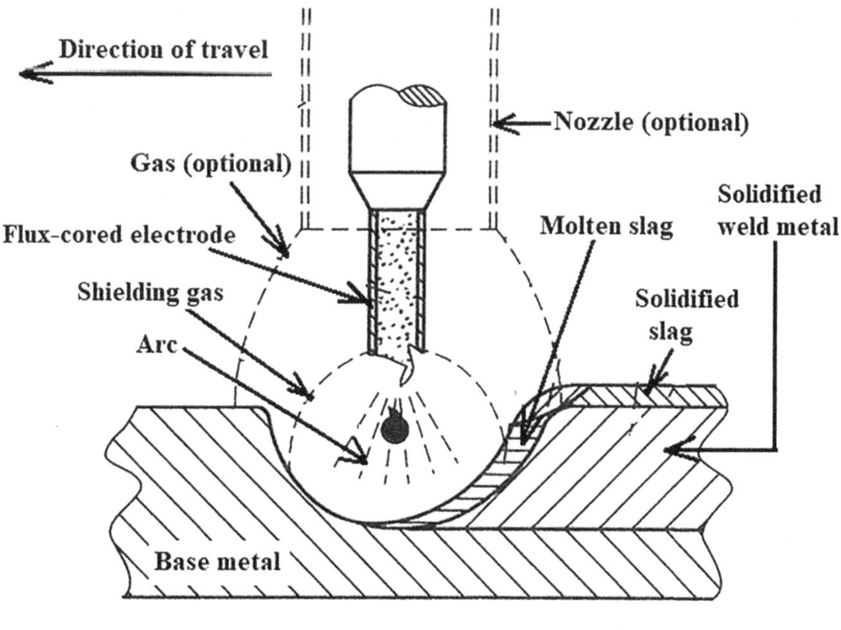

Fig. 4.9 Flux-cored arc welding

The electrode, in FCAW, contains flux (compounds that do not become part of the weld bead) (see Fig. 4.9). The cross-sectional area of the metal in the electrode can be calculated by:

$$\text{Area of the metal in FCAW electrode} = A_m = A_e — A_f = \frac{\pi}{4} * \left(D_e^2 - D_f^2\right) \tag{4.23}$$

where A_e and A_f are the cross-sectional areas of the electrode and flux, respectively, and D_e and D_f are the diameters of the electrode and flux, respectively. The feed rate of the electrode, in FCAW, can be determined by modifying Eq. 4.15, as follows:

$$\text{Feed rate of electrode} = \frac{Volume\ rate\ of\ metal\ welded \times \%electrode\ metal\ in\ the\ weld\ bead}{Cross-sectional\ area\ of\ the\ metal\ in\ the\ electrode} \tag{4.24}$$

The significance of Eqs. 4.23 and 4.24 is illustrated in Examples 4.18 and 4.19.

4.7 Plasma Arc Welding

Plasma Arc Welding (PAW) is a special form of GTAW process that employs a high-temperature, constricted plasma gas column directed at the weld area. Plasma is a gas that has been superheated to a point where it becomes highly conductive. The normal combination of gases, in PAW, is argon for the plasma gas, with argon plus 2–5% hydrogen for the shielding gas. A tungsten electrode is contained in a specially designed nozzle that focuses a high-velocity stream of the inert gas into the region of the arc to form a high-velocity extremely-hot plasma arc stream (see Fig. 4.10). Temperatures in PAW may exceed 17,000 °C (30,000 °F), which are high enough to melt any known metal.

4.8 Electroslag Welding

Electroslag welding (ESW) involves the use of a flux-coated filler metal and the heat generated between two materials from an arc and electrical resistance for joining thick materials in a vertical or near-vertical position. The heat generated by the arc initially melts the filler metal (to create a conductive molten slag) as well as the surfaces of the metals to be joined. The arc is then extinguished; the molten sag provides heat to melt the filler metal and base metal, as well as protection from

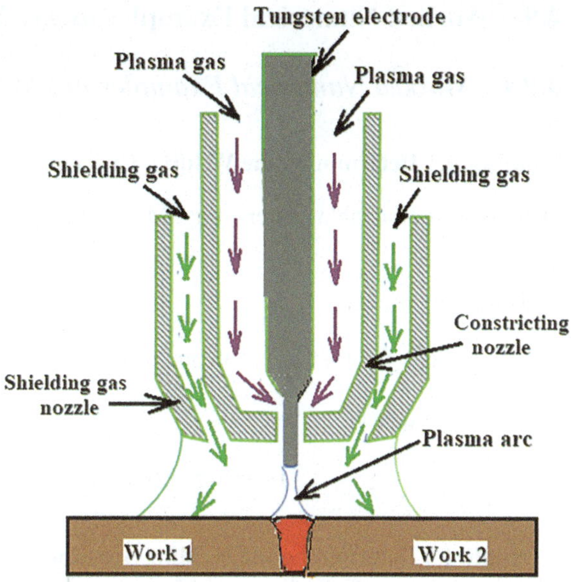

Fig. 4.10 Plasma arc welding

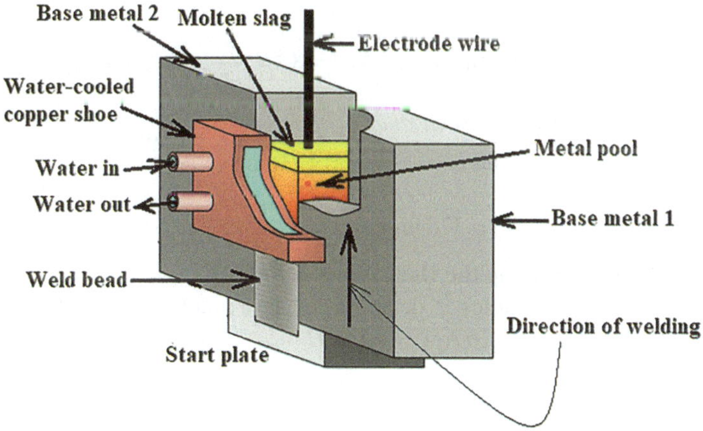

Fig. 4.11 Electroslag welding

atmospheric contamination as it travels along the weld joint. The ESW is used to butt-weld together thick sections (thickness >20 mm). The sections to be joined are fixed in the vertical plane. Water-cooled copper shoes are pressed tightly against each side of the joint (at the bottom position of the start of welding) to prevent leakage of weld metal and slag (see Fig. 4.11).

4.9 Worked Numerical Examples in Arc Welding

4.9.1 Worked Numerical Examples in SMAW

Example 4.1 Determining the Welding Current in SMAW

Determine the suitable current for SMAW using a $\frac{3}{32}$ in. diameter mild-steel electrode.

Solution

$d = \frac{3}{32}$ in. = 0.09375 in = 2.38 mm, I =?

By using Eq. 4.1,

$$I = 40d = 40 \times 2.38 = 95.25\,A$$

Example 4.2 Determining the Welding Speed in SMAW

The electrode travelled a distance of 50 mm in 8 s during an SMAW operation. The welding current and arc voltages were 120 A and 25 V, respectively. Determine the welding speed.

Solution

Distance travelled = 50 mm, time taken = 8 s = $\frac{8}{60}$ = 0.133 min, v_w =?

$$v_w = \frac{Distance\,travelled}{time\,taken} = \frac{50mm}{0.133min} = 376mm/min$$

The welding speed = v_w = 376 mm/min.

Example 4.3 Determining the Heat Input in SMAW

By using the data in Example 4.2, determine the heat input for the welding operation.

v_w = 376 mm/min, I = 120 A, U = 25 V, H_i =?

By using Eq. 4.2,

$$H_i = \frac{0.06U * I}{v_w} = \frac{0.06 \times 25 \times 120}{376} = 0.478\,kJ/mm$$

The heat energy input = 0.478 kJ/mm = 478 J/mm.

Example 4.4 Determining the Flux Coating Thickness in SMAW

A coating factor of 1.4 is recommended for an electrode with a core diameter of 1.6 mm. Determine the flux covering thickness.

Solution

CF = 1.4, d = 1.6 mm, coating thickness = t =?

By using the modified form of Eq. 4.3,

$$D = (CF) * d = 1.4 \times 1.6 = 2.24\text{mm}$$

$$d + 2\,t = D$$

$$t = \frac{D - d}{2} = \frac{2.24 - 1.60}{2} = 0.32\text{mm}$$

The flux coating thickness = 0.32 mm.

Example 4.5 Determining the Electrode Melting Rate in SMAW

A welding current of 120 A is passed through a 1.6-mm-diameter mild-steels electrode with DECP during an SMAW operation. The electrode extension is 15 mm. Determine the electrode melting rate in the welding operation. Hint: refer to Table 4.3.

Solution

I = 120 A, $a = 8.6 \times 10^{-3}$ kg/(h•A), $b = 2.5 \times 10^{-5}$ kg/(h•A^2•mm), L = 15 mm, MR =?

By using Eq. 4.3,

$$\text{MR} = a * I + b * L * I^2 = \left(120 \times 8.6 \times 10^{-3}\right) + \left(2.5 \times 10^{-5} \times 15 \times 120^2\right)$$

$$MR = 1.032 + 5.400 = 6.43\ \text{kg} / \left(\text{h} \cdot \text{mm}\right)$$

The electrode melting rate = 6.43 kg/(h •mm).

4.9.2 *Worked Numerical Examples in GMAW*

Example 4.6 Determine the Rate of Heat Energy Input for GMAW with Short-Circuit Transfer

A GMAW process with short-circuiting transfer was performed using an arc voltage of 25 V and a current of 300 A. The arc is on for 80% of the total given time. Determine the rate of heat energy input.

Solution

U = 25 V, I = 300 A, C = 80% = 0.80, Q =?

By using Eq. 4.5,

$$Q = C * U * I = 0.8 \times 25 \times 300 = 6000\ \text{W}$$

The rate of heat energy input = 6000 J/s = 6000 W.

Example 4.7 Determining the Power Delivered for Melting for the Welding Operation

The power source in a GMAW setup generates 3200 W that can be transferred to the work surface with a heat transfer factor of 0.6. The metal to be welded is low-carbon steel; the unit melting heat energy for the steel is 10 J/mm^3. The melting factor in the welding operation is 0.48. A continuous fillet weld is to be made with a cross-sectional area of 22 mm². Determine the power or the rate of heat energy delivered for melting for the welding operation.

Solution

$R_H = 3200$ W, $U_m = 10$ J/mm³, $f_1 = 0.6$, $f_2 = 0.48$, $A_w = 22$ mm², $R_{Hw} =?$

By using Eq. 4.13,

$$R_{Hw} = f_1 * f_2 * R_H = 0.6 \times 0.48 \times 3200 = 921.6\,\text{J/s}$$

The rate of heat energy delivered for melting for the welding operation = R_{Hw}= 921.6 J/s.

Example 4.8 Determining the Welding Speed in GMAW

By using the data in Example 4.7, determine the welding speed for the operation.

Solution

R_{Hw}= 921.6 J/s, $U_m = 10$ J/mm³, $A_w = 22$ mm², $v_w =?$

By using Eq. 4.13,

$$R_{Hw} = U_m * A_w * v_w$$

$$v_w = \frac{R_{Hw}}{U_m * A_w} = \frac{921.6}{10 \times 22} = 4.19\,\text{mm/s}$$

The welding speed (or travel velocity) = $v_w = 4.19$ mm/s.

Example 4.9 Determining the Power Required to be Generated by the Welding Source

It is required to fillet weld two medium-carbon steel plates by GMAW, as shown in Fig. 4.12. The unit melting heat energy for the steel is 9.6 J/mm³. The travel velocity of the welding head is 7 mm/s. The heat transfer factor is 0.75, and the melting factor is 0.60. Determine the power required to be generated by the welding source (electrode tip).

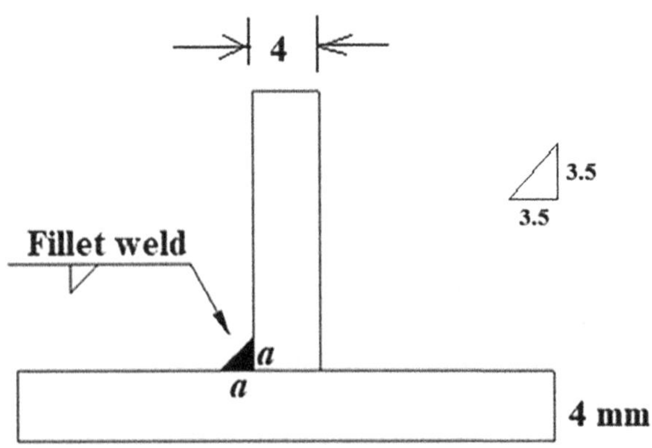

Fig. 4.12 Fillet weld design for Example 4.9

Solution

Weld metal (fillet) area = A_w = ½ (base)∗(height) = ½ × 3.5 × 3.5 = 6.125 mm^2.
$U_m = 9.6$ J/mm^3, $f_1 = 0.75$, $f_2 = 0.60$, $A_w = 6.125$ mm^2, $v_w = 7$ mm/s, R_H =?
By using the modified form of Eq. 4.13,

$$R_H = \frac{U_m * A_w * v_w}{f_1 * f_2} = \frac{9.6 \times 6.125 \times 7}{0.75 \times 0.60} = 914.66 \text{ J/s} \cong 915 \text{ W}$$

Example 4.10 Determining the Cross-sectional Area and Volume of the Weld with U-groove

Two titanium plates are to be butt-welded using a U groove by GMAW, as shown in Fig. 4.13. The length of the weld is 180 mm. The heat transfer factor is 0.80, and the melting factor is 0.55. The unit melting heat energy for titanium is 14.3 J/mm^3. Determine the cross-sectional area and volume of the weld (including the weld metal and base metal).

Solution

With reference to Fig. 4.13, the final cross-sectional area of the weld can be approximated by a semicircle with a radius of 5 mm. Accordingly,

$$\text{Cross-sectional area of weld} = A_w = \tfrac{1}{2}(\text{area of circle})$$
$$= \tfrac{1}{2}(\pi r^2) = \tfrac{1}{2}(\pi \times 5^2) = 39.27\text{mm}^2$$
$$\text{Volume of weld} = \text{V} = A_w * \text{L} = 39.27 \times 180 = 7{,}069 \text{ mm}^3$$

Example 4.11 Determining the Actual Heat Available for Welding to Melt the Joint

By using the data in Example 4.10, determine the actual heat available for welding to melt the Joint (including weld metal and base).

Solution

$U_m = 14.3$ J/mm^3, $f_1 = 0.80$, $f_2 = 0.55$, $A_w = 39.27$ mm^2, V = 7069 mm^3.
By using Eq. 4.11,

$$H_w = U_m * \text{V} = 14.3 \times 7{,}069 = 101081 \text{ J} = 101.08 \text{ kJ}$$

The actual heat available for welding to melt the joint = H_w = 101.08 kJ.

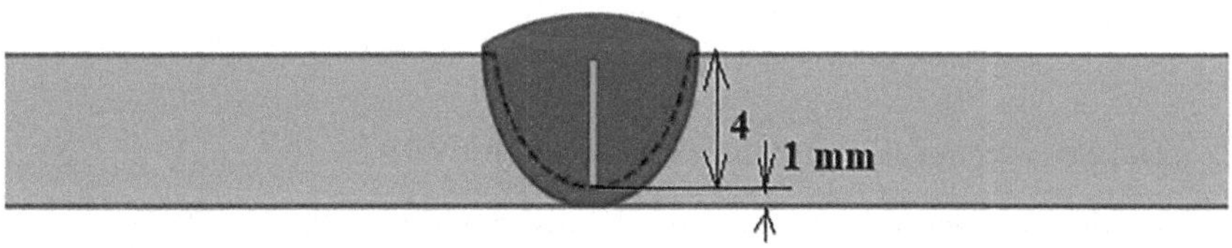

Fig. 4.13 Butt weld with U-groove for Example 4.10

Example 4.12 Determining the Heat Input Generated at the Heat Source in GMAW
By using the data in Examples 4.10 and 4.11, determine the heat input *i.e.*, the total heat generated at the heat source (electrode tip).

Solution

$H_w = 101.08$ kJ, $f_1 = 0.80, f_2 = 0.55$, $H_i = ?$

$$H_w = f_1 * f_2 * H_i \quad (\text{Eq.4.9})$$

$$101.08 = 0.80 \times 0.55 \times H_i$$

$$H_i = \frac{101.08}{0.80 \times 0.55} = 229.72 \text{ kJ}$$

The heat input (or the total heat generated at the heat source) = 229.72 kJ.

4.9.3 Worked Numerical Examples in GTAW, SAW, and FCAW

Example 4.13 Determining the Cross-sectional Area of Weld Bead in GTAW
A GTAW operation is performed on low-carbon steel with unit melting energy of 10.25 J/mm^3. The arc voltage is 18 V and the welding current is 140 A. The heat transfer factor and the melting factor are 0.72 and 0.68, respectively. The diameter of the filler metal is 3.7 mm. The final weld metal comprises of 65% volume of filler metal and 35% volume of base metal. The welding speed is 4 mm/s. Determine the cross-sectional area of the weld metal (bead).

Solution

U_m= 10.25 J/mm^3, U = 18 V, I = 140 A, v_w = 4 mm/s, $f_1 = 0.72, f_2 = 0.68$, filler dia. = 3.7 mm.

Solution

Power generated by the heat source = R_H = Arc voltage × current = $U*I$ = 18 × 140 = 2520 W.

By using Eq. 4.13,

$$U_m * A_w * v_w = f_1 * f_2 * R_H$$

$$10.25 \times A_w \times 4 = 0.72 \times 0.68 \times 2520$$

$$A_w = \frac{0.72 \times 0.68 \times 2520}{10.25 \times 4} = 30.1 \text{mm}^2$$

The cross-sectional area of the weld metal (bead) = A_w = 30.1 mm^2.

Example 4.14 Determining the Feed Rate of the Filler Wire in GTAW
By using the data in Example 4.13, determine the feed rate of the filler wire.

Solution

$$A_w = 30.1\text{mm}^2, \text{vol.filler metal in the weld metal}$$
$$= 65\% = 0.65, v_w = 4\text{mm}/\text{s}, \text{d} = 3.7\text{mm}.$$

$$\text{Cross-sectional area of filler wire} = \frac{\pi}{4} d^2$$

$$= 0.785 \times 3.7^2 = 10.75\text{mm}^2$$

By using Eq. 4.14,

$$\text{Volume rate of metal welded} = A_w * v_w = 30.1\text{mm}^2 \times 4\text{mm}/\text{s} = 120.4\text{mm}^3/\text{s}$$

By using Eq. 4.15,

Feed rate of filler wire

$$= \frac{\textit{Volume rate of metal welded} \times \% \textit{volume filler metal}}{\textit{cross-sectional area of filler wire}}$$

$$= \frac{120.4 \times 0.65}{10.75}$$

Feed rate of filler wire = 7.28 mm/s.

Example 4.15 Determining the Heat Input for Pulsed Current GTAW
In a DC-pulsed-current GTAW process, the pulsed welding current and the background current are 150 and 80 A, respectively. The pulsed current is applied for 50% of the total time. The arc voltage is 12 V, and the travel speed is 128 mm/min. The heat transfer factor is 60%. Determine the heat input for the process.

Solution

I_p = 150 A, I_b = 80 A, t_p= 0.50, t_b = 0.50, U = 12 V, v_w= 128 mm/min, f_1 = 60%, $H_i^{pc-GTAW}$ =?

By using Eq. 4.17,

$$I_m = \frac{(I_p{}^* t_p) + (I_b{}^* t_b)}{t_p + t_b} = \frac{(150 \times 0.50) + (80 \times 0.50)}{0.5 + 0.5} = 115\,\text{A}$$

By using Eq. 4.16,

$$H_i^{pc-GTAW} = f_1 * \left(\frac{0.06 * U * I_m}{v_w} \right) = \frac{0.6 \times 0.06 \times 12 \times 115}{128} = 0.388\,\text{kJ}/\text{mm}$$

The heat input for the pulsed-current GTAW process = 0.388 kJ/mm = 388 J/mm.

Example 4.16 Determining the Rotational Speed of the Tube for Pressure Vessels by SAW

A thin-walled pressure vessel is to be manufactured by SAW (under automatic control) by making a continuous weld around the circumference of a round steel tube with a diameter of 1.8 m. The tube is slowly rotated under a stationary welding head. The arc voltage is 22 V and the welding current is 300 A. The heat transfer factor for SAW is 0.93 and the melting factor is assumed to be 0.68. The cross-sectional area of the weld bead is 77 mm^2. The unit melting energy for the steel is 9.6 J/mm^3. Determine the rotational speed of the tube.

Solution

D = 1.8 m = 1800 mm, U = 22 V, I = 300 A, $f_1 = 0.93$, $f_2 = 0.68$, $A_w = 77$ mm^2, U_m= 9.6 J/mm^3.

By using Eq. 4.21,

$$v_w = \frac{f_1 * f_2 * \mathrm{U} * \mathrm{I}}{U_m * A_w} = \frac{0.93 \times 0.68 \times 22 \times 300}{9.6 \times 77} = 5.64\,\mathrm{mm/s} = 338.4\,\mathrm{mm/min}$$

By using Eq. 4.19,

$$\mathrm{N} = \frac{v_w}{\pi * D} = \frac{338.4}{\pi \times 1800} = 0.0598\,\mathrm{rev/min}$$

The rotational speed of the tube = N = 0.06 rev/min.

Example 4.17 Determining Time to Weld Around Circumference for Pressure Vessel by SAW

By using the data in Example 4.16, determine the time to weld around circumference of the tube.

Solution

Circumference = C = $\pi * D = \pi \times 1800 = 5655$ mm^2, v_w = 338.4 mm/min

$$\text{Time to weld around circumference} = \frac{C}{v_w} = \frac{5655}{338.4} = 16.71\,\mathrm{min}$$

Example 4.18 Determining the Volume Rate of Metal Welded in FCAW

A FCAW process is used to fillet-weld two low-alloy steel plates that are oriented at right angle. The steel plates are 10-mm thick, and the unit melting energy of the steel is 8.85 J/mm^3. The heat transfer factor is 0.85 and the melting factor is 0.65. The welding current is 80 A and the arc voltage is 20 V. The travel velocity of the welding head is 1 m/min. The electrode diameter is 2.5 mm. The flux core, running through the center of the electrode, has a diameter of 1.3 mm; the flux contains compounds that do not become part of the weld bead. The weld bead consists of 60% metal from the electrode and the remaining 40% from the steel plates. Determine the volume rate of metal welded.

Solution

$f_1 = 0.85, f_2 = 0.65$, I = 80 A, U = 20 V, v_w = 1 m/min = 16.67 mm/s, U_m = 8.85 J/mm^3.
By using Eq. 4.20,

$$U_m * A_w * v_w = f_1 * f_2 * \mathrm{U} * \mathrm{I}$$
$$8.85 \times A_w \times 16.67 = 0.85 \times 0.65 \times 20 \times 80$$
$$A_w = \frac{0.85 \times 0.65 \times 20 \times 80}{8.85 \times 16.67} = 6 \text{ mm}^2$$

By using Eq. 4.14,

$$\text{Volume rate of metal welded} = A_w * v_w = 6 \times 16.67$$
$$= 100 \,\text{mm}^3 / \text{s} = 6000 \,\text{mm}^3 / \text{min}$$

Example 4.19 Determining the Feed Rate of the Electrode for FCAW Operation

By using the data in Example 4.18, determine the feed rate of the electrode for the operation.

Solution

D_e = 2.5 mm, D_f = 1.3 mm, volume rate of metal welded = 6000 mm^3/min, Electrode metal in weld bead = 60% = 0.60.
By using Eq. 4.23,

$$\text{Cross-sectional area of the metal in the electrode}$$
$$= \frac{\pi^*}{4}\left(D_e^2 - D_f^2\right) = 0.785 \times \left(2.5^2 - 1.3^2\right)$$
$$\text{Cross-sectional area of the metal in the electrode}$$
$$= 0.785 \times (6.25 - 1.69) = 3.58 \text{mm}^2$$

By using Eq. 4.24,

$$\text{Feed rate of electrode}$$
$$= \frac{\textit{Volume rate of metal welded} \times \textit{\% electrode metal in the weld bead}}{\textit{Cross-sectional area of the metal in the electrode}}$$
$$= \frac{6000 \times 0.60}{3.58}$$
$$\text{Feed rate of the electrode} = 1005.58 \,\text{mm} / \text{min} \cong 1 \text{m} / \text{min}$$

Questions and Problems

4.1.(MCQs). Encircle the most appropriate answers for each of the following statements.

(1) Which one of the following arc-welding processes is the most suitable for health and safety as well as for longitudinal and circumferential butt welds required in pressure vessels?

(a) SMAW, (b) SAW, (c) FCAW, (d) ESW, (e) PAW, (f) GMAW, (g) GTAW.

(2) Which arc-welding process is the most suitable for melting/welding any known metal?

(a) SMAW, (b) SAW, (c) FCAW, (d) ESW, (e) PAW, (f) GMAW, (g) GTAW.

(3) Which arc-welding process is unsuitable for welding aluminum alloys and stainless steels?

(a) SMAW, (b) SAW, (c) FCAW, (d) ESW, (e) PAW, (f) GMAW, (g) GTAW

(4) Which arc-welding is the most suitable to butt-weld together thick sections (>20 mm)?

(a) SMAW, (b) SAW, (c) FCAW, (d) ESW, (e) PAW, (f) GMAW, (g) GTAW

(5) Which arc-welding process is unsuitable for welding in vertical and overhead positions?

(a) SMAW, (b) SAW, (c) FCAW, (d) ESW, (e) PAW, (f) GMAW, (g) GTAW

(6) Which arc-welding process is the simplest, cheapest, and suitable for general fabrication?

(a) SMAW, (b) SAW, (c) FCAW, (d) ESW, (e) PAW, (f) GMAW, (g) GTAW

(7) Which of the following arc-welding processes is capable of producing the highest quality welds for joining stainless steels, aluminum, and other non-ferrous metals?

(a) SMAW, (b) SAW, (c) FCAW, (d) ESW, (e) PAW, (f) GMAW, (g) GTAW.

(8) Which of the following metal-transfer modes in GMAW results in high spatter formation, irregular bead shapes and the formation of numerous ceramic/slag islands?

(a) GMAW-S, (b) globular transfer GMAW, (c) spray transfer GMAW, (d) GMAW-P.

(9) Which modes in GMAW is the best for welding in vertical and overhead positions?

(a) GMAW-S, (b) globular transfer GMAW, (c) spray transfer GMAW, (d) GMAW-P.

(10) Which of the following modes in GMAW is the most useful for welding thin-to-medium thickness materials, providing good control and reduced distortion?

(a) GMAW-S, (b) globular transfer GMAW, (c) spray transfer GMAW, (d) GMAW-P.

4.2. (a) Explain the working principle of GMAW with the aid of a sketch.
(b) Diagrammatically illustrates the various metal-transfer modes in GMAW.
(c) What is the advantage of short-circuiting GMAW (GMAW-S)?
4.3. (a) Draw a sketch showing the equipment set-up for SMAW.
(b) What are the functions of flux used in SMAW?
(c) Differentiate between DCEP and DCEN polarities used in SMAW.
4.4. Describe FCAW with the aid of a sketch.
4.5. (a) Explain the working principle of GTAW with the aid of a sketch.
(b) Why is GTAW the best arc-welding process for producing the highest quality welds?
(c) What is pulsed-current GTAW? And what is its advantage?
4.6. Draw a sketch illustrating plasma arc welding (PAW).
4.7. Briefly explain electro-slag welding (ESW).
P-4.8. The electrode travelled a distance of 80 mm in 10 s during an SMAW operation. The welding current and arc voltages were 200 A and 20 V, respectively. Determine the heat input for the welding operation.
P-4.9. A welding current of 150 A is passed through a 1.6-mm-diameter mild-steels electrode with DECN during an SMAW operation. The electrode extension is 17 mm. Determine the electrode melting rate in the welding operation. Hint: refer to Table 4.3.
P-4.10. The power source in a GMAW setup generates 3500 W that can be transferred to the work surface with a heat transfer factor of 0.70. The metal to be welded is low-carbon steel; the unit melting heat energy for the steel is 10 J mm^3. The melting factor in the welding operation is 0.50. A continuous fillet weld is to be made with a cross-sectional area of 25 mm^2. Determine the welding speed for the operation.
P-4.11. Two titanium plates are to be butt-welded by GMAW, as shown in Fig. 4.14. The length of the weld is 200 mm. The heat transfer factor is 0.80, and the melting factor is 0.55. The heat energy required to melt titanium is 14.3 J/mm^3. Determine the heat input *i.e.*, the total heat generated at the heat source (electrode tip).
P-4.12. In a DC-pulsed-current GTAW process, the pulsed welding current and the background current are 175 and 85 A, respectively. The pulsed current is applied

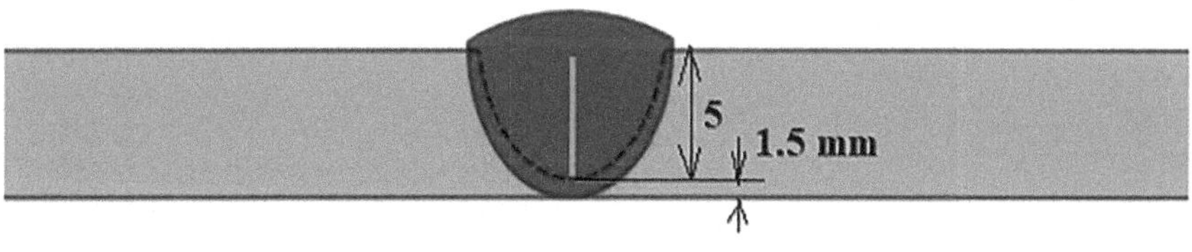

Fig. 4.14 Butt weld with U-groove for Problem P-4.11

for 55% of the total time. The arc voltage is 15 V, and the travel speed is 130 mm/min. The heat transfer factor is 65%. Determine the heat input for the process.

P-4.13. A GTAW operation is performed on low-carbon steel with unit melting energy of 10.25 J/mm^3. The arc voltage is 16 V and the welding current is 130 A. The heat transfer factor and the melting factor are 0.72 and 0.68, respectively. The diameter of the filler metal is 3 mm. The final weld metal comprises of 70% volume of filler metal and 30% volume of base metal. The welding speed is 3 mm/s. Determine the feed rate of the filler wire.

P-4.14. A FCAW process is used to fillet-weld two low-alloy steel plates oriented at right angle. The steel plates are 8-mm thick, and the unit melting energy of the steel is 8.85 J/mm^3. The heat transfer factor is 0.85 and the melting factor is 0.65. The welding current is 100 A and the arc voltage is 22 V. The travel velocity of the welding head is 800 mm/min. The electrode diameter is 2.7 mm. The flux core, running through the center of the electrode, has a diameter of 1.4 mm; the flux contains compounds that do not become part of the weld bead. The weld bead consists of 65% metal from the electrode and 35% from the steel plates. Determine the volume rate of metal welded.

References

AWS (2004). Welding Processes-AWS Handbook, 9th edition, *American Welding Society Committee,* Miami, USA

Baghel, P.K. (2022). Effect of SMAW process parameters on similar and dissimilar metal welds: An overview, *Heliyon*, Vol. 8, e12161

Hoffman, D.J., Dahle, K.R., Fisher, D.J. (2017). Welding, *Pearson India Education Services (Pvt) Ltd.* India

Houle, M.J. (1985). Low Voltage Short Circuiting-GMAW (**Category: Design/Fabrication**), *The National Board of Boilers and Pressure vessel Inspector,* Columbus, Ohio, USA

Khan, M.I. (2007). Welding Science and Technology, *New Age International (Pvt) Ltd*, India

Kumar, M.D.B., Arivazhagan, N., Tofil, S., *et al.* (2024). Influence of pulsed current GTAW-WAAM process parameters on the single layer bead geometry and multi bead multi-layer deposition of a nickel-based superalloy. *materialstoday,* Vol. 39, 108824

Phillips, D.H. (2023). Welding Engineering – An Introduction, 2nd Ed., Wiley, New York, USA

Srinivasan, S., Ravi, B.R., Atrens, A., Bala, S.P. (2024). Effect of pulse gas tungsten arc welding on the metallurgical characteristics & microstructural evolution in AZ31B magnesium alloy welds, *Journal of Materials Research and Technology,* Vol. 29, pp. 5412-5424

TWI (2025). What is Gas Metal Arc Welding? MIG Welding/MAG Welding, *The Welding Institute*, UK. Online: https://www.twi-global.com/technical-knowledge/faqs/faq-what-is-mig-mag-welding

Chapter 5
Resistance Welding Processes

Abstract This chapter begins with an introduction to resistance welding processes, followed by detailed descriptions of resistance spot welding (*RSW*), resistance seam welding (*RSeW*), and resistance projection welding (*RPW*). The *RSW* process is explained with reference to its working principle, equipment, steps of operation, applications, advantages and limitations, process parameters (current control, time, and electrode force) and their engineering analyses, and calculations of weld-time and transformer's duty cycle in RSW. The discussion on the recent advances in RSW includes laser beam spot welding (LBSW), the selection of ultra-high-strength steels (UHSS) for RSW of car bodies, and achieving high precision in RSW. The effects of weld-nugget geometry on the heat energy required for spot welding are also quantitatively explained. The explanation of the *RSeW* includes its working principle, operating data, and engineering analysis. The RPW process is also described with the aid of a sketch. This chapter includes 7 diagrams, 12 worked numerical examples, 2 data tables, and 6 MCQs with their answers given at the end of this book.

Keywords Resistance spot welding · Resistance seam welding · Resistance projection welding · Laser beam spot welding · Weld nugget · Transformer's duty cycle

Nomenclature

Q	Heat energy required for welding (J)
I	Current (A)
R	Resistance of base metals and contact interfaces (**Ω**)
t	Time current flows (s)
$2a$	Nugget diameter
2b	Nugget thickness
V_{nugget}	Nugget volume
I_p	Current in the primary coil

Z. Huda, *Welding and Joining Engineering*,
https://doi.org/10.1007/978-3-032-20462-2_5

I_s	Currents in the secondary coil
N_p	Number of turns in the primary coil
N_s	Number of turns in the secondary coil
F	Electrode force (N)
D	Diameter of the piston (mm)
p	Air or fluid pressure (MPa)
F_{max}	Maximum load or the load-carrying capacity of the weld-joint (N)

5.1 Resistance Welding Fundamentals

Resistance welding (or electric resistance welding) involves the joining of two plain metal work-pieces using pressure and electrical current, for a specified time duration, through the area of the metal to be joined. The key advantage of resistance welding is that no other materials such as shielding gasses, fluxes or filler rods are required in the union of the metals. In resistance welding (RW), an electric current is delivered to the metal sheets (or any work pieces being joined) through weld electrodes which apply force to the work. This force (mechanical energy) is then converted to heat energy; the heat, so generated, melts the metal at the point of contact—the point of 'resistance' between the faying surfaces. The electrode then extracts heat from the molten weld area which forms a weld 'nugget' at the point where it solidifies. The RW process involves the application of force before, during, and after a current is applied; this technique confines the contact area, resulting in a sound weld.

There are three principal types of resistance welding processes: (a) resistance spot welding (*RSW*), resistance seam welding (*RSeW*), and resistance projection welding (*RPW*); these processes are explained in the following sections.

5.2 Resistance Spot Welding (RSW)—*Welding Automation*

5.2.1 Working Principle and Equipment

Resistance spot welding (*RSW*) is commonly used to join two or more metal sheets by applying pressure and current using two copper cylindrical electrodes; the heat from electric current forms a spot-weld nugget (see Fig. 5.1). The *RSW* process works on the principle of joining whereby pressure and electric current are applied and heat is generated by the passage of current through resistive materials, such as low-carbon steels. In the *RSW process*, electric current is switched on (for a very short duration) between the welding electrodes to flow through the overlapping metal sheets under pressure. The metal fuses in the central area of the interface between the two sheets and then solidifies in the form of a *nugget*, thus providing

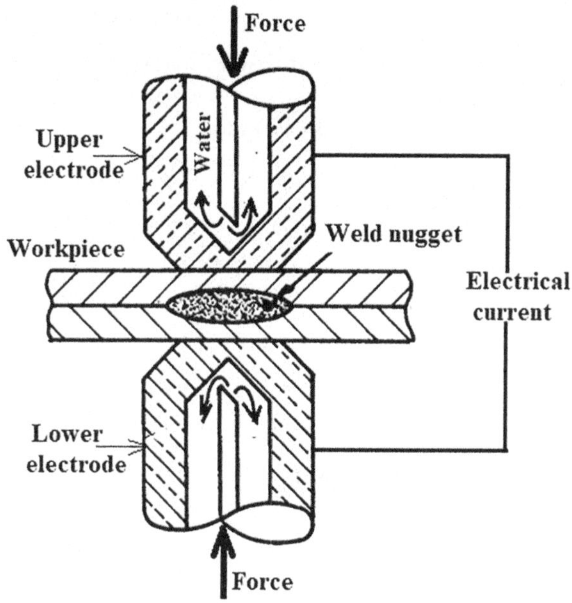

Fig. 5.1 Resistance spot welding—*working principle*

the weld joint. The *spot* welds are made at regular intervals on overlapping metal sheets.

A spot-resistance weld forms due to the resistance heating that occurs when current passes through the resistive metallic work-pieces being welded. For this reason, metals with higher resistivity (such as steel) are easier to weld than those that are better conductors of electricity such as aluminum (Kimchi and Phillips 2023). The heat energy generated in resistance spot welding is governed by the following formula:

$$Q = I^2 * R * t \tag{5.1}$$

where Q is the heat energy required for welding (J), I is the amperage, R is the resistance of base metals and contact interfaces (**Ω**), and t is the time current flows (s) (see Example 5.1).

The equipment used in the spot welding consists of tool-holders, electrodes, pressure mechanisms, current control switches, and other related parts (see Fig. 5.2). The tool-holders function as a mechanism to hold the electrodes firmly in place as well as support optional water hoses that cool the electrodes during welding. The electrode force is applied by means of a pneumatic or hydraulic cylinder with a moveable piston. The electrodes are generally made of a highly-conductive alloy, usually copper, and are designed in many different shapes and sizes depending on the application needed. The mechanical and electrical aspects of welding machines strongly influence the quality of welds in resistance welding, particularly spot welding (Zhang and Senkara 2011).

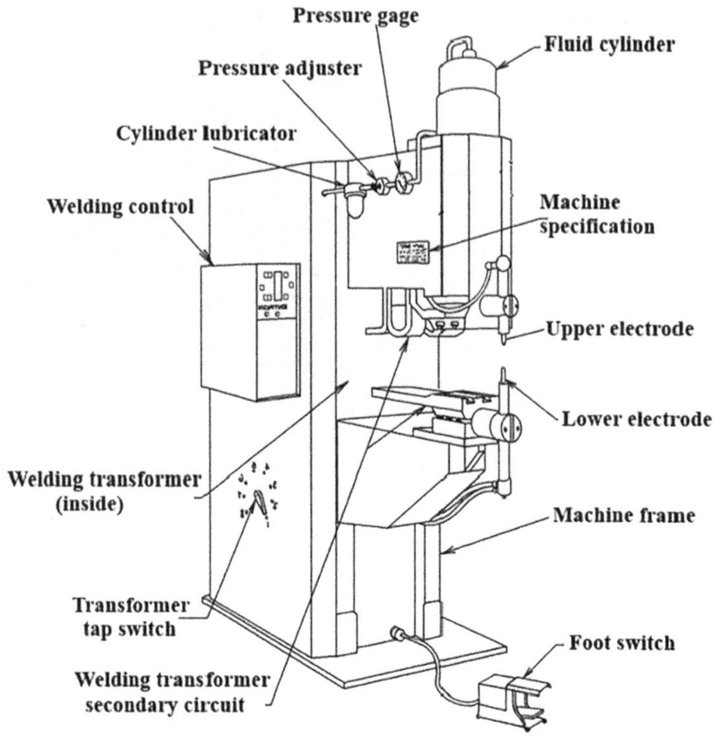

Fig. 5.2 Spot resistance welding machine—*schematic illu*

5.2.2 Steps of Operation in RSW

In the RSW operation, it is important to select a work-sheet with a thickness in the range of 0.50–3.0 mm. The RSW operation involves the following steps.

I. The overlapping parts (metal sheets) are inserted between the electrodes.
II. A suitable pressure is applied between the electrodes to ensure intimate contact between the workpieces.
III. The current is switched on for a short duration of time (usually in the range of 0.1–0.6 s), causing heating at the contact surfaces. The applied voltage is in the range of 1–20 V. The peak resistance near the faying surface results in melting of nuggets.
IV. The pressure is maintained until the nugget solidifies, and
V. The electrodes are opened, and the welded part (weldment) is removed.

5.2.3 *Applications, Advantages, and Limitations of RSW*—Automation

Spot resistance welding has applications in a number of industries, including automotive, aerospace, rail, metal furniture, electronics, and the like. The *RSW* has been the main joining process in the construction of steel cars for over 100 years. In particular, essential car parts (e.g. door frames, fenders, car doors, hoods, etc.) are frequently assembled using this technique. There are several advantages of the *RSW*. This process can be easily automated when combined with robots and manipulation systems. The RSW is the most common joining process in high-volume manufacturing lines, since it is a fast and efficient process. An ordinary car has about 5000 spot welds for joining different metal sheets and add-ons; this objective is achieved by the use of welding robots (see Chap. 1, Fig. 1.8). The weld joint produced by RSW is strong and durable.

The main limitations of the RSW process include the work-piece thickness limits and the voltage-range limit. Additionally, spot welding can affect the chemical and physical properties of the metals being fused, *i.e.* the resistance to corrosion may be reduced in stainless steel, aluminum and other metals. It is difficult to control the heat and pressure in spot welding.

5.2.4 *Engineering Analysis of RSW*

5.2.4.1 Determination of Nugget Volume in RSW

In resistance spot welding, the fused weld metal (nugget) (Fig. 5.3a) has a shape that approximate to an spheroid or an ellipsoid of revolution with a pair of equal semi-axes, *a*, and a distinct third semi-axis, b, which is an axis of symmetry (see Fig. 5.3b).

By reference to Fig. 5.3, the parameter 2*a* or the nugget diameter *D* can be determined by:

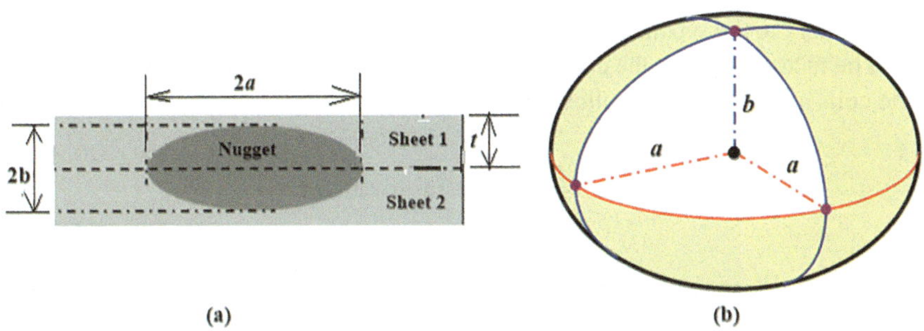

Fig. 5.3 Weld nugget geometric parameters (**a**), and a 3-D sketch of spheroid (**b**)

$$2a = \mathrm{D} = 5\sqrt{t} \tag{5.2}$$

The nugget parameter 2*b* or the nugget thickness can be calculated by (Huda 2018):

$$2\mathrm{b} \cong 0.8(2t) = 1.6t \tag{5.3}$$

where *t* is the thickness of each sheet. Once the values of *2a* and *2b* have been determined, the nugget volume can then be computed by (Schwedler et al. 2014):

$$V_{nugget} = \frac{4}{3}\pi * a^2 * \mathrm{b} \tag{5.4}$$

The significance of Eqs. 5.2–5.4 is illustrated in Examples 5.2–5.4.

5.2.4.2 Process Parameters in RSW

The following are the three principal process parameters in RSW: (a) current, (b) time duration, and (c) electrode force. The optimum value of these parameters varies with the type of metal and its thickness. For commonly used low-carbon steel sheets with a thickness of 1.6 mm, a typical value of current is 10,000 A, which is applied for a time-duration of 0.25 s and an electrode force of 2700 N.

5.2.4.3 Current Control by Transformer in RSW—*Engineering Analysis*

It is mentioned in the preceding subsection that a current of around 10,000 A is required for RSW; this high value of amperage is not readily available from any standard electrical outlet. A typical 200-A current is usually available from electrical distribution circuits in factories, which is not enough for RSW operation (see Example 5.5). Therefore, to get the 10,000 A needed for resistance welding, it is necessary to use a step-up transformer. A transformer consists of two coils of wire (primary and the secondary) wound around an iron core; power is transferred from the primary to the secondary via the magnetic flux of the core, as shown in Fig. 5.4.

The factor by which the current is stepped up is related to the number of turns in the coils according to the following formula:

$$\frac{I_s}{I_p} = \frac{N_p}{N_s} = \text{turns ratio} \tag{5.5}$$

where I_p and I_s are currents in the primary and secondary coils, respectively; and N_p and N_s are the number of turns in the primary and secondary coils, respectively. In order to achieve the 10,000 amperage (needed for resistance welding), the engineer would use a transformer with a *turns- ratio* of 50 (see Example 5.6).

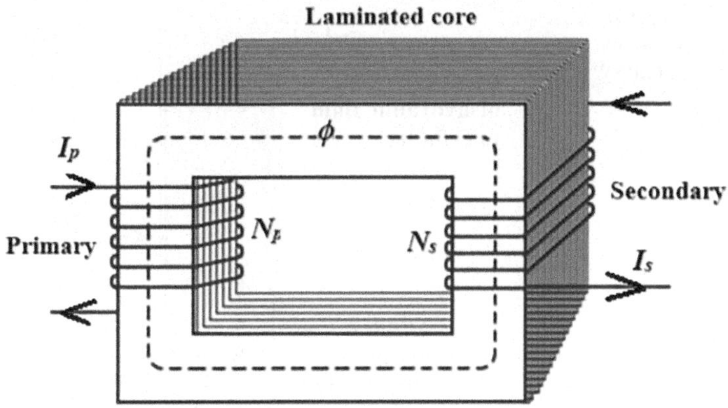

Fig. 5.4 A transformer's working principle

5.2.4.4 Controlling Time in RSW

The flow of current, through the work-pieces of metal to be welded, can be manually controlled by a foot switch (see Fig. 5.2). However, the length of time during which the welding-current flows is very short (0.1–0.6 s). This precision requires the use of some electronic device to control the flow of current. In industrial practice, two devices are used to precisely control current: (a) ignition tube and (b) silicon-controlled rectifier (SCR). These electronic devices work on the principle that a small electrical signal applied to the device allows it to turn on in a small fraction of a second and conduct a large amount of current. The removal of the electrical signal allows the device to turn off again. Fast turn-ons and turn-offs are possible because there are no mechanical moving parts in the two electronic devices.

5.2.4.5 Determination of Electrode Force in RSW

A reasonably high electrode force is necessary to assure good electrical contact between the parts being welded in RSW. The application of electrode force helps to hold the parts steady until the molten metal forming the welded joint has time to solidify. In general, the electrode force is developed by the use of compressed air in a cylinder and piston arrangement (see Fig. 5.5). The cylinder is rigidly attached to the welding-machine frame, and the moveable piston is connected to the upper electrode. The introduction of compressed air into the cylinder develops a force on the piston which, in turn, pushes the electrode down against the metal being welded.

The magnitude of the force applied depends on the cross-sectional area of the piston and the pressure of the compressed air, according to the following formula:

$$\mathrm{F} = \frac{\pi}{4} * D^2 * \mathrm{p} \tag{5.6}$$

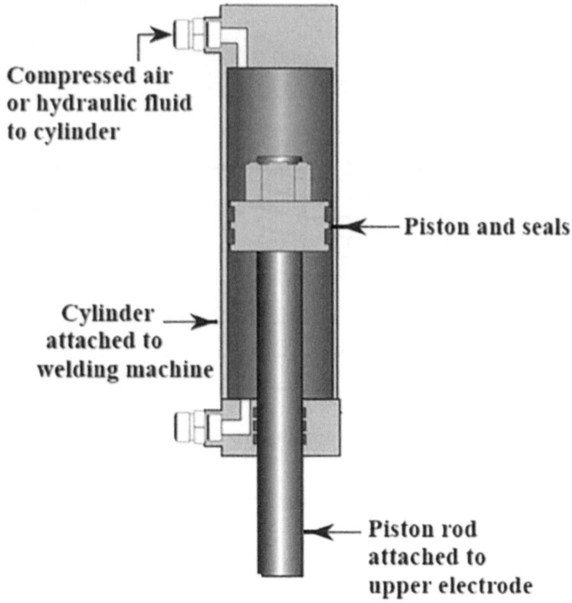

Fig. 5.5 Pneumatic/hydraulic cylinder for welding force in RSW

where F is the electrode force (N), D is the diameter of the piston (mm), and p is the air or fluid pressure (MPa) (see Example 5.7).

5.2.4.6 Calculations for Weld-Time and Transformer's Duty Cycle in RSW

The first step in calculating the duty cycle is to determine the weld time, which requires the value of utility frequency. In North America, the line frequency is 60 Hz, whereas in many other parts of the world, the utility frequency is 50 Hz. Single-phase AC spot-welding machines generally count the number of cycles of the utility frequency to control the time of a weld. The weld time (in seconds) can be determined by:

$$\text{Weld time on} = \frac{\text{Number of cycles}}{\text{Utility frequency}} \tag{5.7}$$

A duty cycle is the *fraction of a given period over which a signal or machinery is active*. If the duty cycle of a machine is 100%, it means that the machine runs for 10 out of 10 min (say), which will be very expensive. In North America, it is generally standard to rate welding transformers at 50%. In welding, the duty cycle can be calculated by:

$$\text{Duty cycle} = \frac{\text{No. of welds produced} \times \text{Weld time on}}{\text{Elapsed time}} \tag{5.8}$$

The elapsed time is taken as either 2 s or 1 min (see Examples 5.8 and 5.9).

5.2.5 Recent Advances in RSW

5.2.5.1 Laser Beam Spot Welding

Laser-beam spot welding (LBSW) is a precise RSW method that uses slow pulse beams to bond a material to a specific area; this process has been developed in recent years. The LBSW refers to welding processes applied to single points. By configuring the laser welding system with high speed and extremely narrow weld seam geometries, laser welding machines can produce extremely precise spot welds. LBSW is a high-energy-density process mainly featured by the low heat input in which laser beams are used to develop the spot welds (Zhang et al. 2020). This technology significantly reduces the weldability-related problems in the joining of ultra-high-strength steel (UHSS), such as cracking in fusion zone, cracking and softening in HAZ; LBSW also results in lower residual stresses, thereby ensuring better-quality spot welds.

5.2.5.2 Selection of High-Strength Steels for RSW for Car Bodies

Resistance spot welding (RSW) process is the primary joining method used in automotive production, since it enables the handling of a wide variety of steel grades and sheets up to 3 mm in thickness. In the last decade, there has been an intensifying customer demand for automobiles with weight reduction for energy saving as well as enhancement of passenger safety. These requirements can be fulfilled by using various combinations of materials, such as conventional deep-drawn steel sheet and high-strength steel sheet, such as TRIP and dual-phase (DP) steels (Spišák and Kaščák 2011). In particular, the high-strength properties of these steels allow us to reduce the thicknesses of particular segments of a car body as well as the amperage to be applied in the RSW process. In modern steel-car manufacturing, DP steels amount to more than 75% of the car body and 5% of TRIP steel sheets in the production of safety elements in the car. The DP1000 steel is a special DP grade of UHSS preferentially employed in the manufacture of next-generation electric vehicles; it exhibits a tensile strength of 1000 MPa (Rajalingam et al. 2024).

Spišák and co-researchers have studied the mechanical and metallurgical behaviors of spot-welded TRIP steel sheet (0.77-mm thickness) and DP steel sheet (1-mm thickness); they have reported that fusion weld joints occurred with all used parameters of RSW, even if the lowest welding current of 5.5 kA was used (Spišák et al. 2015). In the same research findings, it was concluded that the highest values of tensile strength were observed in the samples made with a welding current of 6.0 kA and the lowest values of load-carrying capacity of the weld-joint were observed in samples made with a welding current of 7 kA. The dependence of load-carrying capacity of the weld-joint on the current was expressed as (Spišák et al. 2015):

$$F_{\max} = -1666.7 * I^2 + 20657 * I - 56449 \quad (5.9)$$

where F_{max} is the maximum load or the load-carrying capacity of the weld-joint (N), and I is the welding current (kA) (see Examples 5.10 and 5.11).

5.2.5.3 Achieving High Precision in RSW

In recent years, there has been an increasing demand for high-precision welding/RSW, which requires advanced systems with real-time feedback control. Very recently (2025), Ozensoy and Uyar have reported a medium-frequency direct current (MFDC)-RSW system integrated with a Python-based welding control interface (WCI) to enhance precision and adaptability. Experimental validation of the data showed that the root-mean-square errors (RMSE) improved from 0.178 kA (open-loop) to 0.109 kA (closed-loop), indicating a substantial enhancement in stability, precision, and reliability under varying load conditions (Ozensoy and Uyar 2025).

5.3 Resistance Seam Welding

5.3.1 Working Principle

Resistance Seam Welding (RSeW) is a resistance welding process similar to the RSW, but here circular rotating electrodes are used. This process results in a continuous weld (series of overlapping welds), which is normally gas-tight or liquid-tight. In *RSeW*, two or more sheets of base metal are usually passed between disc-shaped electrode rollers. The electrodes transmit the current and also the mechanical pressure required for producing a welded seam, as shown in Fig. 5.6. The electrodes rotate as the material passes between them; this allows the electrodes to stay in constant contact with the material to make long continuous welds.

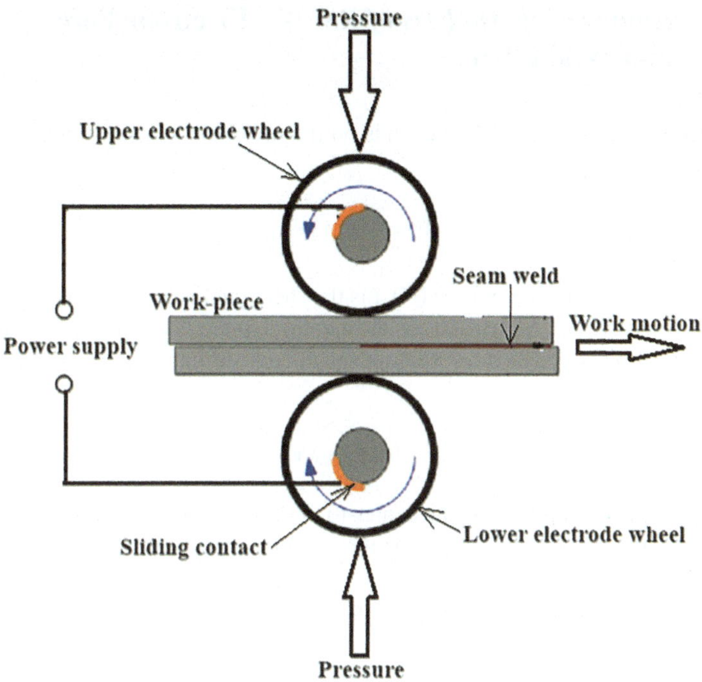

Fig. 5.6 Resistance seam welding—*a schematic illustration*

Table 5.1 Recommended data for RSeW of stainless steels (AWS 2022)

Sheet thickness	Electrode thickness	Net electrode Force	On time (cycles)	On time (ms)	Off time (cycles)	Off time (ms)	Max. weld speed
0.53 mm	6.0 mm	3.11 kN	3	50	2	33	1.4 m/min
1.98 mm	16 mm	10.23 kN	4	66	6	100	1.0

5.3.2 *Operating Data for RSeW*

The data for recommended practices for resistance welding are available in AWS Standard C1.1 for many of the most frequently used materials. Table 5.1 presents the data for the recommended data for the *RSeW* of stainless steels (300 series) with thicknesses in the range of 0.5–2.0 mm with the use of RWMA Class 3 electrode material.

5.3.3 *Engineering Analysis of RSeW*—Electrode Force and Weld Pitch

The electrode force, in *RSeW*, depends on the material's tensile strength and sheet thickness; it is expressed as:

$$F = k * (2t) * \sigma_{UTS} \tag{5.10}$$

where *F* is the electrode force (N), *t* is the sheet thickness (mm), σ_{UTS} is the material's tensile strength (MPa), and *k* is the electrode force factor (k = 1 for mild steel).

The weld pitch depends on welding speed and welding time per weld; it can be determined by:

$$\text{Pitch} = v * t' \tag{5.11}$$

where pitch is weld pitch, *i.e.* the center-to-center distance between two successive nuggets in the seam weld (mm), *v* is the welding speed (mm/s), and *t'* is the time per weld.

The significance of Eqs. 5.10–5.11 is illustrated in Example 5.12.

5.4 Resistance Projection Welding

Projection resistance welding (*RPW*) uses pressure and electrical current to join two or more metal parts that meet at one or more specific points with the maximum amount of contact. In *RPW*, small projections are formed on one or both pieces of the base metal to obtain contact at a point which localizes the current flow and concentrates the heat, as shown in Fig. 5.7. Under the application of force, the heated and softened projections collapse and a weld nugget is formed. Projection on the upper component is pressed against the lower component by electrode force. The

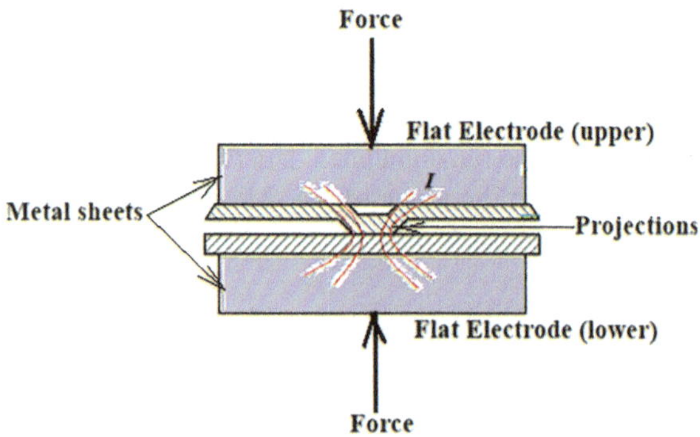

Fig. 5.7 Projection resistance welding—*a schematic*

projection collapses resulting in the formation of a fused weld nugget under the application of current. This technique is of special value in mounting attachments to surfaces of which the back side is inaccessible to a welding operator. *RPW* allows for efficient welding with less energy and greater weld strength.

5.5 Worked Numerical Examples in Resistance Welding

Example 5.1 Determining the Heat Energy Generated in a RSW Operation
The contact resistance of mild steel (MS) in an RSW operation was measured to be 200 $\mu\Omega$. A current of 10,000 A flows for a duration of 0.5 s in the welding process. Determine the heat energy generated for the operation.

Solution

I = 10,000 A, R = 200 × 10^{-6} Ω, t = 0.5 s, Q =?

By using Eq. 5.1,

$$Q = I^2 * R * t = 10000^2 \times 200 \times 10^{-6} \times 0.5 = 10^8 \times 100 \times 10^{-6} = 10{,}000\,\text{J}$$

The heat energy generated in the RSW = 10 kJ.

Example 5.2 Determining the Diameter and Thickness of a Spot-Weld Nugget
The RSW was performed to weld two mild-steel (MS) sheets, each with 1.5-mm thickness. By reference to Fig. 5.3, determine the dimensions for 2a and 2b for the weld nugget.

Solution

t = 1.5 mm, Dia. =?, thickness =?

By using Eqs. 5.2 and 5.3,

$$\text{Nugget diameter} = 2a = D = 5\sqrt{t} = 5 \times \sqrt{1.5} = 6.12 \text{ mm}$$
$$\text{Nugget thickness} = 2b = 1.6\ t = 1.6 \times 1.5 = 2.4 \text{ mm}$$

Example 5.3
By using the data in *Example 5.2*, determine the volume and the mass of the nugget. The density of MS is 7860 kg/m^3.

Solution

ρ = 7860 kg/m^3, 2a = 6.12 mm, a = 3.06 mm. 2b = 2.4 mm, b = 1.2 mm.

By using Eq. 5.4,

$$V_{nugget} = \frac{4}{3}\pi * a^2 * \text{b} = 4.1888 \times 3.06^2 \times 1.2 = 47.1\,\text{mm}^3 = 47.1 \times 10^{-9}\,\text{m}^3$$
$$\text{m} = \rho * \text{V} = 7{,}860 \times 47.1 \times 10^{-9} = 370206 \times 10^{-9}\,\text{kg}$$

Mass of the weld-nugget = m = 370.2 ×10^{-6} kg = 370.2 ×10^{-3} g = 0.37 g.

Table 5.2 Thermal properties of mild steel (MS)

Properties	T_m, °C	c, J/kg•°C	L_f, J/kg
Values	1380	510	14×10^5

Example 5.4 Assessing the Successfulness of RSW Operation Using the Given Data

Refer to the data in Examples 5.2 and 5.3. Are the amperage and the amount of heat energy generated in the welding operation enough for a successful RSW operation? The thermal properties of MS is given in Table 5.2.

Solution

Heat energy generated in the RSW operation = Q = 10,000 J (from Example 5.1).
m = 370.2 ×10^{-6} kg (from Example 5.3).
Let the heat energy required to form the nugget at the contact point for RSW = Q'.

$$Q' = m * c * (\Delta T) + m * L_f = m^* \left[c^* (\Delta T) + L_f \right]$$
$$Q' = 370.2 \times 10^{-6} \times \left[510 \times (1380 - 20) + 14 \times 10^5 \right]$$
$$= 370.2 \times 10^{-6} \times (693600 + 1400000)$$
$$Q' = 370.2 \times 10^{-6} \times 2093600 = 775.051\,\mathrm{J}$$

Since Q ≫ Q', the amperage and the amount of heat energy generated is more than enough for a successful RSW operation.

Example 5.5 Assessing the Feasibility of RSW Operation Based on Amperage Applied

Repeat Example 5.1 when a current of 200 A is used (instead of 10,000 A) in the RSW operation. Is the amount of heat generated enough for the welding operation? Hint: refer to the data in Example 5.4.

Solution

I = 200 A, R = 200 × 10^{-6} **Ω,** t = 0.5 s, Q =?, c = 510 J/kg•°C, Q' = 775.051.
By using Eq. 5.1,

$$\text{Heat generated} = Q = I^2 * R * t = 200^2 \times 200 \times 10^{-6} \times 0.5 = 40{,}000 \times 100 \times 10^{-6} = 4\,\mathrm{J}$$

The heat energy required for RSW = Q' = 775.051 J.

Since Q < Q', the amount of heat generated is NOT enough for the welding operation.

Example 5.6 Determining the Turns Ratio and Current Obtained from a RSW Transformer

A current of 200 A is supplied to the 100-turns primary coil of a welding transformer. The secondary coil has 2 turns Determine the *turns ratio* of the transformer, and the current obtained from the secondary coil.

Solution

$N_p = 100$, $N_s = 2$, $I_p = 200$ A, $I_s = ?$

By using Eq. 5.5,

$$\text{Turns ratio} = \frac{N_p}{N_s} = \frac{100}{2} = 50$$

$$I_s = I_p * (\text{turns ratio}) = 200 \times 50 = 10{,}000\,\text{A}$$

Example 5.7 Determining the Electrode Force to Be Applied for RSW

The piston diameter of a hydraulic cylinder, in a RSW machine, is 127 mm. The cylinder pressure-gage is showing a fluid pressure of 200 kPa. Determine the electrode force to be applied for the welding operation.

Solution

D = 127 mm, p = 200 kPa = 0.2 MPa, F =?

By using Eq. 5.6,

$$\text{F} = \frac{\pi}{4} * D^2 * \text{p} = 0.785 \times 127^2 \times 0.2 = 2532.25\,\text{N}$$

The electrode force ≅ 2532 N = 2.53 kN ≅ 570 pound-force.

Example 5.8 Determination of Weld Time-on and Duty Cycle for RSW in North America

A single-phase AC spot-welding machine counts 20 weld cycles in the USA. The machine can produce an average of 1 weld per 2 seconds Determine the: (a) weld time-on, and (b) duty cycle.

Solution

(a) Number of weld cycles = 20, utility frequency = 60 (for USA).

By using Eq. 5.7,

$$\text{Weld time on} = \frac{\text{Number of cycles}}{\text{Utility frequency}} = \frac{20}{60} = 0.33\,\text{s}$$

(b) No. of welds produced = 1, elapsed time = 2 s, weld time-on = 0.33 s, duty cycle =?

By using Eq. 5.8,

$$\text{Duty cycle} = \frac{\text{No. of welds produced} \times \text{Weld time on}}{\text{Elapsed time}} = \frac{1 \times 0.33}{2} = 0.165 = 16.5\%$$

Example 5.9 Determination of Weld Time-on and Duty Cycle for RSW in Europe
A single-phase AC spot-welding machine counts 20 weld cycles in Europe. The machine can produce an average of 6 welds per min. Determine the: (a) weld time-on, and (b) duty cycle.

Solution

(a) Number of cycles = 20, utility frequency = 50 (for Europe)
By using Eq. 5.7,

$$\text{Weld time on} = \frac{\text{Number of cycles}}{\text{Utility frequency}} = \frac{20}{50} = 0.4\,\text{s}$$

(b) No. of welds produced = 6, elapsed time = 60 s, weld time-on = 0.4 s, duty cycle =?

By using Eq. 5.8,

$$\text{Duty cycle} = \frac{\text{No. of welds produced} \times \text{Weld time on}}{\text{Elapsed time}} = \frac{6 \times 0.4}{60} = 0.04 = 4\%$$

Example 5.10 Determining the Load-Carrying Capacity of a RSW Weld Joint
In the experimental study on the RSW of TRIP and DP steels, the maximum load for the weld-joint is expressed by Eq. 5.9. Determine the load-carrying capacity of the weld-joint for a welding current of 6 kA.

Solution
By using Eq. 5.9,

$$F_{\max} = -1666.7 * I^2 + 20657 * I - 56449 = \left(-1666.7 \times 6^2\right) + \left(20657 \times 6\right) - 56449$$

$$F_{\max} = -60001.2 + 123942 - 56449 = 7{,}491.8\,\text{N}$$

Example 5.11 Determining the Welding Current Based on the Load Capacity of Weld Joint
The mechanical behavior of an RSW joint follows Eq. 5.9. It is required to achieve a load-carrying capacity of 7000 N in the weld-joint. Determine welding current to be used in the welding operation.

Solution
By substituting $F_{max} = 7000$ N in Eq. 5.9, we obtain:

$$-1666.7 * I^2 + 20657 * I - 56449 = 7000$$

$$-1666.7 * I^2 + 20657 * I - 63449 = 0$$

By comparing this Equ. with the standard quadratic equation ($ax^2 + bx + c = 0$), we can write:

a = −1666.7, b = 20,657, c = −63,449; thus,

$$I = \frac{-b \mp \sqrt{b^2 - 4ac}}{2a} = \frac{-20657 \mp \sqrt{-20657^2 - 4 \times -1666.7 \times -63449}}{2 \times -1666.7}$$
$$= \frac{-20657 \mp 1926}{-3333.4}$$
$$I_1 = \frac{-20657 + 1926}{-3333.4} = 5.6\,\text{kA}$$
$$I_2 = \frac{-20657 - 1926}{-3333.4} = 6.7\,\text{kA}$$

Thus, the welding current should be either 5.6 kA or 6.7 kA.

Example 5.12 Determining the Electrode Force and Weld Pitch

A RSeW is performed to join two mild-steel (MS) sheets, each of thickness 2 mm by applying a welding current of 10 kA. The welding time per weld is 0.2 s, and the welding speed is 550 mm/min. The tensile strength of MS is 430 MPa. Determine the electrode force, weld pitch, and weld nugget diameter.

Solution

I = 10.000 A, t = 2 mm, t' = 0.2 s, v = 550 mm/min = 9.16 mm/s, σ_{UTS} = 430 N/mm², k = 1.

By using Eqs. 5.10 and 5.11,

$$F = k * (2t) * \sigma_{UTS} = 1 \times 2 \times 2 \times 430 = 1720\,\text{N}$$
$$\text{Pitch} = v * t' = 9.16 \times 0.2 = 1.83\ \text{mm}$$

The electrode force = 1720 N, and weld pitch = 1.8 mm.

Questions and Problems

5.1. Encircle the most appropriate answers for the following statements.

(1) Which of the following welding processes results in gas-tight and liquid-tight weld?

(a) RSW, (b) RSeW, (c) RPW, (d) LBSW

(2) Which of the following resistance welding processes can produce extremely precise spot welds with lower residual stresses and better quality?

(a) RSW, (b) RSeW, (c) RPW, (d) LBSW

(3) Which resistance welding process is most extensively used in automotive industry?

(a) RSW, (b) RSeW, (c) RPW, (d) LBSW

(4) Which metallic material ensures the best passenger safety for car body fabrication?

(a) MS, (b) aluminum, (c) DP steel, (d) high carbon steel

(5) Which of the following resistance welding processes is the most suitable to join two metal parts that meet at one or more specific points with the maximum amount of contact?

(a) RPW, (b) RSeW, (c) RSW, (d) LBSW

(6) Which of the following time-duration ranges for current flow is the best for RSW?

(a) 0.6–0.9 s, (b) 0.1–0.6 s, (c) 1–5 s, (d) 6–9 s.

5.2. Why is it difficult to resistance weld aluminum as compared to MS under the same operating conditions?

5.3. Is an amperage of 300 A suitable for RSW? Explain your answer.

5.4. (a) Illustrate RSW with the aid of a sketch.

(b) List the steps of operation in RSW.

5.5. Explain RSeW with the aid of a sketch.

P5.6. The contact resistance of MS in an RSW operation was measured to be 200 $\mu\Omega$ A current of 8000 A flows for a duration of 0.3 s in the welding process. Determine the heat energy generated for the operation.

P5.7. RSW was performed to weld two mild-steel (MS) sheets, each with 2-mm thickness.

Determine the: (a) weld nugget diameter, (b) nugget volume, (c) mass, and (d) heat energy required for the successful welding operation. The density of MS is 7860 kg/m^3.

Hint: refer to the data in Table 5.2.

P5.8. The piston diameter of a hydraulic cylinder of a spot-welding machine is 120 mm.

The cylinder pressure-gage is showing a fluid pressure of 180 kPa. Determine the electrode force to be applied for welding.

P5.9. In the experimental study on the RSW of TRIP and DP steels, the maximum load for the weld-joint is expressed by Eq. 5.4. Determine the load-carrying capacity of the weld-joint for a welding current of 7 kA.

P5.10. A current of 240 A is supplied to the 80-turns primary coil of a welding transformer. The secondary coil has 2 turns Determine the *turns ratio* of the transformer, and the current obtained from the secondary coil.

P5.11. A single-phase AC spot-welding machine counts 18 weld cycles in Europe. The machine can produce an average of 4 welds per min. Determine the weld time-on, and duty cycle.

P5.12. A RSeW is performed to join two mild-steel (MS) sheets, each of thickness 1.5 mm by applying a welding current of 8 kA. The welding time per weld is 0.1 s, and the welding speed is 600 mm/min. The tensile strength of MS is 430 MPa. Determine the electrode force and weld pitch.

References

AWS (2022). Recommended Practices for Resistance Welding (6th Ed.): AWS C1.1M/C1.1:2019-AMD1, *American Welding Society*, USA

Huda, Z. (2018). Manufacturing: *Mathematical Models, Problems & Solutions. CRC Press,* FL

Kimchi M. and Phillips, D.H. (2023). Introduction to Resistance Spot Welding, *Springer Nature*, Switzerland

Ozensoy, C. and Uyar, M. (2025). Design and implementation of a medium frequency direct current resistance spot welding system integrated with welding control interface. *Ain Shams Engineering Journal,* Vol. 16, Issue 5, 103357

Rajalingam, P., Rajakumar, S., Sonar, T., Kavitha, S. (2024). A comparative study on resistance spot and laser beam spot welding of ultra-high strength steel for automotive applications. *Int. Journal of Lightweight Materials and Manufacture,* Vol. 7, Issue 5, pp. 648–661

Schwedler, O., Zinke, M., Jüttner, S. (2014). Determination of hydrogen input in welded joints of press-hardened 22MnB5 steel. *Welding in the World,* Vol. 58, pp. 339—346

Spišák, E., Kaščák, Ľ. (2011), Joining car body steel sheets using the clinching method. *Acta Mechanica Slovaca.* Vol. 15, No. 1, pp. 28–34.

Spišák, E., Kaščák, Ľ., Viňáš, J. (2015). Application of resistance spot welding in car body production, *Acta Mechanica Slovaca*, Vol. 19, No. 2, pp. 28—34.

Zhang, H. and Senkara, J. (2011). Resistance Welding—*Fundamentals and Applications* (2nd Ed.), CRC Press, FL, USA

Zhang, W., Yang, S., Lin, Z., Tao, W. (2020). Weld morphology and mechanical properties in laser spot welding of quenching and partitioning 980 steel. *Journal of Manufacturing Processes,* Vol. 56, pp. 1136–1145

Chapter 6
Chemical-Based Welding—*OFW & TW*

Abstract This chapter presents both qualitative and quantitative aspects of oxyfuel welding (OFW) and thermit welding processes. The OFW is explained with reference to its working principle, thermochemistry of oxyacetylene welding, equipment, operation, flame adjustment, forehand & backhand welding, and engineering analysis; safety precautions in the handling of acetylene cylinders are also included. Thermit welding is described by taking into consideration its working principle, chemistry, applications, and the steps of operation. This chapter includes 11 worked numerical examples, 8 diagrams, and 8 multiple-choice questions (MCQs); the answers of all MCQs and selected problems are given at the end of this book.

Keywords Oxyfuel welding (OFW) · Thermit welding (TW) · Oxyacetylene welding · Forehand welding · Backhand welding · Thermit mixture · Exothermic reaction · Heat of combustion

Nomenclature

$\dot{H}_i$	Total heat-input rate (kJ/h)
f_1	Heat-transfer factor
$\dot{V}$	Volumetric flow rate of fuel gas (m³/h)
ΔH	Heat of combustion of the fuel gas
V_g	Volume of gas consumed (m³)
$\dot{V}_g$	Volumetric-flow rate of the gas
t_w	Welding time (h)
V_f	Volume of filler rod (m³)
A_{groove}	Cross-sectional area of the weld groove (m²)
l_w	Length of the weld

Z. Huda, *Welding and Joining Engineering*,
https://doi.org/10.1007/978-3-032-20462-2_6

6.1 Oxyfuel Welding (OFW)

6.1.1 Working Principle

Oxyfuel Welding (OFW), also called oxy-acetylene welding, is a fusion welding process that uses a flame produced by burning a fuel gas (usually acetylene) mixed with oxygen to fuse and join metals. It is one of the oldest and most versatile welding processes, still used today for specific applications. In the OFW process, oxygen and a fuel gas/acetylene are combined in an appropriate proportion in a torch. The mixture of gases (C_2H_2 + O_2) (in oxy-acetylene welding) is ignited to produce a high-temperature flame, which is directed at the metal workpieces; the flame melts the work edges.

A filler rod is generally added to help join the metals, as shown in Fig. 6.1. The OFW can be performed by the use of a variety of mixtures of gases, such as oxygen-acetylene, oxygen-propane, oxygen-hydrogen, etc. However, oxyacetylene welding is the commonest method, since an oxyacetylene torch produces the hottest flame with a temperature exceeding 3400 °C.

6.1.2 Thermo-Chemistry of Oxy-Acetylene Welding

The chemistry of oxy-acetylene welding involves two-stage reactions. The first-stage reaction results from the combustion of acetylene with oxygen in which equal volumes of acetylene and oxygen react to produce carbon monoxide and hydrogen, according to the following reaction:

$$C_2H_2 + O_2 = 2CO + H_2,\ \Delta H = -1299 kJ/mol \tag{6.1}$$

The exothermic reaction with the heat of combustion of 1299 kJ/mol liberates a large amount of heat, resulting in melting of the metal. An alternative value of the heat of combustion of acetylene is $\Delta H = -49.37 \times 10^6\ J/m^3$ (obtained using the ideal gas law) (Lide 2005). In the second stage, carbon monoxide burns with oxygen to form carbon dioxide, while the hydrogen (formed in stage 1) combines with oxygen to form water, according to the following reaction:

$$\mathrm{CO + H_2 + O_2 = CO_2 + H_2O} \tag{6.2}$$

Thus, the combustion is complete and carbon dioxide and water (turned to steam) are the chief products of combustion. The overall reaction produces a flame temperature of around 3300 °C. The total heat liberated in the overall combustion reactions in oxy-acetylene welding is 55 × 10^6 J/m^3; however, because of the temperature distribution in the flame, the way in which the flame spreads over the

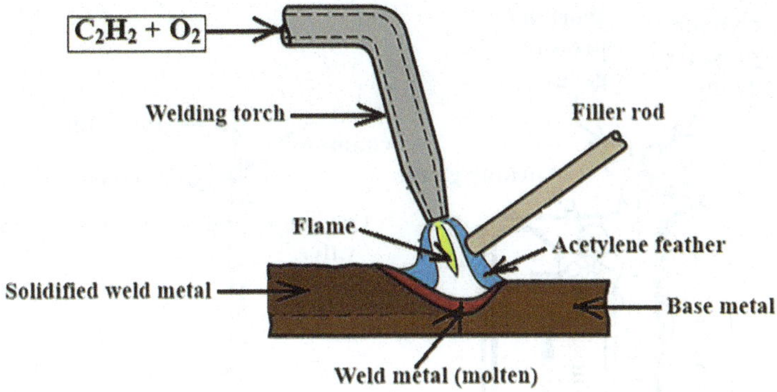

Fig. 6.1 Oxy-acetylene welding—*working principle*

work surface, and losses to the air, power densities and heat transfer factors in oxyacetylene welding are relatively low, *i.e.*; f_l in the range of 0.1–0.30 (Groover 2020).

6.1.3 OFW Equipment and its Operation—Flame Adjustment

The equipment for the OFW basically consists of the following components: (a) cylinders: steel pressurized cylinders containing oxygen and fuel gas, (b) regulators: oxygen regulator and fuel-gas regulator, (c) cylinder pressure gauges, (d) working pressure gauges, (e) hoses: oxygen hose and fuel-gas hose, (f) safety valves, (g) torch, and (h) torch tip (see Fig. 6.2). Oxygen cylinders are usually painted green, while fuel-gas cylinders are painted red or orange. Prior to igniting a flame in the OFW, it is important to check the equipment for safety by following the three steps by: (I) ensuring that hoses are undamaged and connections are tight, (II) opening oxygen and acetylene cylinder valves slowly, and (III) checking for leaks using soapy water at connections (no bubbles means no leaks).

In order to ignite a flame in oxy-acetylene welding, firstly, the acetylene valve on the torch handle is opened to set acetylene pressure in the range of 7–10 psi (50–70 kPa). Secondly, the gas is lit with a spark lighter; then the oxygen valve is slowly opened to set oxygen pressure in the range of 20–40 psi (135–275 kPa). Finally, the flame is adjusted (Fig. 6.3) by watching the flame color change, as follows. Figure 6.3a shows the carburizing flame indicating an excess of acetylene; this flame (also called a reducing flame) is used for the welding of aluminum and its alloys, as well as for soft soldering. The neutral zone is shown in Fig. 6.3b; here, two distinct zones may be seen. The inner core of the flame has a bright white-bluish light and extends only a short distance from the tip; around this inner cone is the blue flame envelope. This flame is metallurgically neutral, and is used for the welding of steel. Fig. 6.3c illustrates an oxidizing flame, which is obtained by slightly increasing the oxygen flow. This flame is shorter and sharper than the neutral flame, with a shorter, more pointed inner cone. This flame is used for welding cast iron, brass, bronze and zinc alloys, and for some brazing alloys.

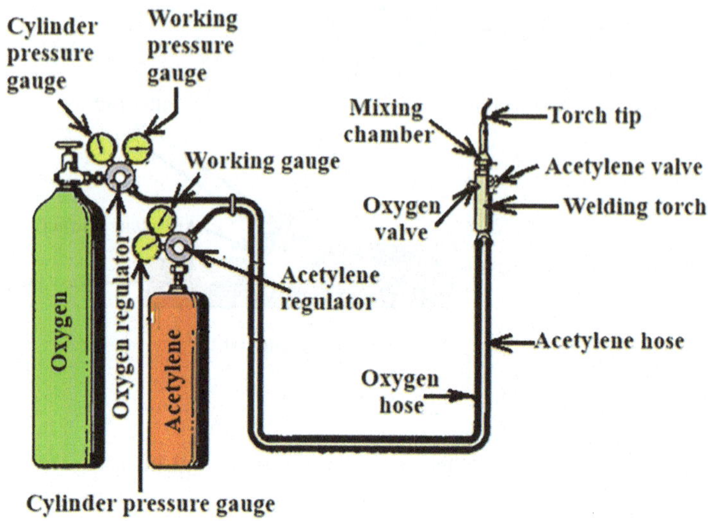

Fig. 6.2 Oxy-acetylene welding equipment

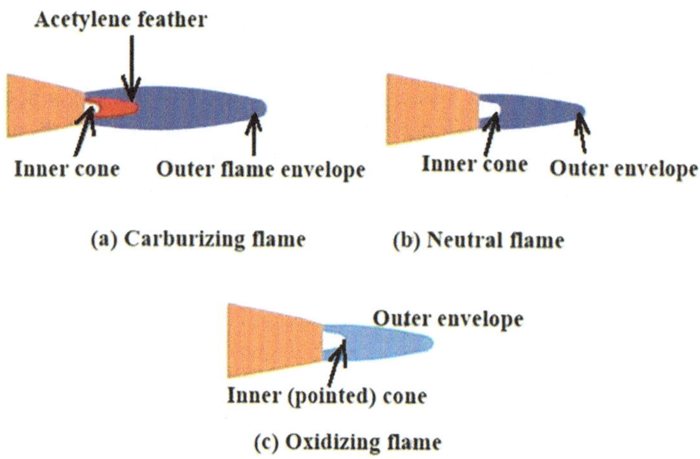

Fig. 6.3 Flame adjustment in oxy-acetylene welding; (**a**) carburizing flame, (**b**) neutral flame, (**c**) oxidizing flame

6.1.4 Safety Precautions for Acetylene Cylinders

The use of acetylene cylinders requires the following safety precautions. It is important that acetylene cylinders are secured **upright**. When not in use, acetylene cylinders (like other gas cylinders) must be stored in a safe, well-ventilated area that is secure and lockable. Additionally, the cylinders should be stored away from heat/ignition sources, and flammable or corrosive materials. In the case of

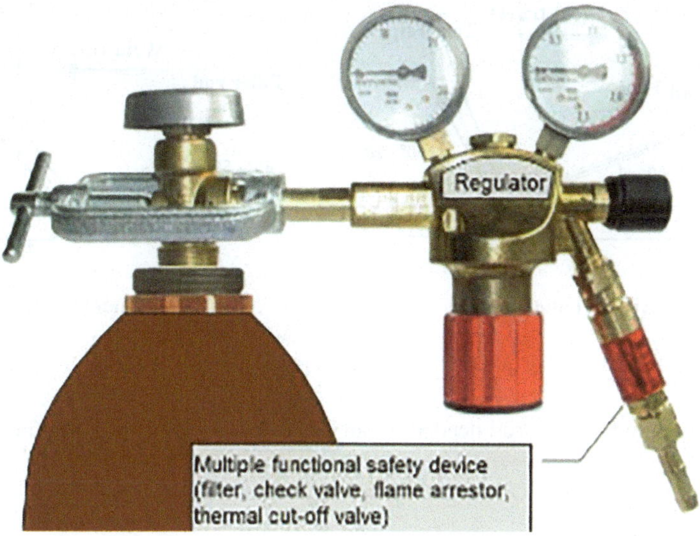

Fig. 6.4 Acetylene cylinder with multiple functional safety device (Courtesy: *European Industrial Gases Association*, Belgium)

leakage in an acetylene cylinder or valve, immediately close the cylinder valve (see Fig. 6.4). In the case of flashback, all valves/blow pipes must be closed (EIGA 2025).

6.1.5 *Engineering Analysis of OFW*

The determinations of heat input and the volume consumptions of gases and filler rod are useful for an OFW operation. The heat input rate to the work in the OFW can be calculated by:

$$\dot{H}_t = f_1 * \dot{V} * (\Delta H) \tag{6.3}$$

where $\dot{H}_i$ is the total heat-input rate (kJ/h), f_l is the heat-transfer factor, $\dot{V}$ is the volumetric flow rate of fuel gas (m^3/h), and ΔH is the heat of combustion of the fuel gas (see Examples 6.1–6.3). The volume of a gas consumed during OFW can be determined by:

$$V_g = \dot{V}_g * t_w \tag{6.4}$$

where V_g is the volume of a gas consumed (m^3), $\dot{V}_g$ is the volumetric-flow rate of the gas, and t_w is the welding time (h) (see Example 6.4).

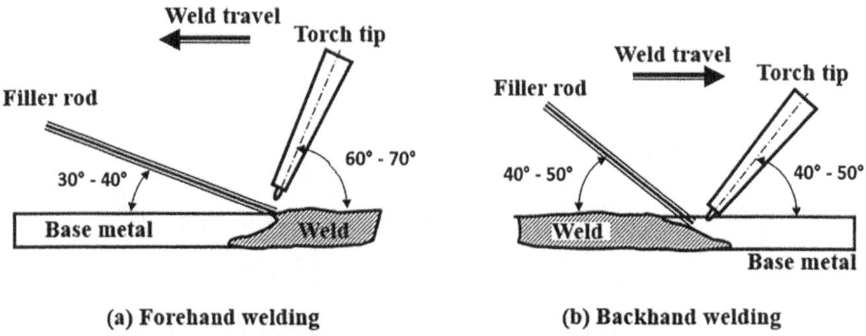

Fig. 6.5 Forehand welding (**a**), and backhand welding (**b**)

The volume of filler rod needed for an OFW operation can be computed by:

$$V_f = A_{groove} * l_w \tag{6.5}$$

where V_f is the volume of filler rod (m^3), A_{groove} is the cross-sectional area of the weld groove (m^2), and l_w is the length of the weld (seam) (see Example 6.5).

6.1.6 Forehand and Backhand Welding in OFW

Based on the relative position of the weld bead and flame, OFW technique can be either forehand welding or backhand welding. **Forehand welding** (or **forward welding**) is the technique in which the flame is pointed towards the direction of travel (weld progression), and the filler rod is applied ahead of the torch (see Fig. 6.5a). Forward welding is generally used for thinner sheets (Bowditch and Bowditch 2021). In forward welding, the welding torch remains just above the completed weld bead (see Fig 6.5a). In backhand welding (or backward welding), the flame is pointed away from the direction of travel, and the filler rod is applied behind the torch (see Fig. 6.5b). Backhand welding is mostly used for thicker work-pieces. Table 6.1 presents the differences between forehand welding and backhand welding.

6.2 Thermit Welding

6.2.1 Working Principle and Applications

Thermit welding was developed in the mid 1890s for joining railroad rails, steel pipes, steel wires, large cast, and forged components; this process is now used in the repair of steel casting and forging. Thermit welding is a chemical-based fusion

Table 6.1 Forehand welding and backhand welding

#	Forehand welding	Backhand welding
1.	The flame is pointed towards the direction of weld travel	The flame is pointed away from the direction of weld travel
2.	The torch is positioned above the deposited weld bead	The torch remains above the unfilled root gap
3.	Filler metal is applied ahead of the torch	Filler metal is applied behind the torch
4.	No post-heating occurs here	This technique facilitates post-heating of the deposited weld bead
5.	It is preferred for joining thin sheets (up to 3-mm thickness)	It is preferred for joining thicker plates

welding process, where an exothermic reaction results in providing the needed heat energy. This process is so named because it involves the use of *thermit*—a mixture of aluminum powder and iron oxide. The pre-heating of *thermit* up to 1300 °C is essential to initiate the exothermic reaction. The *thermit* mixture reacts according to the following chemical reaction:

$$\mathrm{Fe_2O_3} + 2Al = 2\mathrm{Fe} + \mathrm{Al_2O_3}, \Delta H = -851.5\ \mathrm{kJ/mol} \tag{6.6}$$

The liberation of a large amount of heat in the exothermic reaction generates a temperature of around 3000 °C. It is obvious in Eq. 6.6 that 1 mole of iron(III) oxide produces 2 moles of iron (see Examples 6.6–6.8). On completion of the reaction, the product is in molten form—the pure iron (steel) settles down, while aluminum oxide floats as slag.

6.2.2 *Steps of Operation in Thermit Welding*

The performance of a thermit-welding operation requires facilities/equipment similar to those in a foundry, since a thermit reaction produces the metal to be poured into the obstinate cavity created around the joint. The process involves the following steps:

I. It is mandatory to properly clean and prepare the edges of the metal parts to be welded.
II. Hot wax is poured into the joint to be fused to create the wax pattern.
III. A molding box is placed around the joint and the sand is loaded all over the wax pattern.
    It provides the necessary pouring basin, riser, sprue, and gating system (see Fig. 6.6).
IV. The bottom opening is used to drain the molten wax, which helps to preheat the joint and render it ready for welding.
V. *Thermit* is mixed in a crucible, which is made from refractory material to withstand the extreme heat and pressure which is produced during this exothermic reaction.

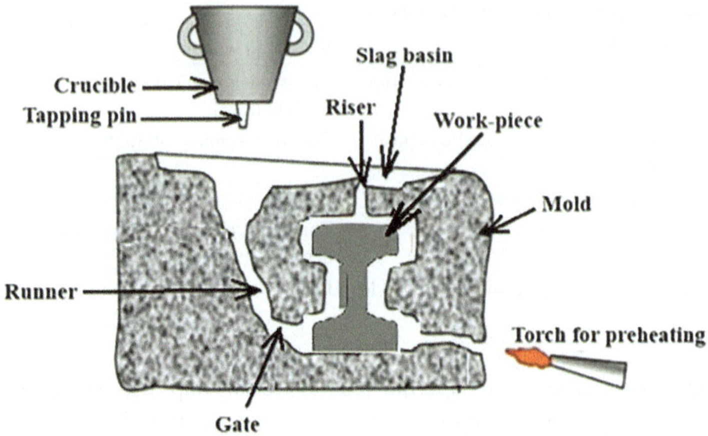

Fig. 6.6 Thermit welding—*working principle*

VI. The igniter (*e.g.* barium oxide or magnesium) is placed on the head of the mixture; it is lighted with red-hot metal rods.
VII. The reaction (Eq 6.6) completes in a short time; a highly heated molten iron flows into the mold cavity which had been created around the joint to be fused.
VIII. The heated molten metal fills the cavity and joins the parent metals; it solidifies to a homogenous, strong joint.

6.3 Worked Numerical Examples

Example 6.1 Determining Rate of Heat Generated in Combustion in Oxyacetylene Welding
An oxyacetylene welding operation is performed on a 5-mm-thick steel cylindrical work-piece (diameter = 8 mm). The welding torch supplies 0.25 m^3 of acetylene per hour and an equal volume rate of oxygen. Determine the rate of heat liberated during overall combustion (in J/h).

The overall heat of combustion of acetylene = 55 × 10^6 J/m^3.

Solution

$$\text{Rate of heat generated} = (\text{volume flow rate of acetylene}) \times (\text{heat of combustion})$$

$$\text{Rate of heat energy generated} = 0.25\frac{m^3}{h} \times 55 \times 10^6 \frac{J}{m^3} = 13.75 \times 10^6 \text{ J/h}$$

Example 6.2 Determining the Rate of Heat Input in Oxyacetylene Welding
By using the data in Example 6.1, determine the rate of heat input (power input) to the work surface, if the heat transfer factor $f_l = 0.20$.

Solution

$\dot{V} = 0.25$ m³/h, $\Delta H = 55 \times 10^6$ J/m³, $f_1 = 0.20$, $\dot{H}_i = ?$

By using Eq. 6.3,

$$\dot{H}_i = f_1 * \dot{V} * (\Delta H) = 0.20 \times 0.25 \frac{m^3}{h} \times 55 \times 10^6 \frac{J}{m^3} = 2.75 \times 10^6 \text{ J/h}$$

The power input is the rate of heat input ($\dot{H}_i$), expressed in J/s or W.

$$\text{Power input} = 2.75 \times 10^6 \text{ J/h} = \frac{2.75 \times 10^6}{3{,}600} \text{ J/s} = 763.89 \text{ J/s} = 763.89 \text{ W}$$

Example 6.3 Determining Power Density in the Area where Heat Is Concentrated

Refer to the data in Examples 6.1 and 6.2. Determine the average power density in the circular area, if 80% of the heat from the flame is concentrated in a circular area on the work surface.

Solution

Power input = 763.89 W, diameter of work = D = 8 mm, thickness = 5 mm.

$$\text{Area} = \text{A} = \frac{\pi}{4} D^2 = \frac{\pi}{4} 8^2 = 50.24 \text{ mm}^2$$

Power input to the work surface area to which heat is concentrated = 80% × 763.89 = 611.112 W.

$$P_i = 611.112 \text{ W}$$

By using Eq. 2.12,

$$\text{PD} = \frac{P_i}{A} = \frac{611.112}{50.24} = 12.16 \text{ W/mm}^2$$

Example 6.4 Determining the Volumes of Acetylene and Oxygen Consumed

By using the data in Example 6.1, determine the volumes of acetylene and oxygen consumed, if the OFW operation is performed for 20 min.

Solution

Volume flow rate of acetylene = $\dot{V}_a = 0.25$ m³/h, Volume flow rate of oxygen = $\dot{V}_o = 0.25$ m³/h

$t_w = 20$ min = 0.33 h

By using Eq. 6.4,

$$V_o = \dot{V}_o * t_w = 0.25 \times 0.33 = 0.0825\text{m}^3$$
$$V_a = \dot{V}_a * t_w = 0.25 \times 0.33 = 0.0825\text{m}^3\text{m}^3$$

The volume of oxygen consumed = 0.0825 m^3, Volume of acetylene consumed = 0.0825 m^3.

Example 6.5 Determining Volume and Length of Filler Rod for OFW
It is required to produce 70-cm long seam by OFW on V-grooved work-piece, as shown in Fig. 6.7. (a) Determine the volume of the filler rod that will be consumed. (b) What length of the filler rod is needed, if its diameter is 1 mm?

Solution

(a) Groove width = 4 mm, groove depth = 3 mm, l_w= 70 cm = 700 mm

Since the V-groove is triangular, its cross-sectional area can be determined by:

$$A_{groove} = \tfrac{1}{2}\text{base} \times \text{height} = \tfrac{1}{2} \times 4 \times 3 = 6\text{mm}$$

By using Eq. 6.5,

$$V_f = A_{groove} * l_w = 6 \times 700 = 4{,}200\,\text{mm}^3 = 4.2\,\text{cm}^3$$

The volume of filler rod needed for the OFW operation = 4.2 cm^3

$$\frac{\pi}{4} d^2 * l = V_f$$

(b) $$\frac{\pi}{4} (0.1)^2 * l = 4.2$$

$$l = \frac{4.2}{0.00785} = 535 \text{ cm} = 5.35 \text{ m}$$

The length of filler rod needed = 5 m 35 cm.

Example 6.6 Determining the Heat Energy Liberated in Thermit Welding
A mixture of 800 grams of iron(III) oxide and sufficient aluminum powder is ignited for a thermit-welding reaction. Determine the amount of evolved heat energy. Take the molar mass of Fe_2O_3 = 160 **g/mol.**

Solution

Mass of Fe_2O_3 = 800, $\Delta H = -851.5$ kJ/mol (from Eq. 6.6)

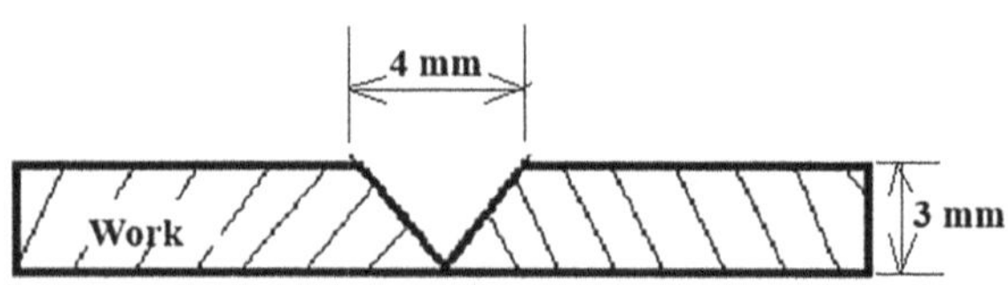

Fig. 6.7 Work dimensions for OFW in Example 6.5

$$\text{Heat energy evolved in the reduction of 1 mole } Fe_2O_3 = 851.5 \text{ kJ}$$

$$\text{No.of moles of } Fe_2\ O_3 \text{ in the thermit reaction} = \frac{\textit{Mass in grams}}{\textit{Molar mass}} = \frac{800}{160} = 5$$

Heat energy evolved in the reduction of 5 mole Fe_2O_3 = 851.5 kJ × 5 = 4257.5 kJ = 4.257 MJ

Example 6.7 Determining Mass of Iron Produced in Thermit Welding
By using the data in Example 6.6, determine the mass of iron produced in the thermit reaction.

Atomic mass of Fe = 55.85 g/mole.

Solution

No. of moles of Fe_2O_3 in the thermit reaction = 5.

By reference to Eq. 6.6, 1 mole of Fe_2O_3 produces 2 moles of iron (Fe).

$$\text{No.of moles of iron } (\text{Fe}) \text{ produced} = 2 \times 5 = 10$$

$$\begin{aligned}\text{Mass of iron produced} &= \text{No.of moles} \times \text{atomic mass} \\ &= 10 \times 55.85 = 559.5 \text{ grams} \cong 560 \text{ g}\end{aligned}$$

Example 6.8 Determining the Mass of Iron(III) Oxide Needed for Thermit Welding
It is required to produce **3 kg of molten iron** for a thermitl welding operation. Determine the mass of Fe_2O_3 needed for the thermit mixture. The molar mass of Fe_2O_3 = 160 **g/mol.**

Solution

$$\text{No.of moles of Fe} = \frac{\textit{Mass of Fe in grams}}{\textit{Atomc mass of Fe}} = \frac{3{,}000}{55.85} = 53.71$$

$$\text{2 moles of Fe requires 1 mole of } Fe_2O_3$$

$$\begin{aligned}&\text{53.71 moles of Fe will requires no.of moles of } Fe_2O_3 \\ &= \frac{1}{2} \times 53.71 = 26.85\end{aligned}$$

$$\begin{aligned}\text{Mass of } Fe_2O_3 \text{ needed} &= \text{No.of moles} \times \text{molar mass} \\ &= 26.85 \times 160 = 4297.22\ g = 4.297 \text{ kg}\end{aligned}$$

The mass of Fe_2O_3 needed for the thermit reaction = 4.3 kg.

Example 6.9 Determining the Total Heat Energy Released in Oxyacetylene Welding
An oxyacetylene welding process releases 10.500 kJ of heat energy per m^3 of acetylene consumed. Determine the total heat energy released, if 0.2 m^3 of acetylene is consumed.

Solution

$$1\text{m}^3 \text{ of consumed acetylene releases heat energy} = 10{,}500\text{kJ}$$

$$0.2\text{m}^3 \text{ of consumed acetylene releases heat energy}$$
$$= \frac{10500kJ}{1m^3} \times 0.2\text{m}^3 = 2{,}100\text{kJ}$$

Total heat energy released = 2100 kJ.

Example 6.10 Determining the Time an Oxygen Cylinder Will Last in Oxyfuel Welding

An oxygen cylinder has a volume of 7 m^3. Determine the time the oxygen cylinder will last, if the flow rate is 22 l/min.

Solution

The volume of oxygen gas in the cylinder = 7 m^3 = 7000 l.

$$22\text{liter oxygen gas is consumed in time} = 1\,\text{min}$$

$$7 \text{ liter oxygengasis consumed in time} = \frac{1\,\text{min}}{22\ liter} \times 7 \text{ liter} = 0.32\,\text{min}$$

The time the oxygen cylinder will last = 0.3 min.

Example 6.11 Determining the Volume of Oxygen with a Neutral Flame in OFW

A neutral flame is achieved in an oxyacetylene welding operation. The oxygen-to-acetylene ratio is 1.2:1. Determine the volume of oxygen consumed, if 16 l of acetylene is used.

Solution

$$1 \text{ liter of acetylene requires oxygen} = 1.2 \text{ liter}$$

$$16 \text{ liter of acetylene requires oxygen} = \frac{1.2\ liter}{1\ liter} \times 16 \text{ liter} = 19.2 \text{ liter}$$

The volume of oxygen consumed = 19.2 l.

Questions and Problems

6.1(MCQs). Encircle the most appropriate answers for each of the following statements.

(1) Which of the following welding processes is the lowest cost process?

(a) Thermit welding, (b) diffusion welding, (c) explosion welding, (d) OFW.

(2) Which of the following welding processes involves combustion reactions?

(a) OFW, (b) diffusion welding, (c) explosion welding, (d) thermit welding.

(3) Which of the following welding processes involves the use of aluminum powder?

(a) OFW, (b) diffusion welding, (c) explosion welding, (d) thermit welding.

(4) Which of the following welding processes involves metal casting?

(a) OFW, (b) diffusion welding, (c) explosion welding, (d) thermit welding.

(5) Which of the following types of flame is required in oxy-acetylene welding of steel?

(a) Carburizing flame, (b) neutral flame, (c) oxidizing flame

(6) Which of the following types of flame is required in oxy-acetylene welding of aluminum?

(a) Carburizing flame, (b) neutral flame, (c) oxidizing flame.

(7) Which of the following mixtures of gases produces the hottest flame for OFW?

(a) Propane-oxygen, (b) acetylene-oxygen, (c) hydrogen-oxygen

(8) Which of the following welding processes is generally used to repair railroad rails?

(a) OFW, (b) diffusion welding, (c) explosion welding, (d) thermit welding.

6.2. (a) Draw a labeled sketch showing OFW.
(b) What reactions occur in oxyacetylene welding?

6.3. Draw sketches showing forehand welding and backhand welding

6.4. Differentiate between forehand welding and backhand welding with the aid of a Table.

6.5. Explain thermit welding with reference to thermit (chemical) reaction.

P6.6. A mixture of 1 kg of iron(III) oxide and aluminum powder is ignited for a thermit-welding reaction. Determine the amount of liberated heat energy. **The heat of reaction in the reduction of** Fe_2O_3 **is** 851.5 kJ/mol. The molar mass of Fe_2O_3 = 160 **g/mol.**

P6.7. An oxyacetylene welding operation is performed on a 5-mm-thick steel cylindrical work-piece (diameter = 8 mm). The welding torch supplies 0.25 m^3 of acetylene per hour and an equal volume rate of oxygen. The overall heat of combustion of acetylene = 55×10^6 J/m^3. The heat transfer factor $f_1 = 0.20$. Determine the: (a) rate of heat liberated during overall combustion (in J/h), (b) rate of heat input (power input) to the work surface (in J/s), (c) the average power density in the circular area, if 75% of the heat from the flame is concentrated in a circular area on the work surface, and (d) the volumes of acetylene and oxygen consumed, if the OFW operation is performed for 20 min.

P6.8. By using the data in P*6.6*, determine the mass of iron produced in the thermit reaction.

Atomic mass of Fe = 55.85 g/mole.

P6.9. It is required to produce 60-cm long seam by OFW on V-grooved work-piece, as shown in Fig. 6.8. (a) Determine the volume of the filler rod that will be consumed. (b) What length of the filler rod is needed, if its diameter is 1.2 mm?

Fig. 6.8 Work dimensions for OFW in P6.9

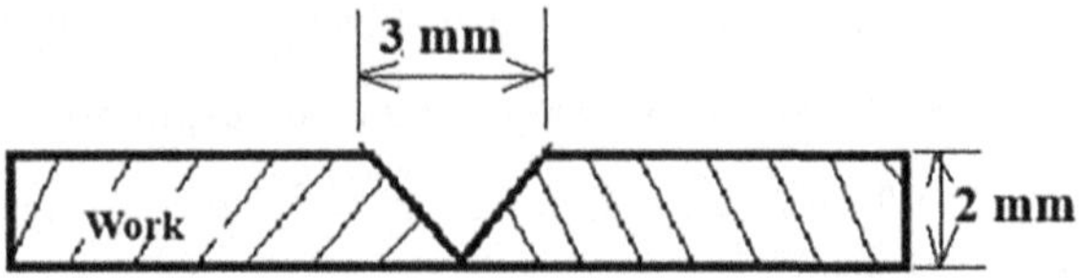

References

Bowditch, K.E., Bowditch, M.A. (2021), Oxyfuel Gas Welding and Cutting (8th Ed.), *Goodheart-Willcox Inc.,* Illinois, USA

EIGA (2025). The Safe Transport, Use, and Storage of Acetylene Cylinders, *European Industrial Gases Association* (EIGA), Brussels—Belgium. https://www.eiga.eu/uploads/documents/SL004.pdf, Accessed on May 22, 2025

Groover, M.P. (2020). Fundamentals of Modern Manufacturing (7th Ed.), *Wiley*, New York

Lide, D.R. (2005). CRC Handbook of Chemistry and Physics (100th Ed.), CRC Press, Boca Raton, FL, USA

Part III
Solid-State Welding Processes

Chapter 7
Cold-Pressure Welding Processes

Abstract This chapter discusses both qualitative and quantitative aspects of cold-pressure welding processes. Cold-pressure processes for welding of sheets and wires are explained with reference to their working principles, operational steps, and engineering analyses, including recent advances in cold welding of nanowires. The need for recrystallization annealing of cold-welded sheets is also justified with the aid of microstructural models. The discussion on ultrasonic welding (USW) includes its working principle, equipment, advantages and limitations, and engineering analysis; here, recent advances in the *USW* of polymers and composites is also explained and analyzed taking into consideration the *energy director* (ED). Explosion welding (EXW) is explained by taking into account its working principle, steps of operation, advantages & limitations, key process parameters, and engineering analysis. The mathematical modeling for *EXW* includes the derivation of workable expressions for flyer-plate velocity, collision-point velocity, and collision angle; here detonation velocities for various explosives are also tabulated. Roll welding (*RoW*) is also defined and illustrated. This chapter contains 14 worked numerical examples, 9 diagrams, 15 formulas, 8 multiple-choice questions (MCQs), and 6 exercise problems. The answers to all MCQs and selected problems are given at the end of this book.

Keywords Cold-pressure welding of sheets and wires · Nanowires · Ultrasonic welding · Energy director · Explosion welding · Flyer-plate velocity · Collision-point velocity · Collision angle · Roll welding

Abbreviations

ED	Energy director
EXW	Explosion welding
HEA	High entropy alloy
LMPAEK	Low-melt poly-aryl-ether-ketone
PEEK	Poly-ether-ether-ketone
RoW	Roll welding
TNT	Tri-nitro toluene
USW	Ultrasonic welding

Z. Huda, *Welding and Joining Engineering*,
https://doi.org/10.1007/978-3-032-20462-2_7

Nomenclature

A	Contact area
D	Contact diameter
F	Punch force
A_v	Amplitude of vibration.
p	Pressure
k	Material constant
σ_{ys}	Yield strength
F_b	Maximum force at break during the tensile test
A_w	Cross-sectional area of the weld
ε_{ED}	*Cyclic strain on the ED*
ΔL	Total deformation on the welding stack or half-peak of the amplitude of vibration
E'_{ED}	Transverse storage modulus of ED
E'_{Ad}	Transverse storage modulus of adherents
L_{ED}	*Thickness of the ED*
ω	Frequency of vibration
E''_{ED}	Loss modulus of the *ED* material
R_e	Critical Reynolds number ($R_e = 8–13$)
v_c	Collision point velocity
ρ_f	Density of flyer plate material
ρ_b	Density of base-plate materials
H_f	Vickers hardness number of the flyer-plate material
H_b	Vickers hardness numbers of the base-plate material

7.1 Cold-Pressure Welding of Sheets and Wires

7.1.1 Working Principle and Steps of Operation in Cold Welding of Sheets

The reader has been introduced to solid-state welding and its advantages in Chap. 1 (Sect. 1.3). Cold-pressure welding is a solid-state welding process that involves just pressure to join two metals with no source of heat. This process is frequently used for joining sheets, wires, and small electrical components. In cold-pressure welding of sheets and wires, a special tool is used to produce localized plastic deformation, which results in coalescence between the two parts. This process works best with soft, ductile metals, such as aluminum, copper, brass, and the like.

Cold-pressure welding of sheets involves the following four steps: (I) surface cleaning, (II) pressure application, (III) coalescence between the two sheets, and (IV) recrystallization annealing. Firstly, the sheet surfaces must be thoroughly cleaned with a wire-brush to remove oxides, dirt, oils, rust, and contaminants; additionally, the sheets must be flat and free from warps to ensure intimate contact. Secondly, a special tool (punch) (Fig. 7.1a) is used to apply pressure to produce

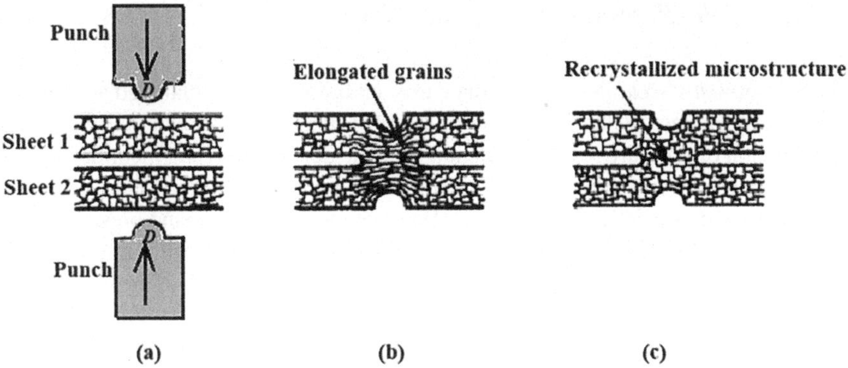

Fig. 7.1 Cold-pressure welding of sheets; (**a**) punches in positions to exert pressure, (**b**) plastically deformed microstructure after the application of pressure, (**c**) recrystallized, refined microstructure with equi-axed grains

localized plastic deformation, which results in coalescence between the two parts; the resulting microstructure with elongated grains indicates reduced ductility (see Fig. 7.1b). In order to restore ductility, the weldment is given a recrystallization-annealing heat-treatment, resulting in a refined, equi-axed microstructure (Fig. 7.1c).

7.1.2 Engineering Analysis of Cold-Pressure Welding of Sheets

The required pressure for cold-pressure welding of sheets can be determined by:

$$\mathrm{p} = k * \sigma_{ys} \tag{7.1}$$

where p is the pressure, k is the material constant ($k = 2.5$ for aluminum), and σ_{ys} is the yield strength of the sheet metal. The contact area of the punch can be calculated by (see Fig. 7.1a):

$$\mathrm{A} = \frac{\pi}{4} * D^2 \tag{7.2}$$

where A is the contact area, and D is the contact diameter. The significance of Eqs. 7.1 and 7.2 is illustrated in Example 7.1. Once the values of pressure and contact areas have been computed, the punch force can be determined by:

$$\mathrm{F} = \mathrm{p} * \mathrm{A} \tag{7.3}$$

where F is the punch force (N), p is the pressure (Pa), and A is the contact area (m^2) (see Example 7.2). In the case of cold-pressure welding of dissimilar metals, σ_y (in Eq. 7.1) would refer to the average of the yields strengths of the two metals (see Examples 7.3 and 7.4).

7.1.3 Cold-Pressure Welding of Wires

In *cold-pressure welding of wires*, the wires' ends are securely clamped and pressed repeatedly against each other by use of a pair of punches, as shown in Fig. 7.2. The application of pressure results in adequate plastic deformation and metallic bonding to create a strong weld joint. Cold welding of wires involves butt welds, because it is easy to trim the ends, get clean metal on the contact surface, and press the wires together (Grill 2025). When making a butt weld for cold welding, it is important that the distance between the clamping points and the contact surface of the work should not be too big, otherwise soft metals may bend sideways instead of joining.

7.1.4 Cold Welding of Nanowires—Recent Advances

Cold welding of nanowires is the solid-state welding of two clean, metallic nanowires (usually gold nanowires) upon contact under slight pressure without the application of heat. At the nanoscale, metals have very high surface-to-volume ratios (especially in ultrathin gold nanowires), which enhances surface-atoms' mobility. The highly-mobile surface atoms diffuse across the interface to form atomic bonds, thereby eliminating the boundary—a distinct advantage. Cold welding of nanowires finds applications in the bottom-up manufacturing of electrical and mechanical nanotechnology devices.

There are generally high-dislocation-density related defects in cold-welded nanowires. This problem has been overcome by recent research. Recently (2024), Nguyen and co-workers have investigated cold welding joints of Co-Ni-Cr-Fe-Mn high-entropy alloy (HEA) nanowires; this study employed molecular dynamics (MD) modeling to investigate the impact of composition, orientation, temperature, and contact pressure on the cold welding mechanism of the HEA nanowires. They

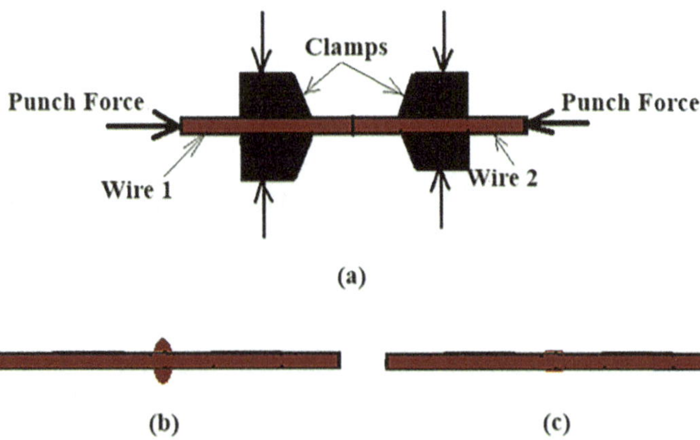

Fig. 7.2 Cold pressure welding of wires (**a**), after welding (**b**), after trimming (**c**)

have found that defect-free cold-welded joints can be successfully formed within a contact pressure range of 10–10,000 Pa (Nguyen et al. 2024).

7.2 Ultrasonic Welding

7.2.1 Working Principle—Advantages and Limitations

Ultrasonic welding (*USW*) is a cold solid-state welding process, whereby high-frequency ultrasonic acoustic vibrations are locally applied to work pieces being held together under pressure. In this process, high-frequency (in the range of 15–40 kHz) low-amplitude (from 20 to 60 microns) lateral vibration is used to create heat owing to friction between the materials to be joined. The generated heat results in plastic deformation and bonding at the works' interface.

The following are the distinct advantages of the *USW*: (a) the cycle times are fast, (b) it is a safe process, (c) it is an energy-efficient process, (d) it results in clean welds, and (e) it is highly capable of welding dissimilar materials. The limitations of USW are as follows: (a) the welding equipment is complex, since there are many variables to control, (b) the equipment is typically costly, and (c) the USW is limited to the welding of small parts.

7.2.2 Ultrasonic Welding Equipment

An ultrasonic welding machine mainly consists of the following components: (a) AC power source, (b) ultrasonic generator, (c) tranducer, (d) coupler, (e) horn, (f) pneumatic system to exert force, (g) reflector sonotrode, (h) anvil, and of course workpiece(s); additionally, there is a control unit that monitors and adjusts the welding parameters (such as frequency, amplitude, and pressure) to ensure consistent and high-quality welds. The ultrasonic generator generates high-frequency electrical energy that is transmitted to a tranducer. The transducer converts the high-frequency electrical energy into low-amplitude mechanical vibrations, which are transmitted to a coupling system (coupler and horn) (see Fig. 7.3). The horn amplifies the vibrations and focuses them onto the materials being welded. The anvil is a flat block that supports the materials being welded and applies pressure to the joint during welding.

7.2.3 Engineering Analysis of Ultrasonic Welding

7.2.3.1 Determinations of Number of Cycles, Total Displacement, and Bonding Strength

The USW is a high-frequency (in the range of 15–40 kHz) low-amplitude (from 20 to 60 microns) welding process. The vibration cycles in USW can be determined by:

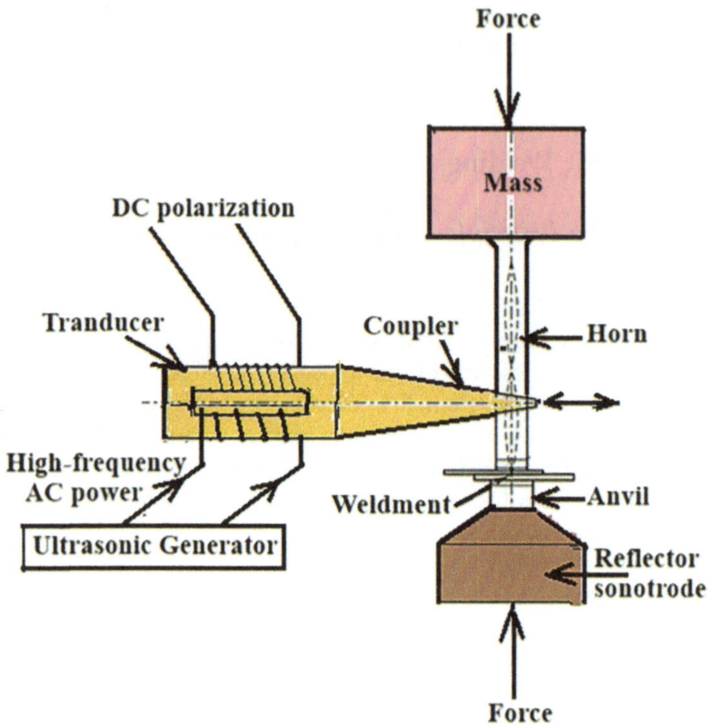

Fig. 7.3 Ultrasonic welding—working principle & equipment

$$\text{No.of cycles} = (\text{frequency}) * (\text{operating time of ultrasonic welder}) \quad (7.4)$$

Since each cycle has double displacements, the total displacement per cycle can be calculated by:

$$\text{Total displacement per cycle} = 2 * A_v \quad (7.5)$$

where A_v is the amplitude of vibration. The total displacement can be determined by:

$$\text{Total Displacement} = (\text{No.of cycles}) * (\text{displacement per cycle}) \quad (7.6)$$

The significance of Eqs. 7.4–7.6 is illustrated in Examples 7.5 and 7.6.

The bonding strength of an ultrasonic weld can be determined by a tensile test on the weldment; this tensile test requires pre-measurement of the cross-sectional area of the weld. The bonding strength is given as follows:

$$\text{Bonding strength} = \frac{F_b}{A_w} \quad (7.7)$$

where bonding strength is in MPa units, F_b is the maximum force at break (in N) during the tensile, and A_w is the cross-sectional area of the weld (mm^2) (see Example 7.7).

7.2.3.2 Ultrasonic Welding of Polymer-Based Composites—*Energy Director*

Ultrasonic welding (USW) of polymer-based composites finds significant application in the modern aerospace industry. In the *USW* of plastics and polymer-based composites, an *energy director* (ED) is created, which is a specific geometric-design feature in the joint that focuses ultrasonic energy to produce a strong weld. The *ED* is typically a small, triangular or other shaped projection on the surface of one of the parts to be joined (Fig. 7.4).

Very recently (2025), Brito and co-researchers have reported a unique USW of polymer-based composite; they used poly-ether-ether-ketone (PEEK) and low-melt poly-aryl-ether-ketone (LMPAEK) reinforced with carbon 5-harness satin to produce the specimens (Brito et al. 2025). In the USW of plastics and polymer-based composites, the cyclic strain on the energy director (ED) can be estimated by (Tateishi, et al., 1992):

$$\varepsilon_{ED} = \frac{\Delta L}{\left[2 * \frac{E'_{ED}}{E'_{Ad}} * L_{Ad} + L_{ED}\right]} \tag{7.8}$$

where ε_{ED} *is the cyclic strain on the ED,* ΔL *is the* total deformation on the welding stack or half-peak of the amplitude of vibration (μm), E'_{ED} is the transverse storage modulus of ED, E'_{Ad} is the transverse storage modulus of adherents, and L_{ED} *is the thickness of the ED. Once the value of the* ε_{ED} *has been computed by using* Eq. 7.8, the viscoelastic heating rate can be determined by (Brito et al. 2025):

$$\dot{Q}_{visc} = \frac{\omega * E''_{ED} * \varepsilon^2_{ED}}{2} \tag{7.9}$$

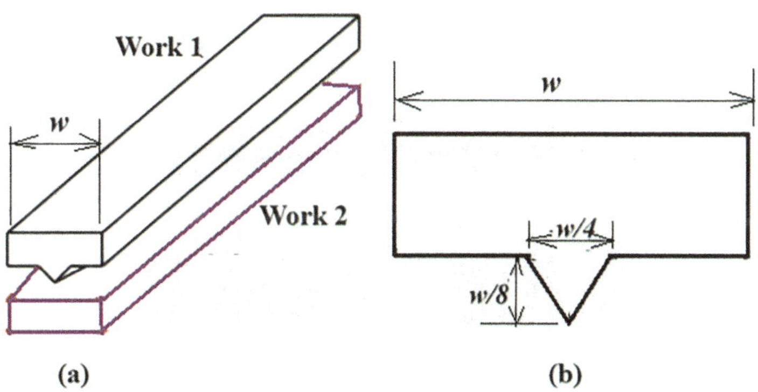

Fig. 7.4 Workpiece(s) with a simple ED for USW of plastics (**a**), a simple energy director's geometric design (**b**)

where ω is the frequency of vibration (Hz), and E''_{ED} is the loss modulus of the energy director (*ED*) material. The significance of Eqs. 7.8 and 7.9 is illustrated in Examples 7.8 and 7.9.

7.3 Explosion Welding

7.3.1 Working Principle—Advantages and Limitation

Explosion welding (EXW) is a cold solid-state welding process that involves a high-velocity oblique collision between two metallic workpieces. The high-velocity collision (impact) is accomplished by high-energy explosives; the impact causes the colliding metal surfaces to flow hydrodynamically when they intimately contact one another to promote solid-state bonding.

Explosive welding has several distinct advantages. Neither melting nor excessive plastic flow are generally essential for this process to succeed, and hence no serious limitation is imposed by large differences in the corresponding metal properties. However, the use of explosives imposes a limitation from the viewpoint of safety regulations (El-Sobky 1983).

7.3.2 Steps of Operation

The *explosion welding* process involves the following steps of operation: (I) plate positioning, (II) explosive distribution, (III) explosive detonation, collision, and bonding; (IV) the final weld. Firstly, the flyer plate (such as clad metal) is placed either parallel or at an angle to the base-plate, while maintaining the stand-off distance (see Fig. 7.5). Then an explosive is distributed over the top surface of the flyer plate. The explosive is detonated resulting in a high-pressure impact between the

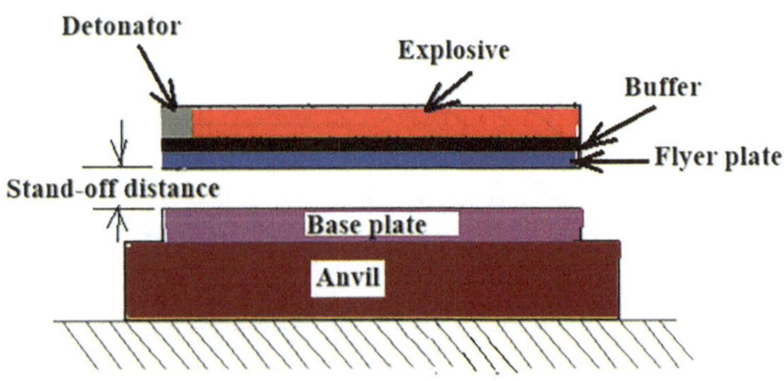

Fig. 7.5 The equipment ready for explosion welding (parallel arrangement)

plates, thereby causing local melting and plastic deformation, leading to bonding at the interface of the plates (see Fig. 7.6). The resulting weld has a wavy appearance.

7.3.3 Engineering Analysis of EXW

The key process parameters in EXW include detonation velocity, flyer-plate velocity, collision point velocity, stand-off spacing, and explosion ratio (Shan et al. 2023). Some of these process parameters are illustrated in Fig. 7.7. The denotation velocity v_d is in the range of 1.5–6.9 km/s, depending on the type of explosive (see Table 7.1).

When the explosive mass and the accelerated mass are known, the flyer-plate velocity can be determined by using the following formula (Salwan et al. 1979):

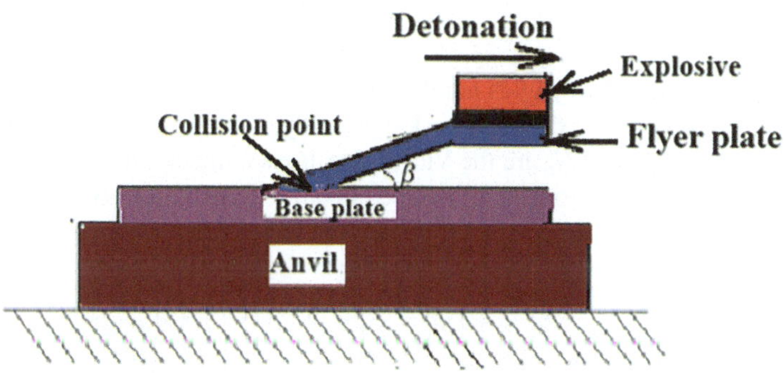

Fig. 7.6 Explosion welding (collision leading to welding) (β = collision angle)

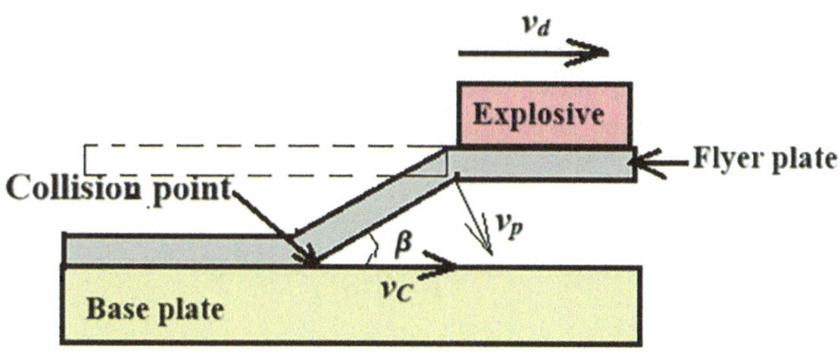

Fig. 7.7 Process parameters in EXW (β = collision angle, v_d = detonation velocity, v_p = flyer-plate velocity, v_c = collision-point velocity)

Table 7.1 Detonation velocities for various explosives

Explosive	Mixture of ammonium nitrate, TNT, and *Al*	Mixture of ammonium nitrate and fuel oil	Tri-nitro toluene (TNT)
Detonation velocity	2.0–3.8 km/s	1.5–2.4 km/s	6.9 km/s

$$\mathrm{v}_p = \mathrm{v}_d * \left[1 + \frac{27}{16} * \frac{C}{m}\left(1 - \sqrt{1 + \frac{32}{27} * \frac{C}{m}}\right)\right] \quad (7.10)$$

where v_p is the flyer-plate velocity, v_d is the detonation velocity, C is the explosive mass, and *m* is the accelerated mass (see Example 7.10).

The collision point velocity v_c is the velocity at the collision point (see Fig. 7.7); it is associated with the flow-transition limit. The flow-transition limit defines the transition of the interfacial material from laminar to turbulent motion, which is the critical value at which the interface changes from smooth to wavy morphology (Chen et al. 2024). Cowan and co-researchers considered welding metal as a fluid and proposed the following flow-transition limit for the collision of metal plates (Cowan et al. 1971):

$$R_e = \frac{\mathrm{v}_c^2 * \left(\rho_f + \rho_b\right)}{2 * \left(H_f + H_b\right)} \quad (7.11)$$

where R_e is the critical Reynolds number ($R_e = 8$–13), v_c is the collision point velocity (m/s), ρ_f and ρ_b are the densities (in kg/m^3) of flyer-plate and base-plate materials, respectively, and H_f and H_b are the Vickers hardness numbers (in IS units) of flyer-plate and base-plate materials, respectively. In order to derive a workable expression for the collision-point velocity, the terms in Eq. 7.11 can be re-arranged as follows:

or $$\mathrm{v}_c = \sqrt{\frac{2 * R_e * \left(H_f + H_b\right)}{\left(\rho_f + \rho_b\right)}} \quad (7.12)$$

It has been experimentally shown that for effective EXW the collision point velocity should be in the range of 1800–3000 m/s (Émurlaeva et al. 2019). The significance of Eq. 7.12 is illustrated in Example 7.11.

Once the values of v_p and v_c have been computed by using Eqs. 7.10–7.12, the collision angle β can be determined by the following formula (Deribas and Zakharenko 1974):

$$\beta = \frac{k_1}{v_c} * \sqrt{\frac{H_v}{\rho}} \quad (7.13)$$

where the angle β is expressed in rad., the constant k_1 is related to the material's surface, with values ranging from 0.6 to 1.2 (from highly clean to less clean surfaces) (for a low-hardness material, $k_1 = 1.14$), H_v is the average of the Vickers hardness numbers of flyer-plate and base-plate materials, and ρ is the average of the densities of the materials (see Example 7.12). For successful EXW, the collision angle should generally be in the range of 5°–20°. The stand-off distance can be related to the thickness of flyer plate by:

$$\text{Stand-off distance} = \mathrm{k} * \mathrm{t} \quad (7.14)$$

where the constant k lies in the range of 0.2–0.5 (see Example 7.13).

7.4 Roll Welding

Roll welding (*RoW*) is a cold solid-state welding process that involves passing stacked sheets or plates through rolls under pressure, causing sufficient deformation to join them (see Fig. 7.8). This technique is commonly used for joining ductile metals and alloys, and for the fabrication of pipes, tubes, and other cylindrical structures.

It is technologically important to determine the effective length of a weld in *RoW*. The effective length, L_{eff}, of a weld can be calculated by:

$$L_{eff} = L_{tot} - 2s \tag{7.15}$$

where L_{tot} is the total or actual length of the weld, and s is the weld size (see Example 7.14).

7.5 Worked Numerical Examples

Example 7.1 Determining the Pressure and Contact Area in Cold-Pressure Welding of Sheets

Cold-pressure welding is performed to join annealed 1060 aluminum sheets with yield strength of 28 MPa. The contact diameter of the punch is 3 mm. Determine the required pressure and the contact area.

Solution

$\sigma_y = 28$ MPa, D = 3 mm, p =?, A =?

By using Eqs. 7.1–7.2,

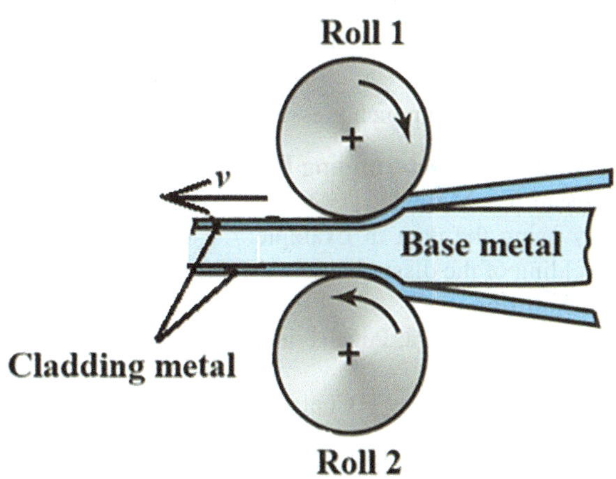

Fig. 7.8 Roll welding—*schematic*

$$p = 2.5 * \sigma_y = 2.5 \times 28 = 70\,\text{MPa}$$

$$A = \frac{\pi}{4} * D^2 = 0.785 \times 3^2 = 7.1\,\text{mm}^2$$

The required pressure is 70 MPa, and the contact area is 7.1 mm^2.

Example 7.2 Determining the Punch Force in Cold Pressure Welding of Sheets

By using the data in Example 7.1, determine the punch force required for the welding operation.

Solution

$p = 70$ MPa $= 70 \times 10^6$ Pa $= 70 \times 10^6$ N/m^2,
A = 7.1 $mm^2 = 7.1 \times (10^{-3}\, m)^2 = 7.1 \times 10^{-6}\, m^2$
By using Eq. 7.3,

$$F = p * A = 70 \times 10^6 \times 7.1 \times 10^{-6} = 497\,\text{N}$$

The required punch force ≅ 500 N.

Example 7.3 Determining the Punch Pressure in Cold Pressure Welding of Dissimilar Metals

Cold-pressure welding is performed to join an annealed 1060 aluminum sheet (yield strength = 28 MPa) with an annealed brass sheet ($\sigma_{y,\,br}$ = 103 MPa). The contact dimensions of the punch are: 15 mm × 20 mm. Determine the required punch pressure. Take $k = 2.8$ (in Eq. 7.1).

Solution

$\sigma_{y,\,Al}$ = 28 MPa, $\sigma_{y,\,br}$ = 103 MPa, k = 2.8, p =?

$$\sigma_y = \frac{\sigma_{y,Al} + \sigma_{y,br}}{2} = \frac{28 + 103}{2} = 65.5\,\text{MPa}$$

By using Eq. 7.1,

$$p = k * \sigma_y = 2.8 \times 65.5 = 183.4\,\text{MPa}$$

The required punch pressure = 183.4 MPa.

Example 7.4 Determining Punch Force for Cold Pressure Welding of Dissimilar Metals

By using the data in Example 7.3, determine the punch force for cold-pressure welding of the dissimilar metals' sheets.

Solution

p = 183.4 MPa, Punch contact dimensions: 15 mm × 20 mm, F =?

$$\text{Contact area} = A = 15 \times 20 = 300\,\text{mm}^2$$

$$F = p * A = 183.4 \frac{N}{mm^2} \times 300\, mm^2 = 55020\,\text{N}$$

The required punch force = 55.02 kN.

Example 7.5 Determining the Number of Cycles in Ultrasonic Welding
A 30-kHz ultrasonic welding machine is operated for 0.4 s. Determine the number of vibration cycles during the welding operation.

Solution

Frequency of vibration = 30 kHz = 30,000 cycles/s, time = 0.4 s.

By using Eq. 7.4,

$$\text{No.of cycles} = (\text{frequency}) * (\text{operating time}) = 30{,}000 \times 0.4 = 12{,}000$$

There are 12,000 vibration cycles during the welding operation.

Example 7.6 Determining Total Displacement per Cycle, and Total Displacement in USW
By using the data in Example 7.5, determine the total displacement per cycle, and the total displacement, if the amplitude of vibration of the horn is 50 μm.

Solution

Frequency = 30,000 cycles/s, A = 50 μm, No. of cycles = 12,000.

(a) By using Eqs. (7.5)–(7.6),
   Total displacement per cycle = 2∗A = 2 × 50 μm = 100 μm.
(b) The total displacement can be determined by:
   Total Displacement = (No. of cycles) ∗ (displacement per cycle) = 12,000 × 100 μm.
   Total Displacement = 12 × 10^5 μm = 1.2 m.

Example 7.7 Determining the Bonding Strength of an Ultrasonic Weld
Two plastic components are welded using ultrasonic welding (USW). The weld area is 22 mm × 10 mm. The weldment was tested by a tensile test. The maximum force at break is 350 N. Determine the bonding strength of the ultrasonic weld.

Solution

A_w = 22 mm × 10 mm = 220 mm², F_b = 350 N.

By using Eq. 7.7,

$$\text{Bonding strength} = \frac{F_b}{A_w} = \frac{350}{220} = 1.59\,\text{MPa}$$

Example 7.8 Determination of the Cyclic Strain in the ED in USW of Polymer Composite
Two poly-ether-ether-ketone (PEEK)-based composite sheets/adherents, each with a thickness of 1.9-mm, are joined by ultrasonic welding (USW); the welding stack is as follows: adherent-ED-adherent. The thickness of the energy director (ED) is 0.25 mm. The frequency of vibration, for the USW operation, is 20 kHz, and the peak-to-peak amplitude of vibration is 86 μm. The transverse storage modulus of ED, at its glass temperature T_g, is 0.97 GPa, while the transverse storage modulus of the adherent, at its T_g, is 7.61 GPa. Determine *the cyclic strain on the ED.*

Solution

L_{Ad}= 1.9 mm, L_{ED} = *0.25 mm*, ΔL = $\frac{1}{2}$ × *86* = *43* μm = 0.043 mm, E'_{ED} = 0.97 GPa,

E'_{Ad} = 7.61 GPa, ε_{ED} =*?*

By using Eq. 7.8,

$$\varepsilon_{ED} = \frac{\Delta L}{\left[2 * \frac{E'_{ED}}{E'_{Ad}} * L_{Ad} + L_{ED}\right]} = \frac{0.043}{\left[2 \times \frac{0.97}{7.61} \times 1.9 + 0.25\right]} = \frac{0.043}{0.73436} = 0.05855$$

The cyclic strain on the ED = ε_{ED} = *0.05855.*

Example 7.9 Determination of the Viscoelastic Heating Rate in the USW Operation

By using the data in Example 7.8, determine the viscoelastic heating rate in the USW operation.

The loss modulus of the *ED* material is 0.226 GPa.

Solution

$$\omega = 20\,\text{kHz} = 20{,}000\,\text{Hz}, \varepsilon_{ED} = 0.05855, E''_{ED} = 0.226 \times 10^9\ \text{Pa}$$

By using Eq. 7.9,

$$\dot{Q}_{visc} = \frac{\omega * E''_{ED} * \varepsilon^2_{ED}}{2} = \frac{20{,}000 \times 0.226 \times 10^9 \times 0.05855^2}{2} = 7.747 \times 10^9\ \text{N/s} \cdot \text{m}^2$$

The viscoelastic heating rate = $\dot{Q}_{visc}$ = 7.747 ×10^9 N/s•m^2

Example 7.10 Determining the Flyer Plate Velocity in Explosion Welding

An aluminum sheet is cladded to a mild-steel plate by EXW using TNT as the explosive; the explosive mass/accelerated mass ratio is 0.128. Determine the flyer-plate velocity for the welding operation. Hint: refer to the data in Table 7.1.

Solution

v_d = 6.9 km/s, C/m = 0.128, v_p =?

By using Eq. 7.10,

$$v_p = v_d * \left[1 + \frac{27}{16} * \frac{C}{m} * \left(1 - \sqrt{1 + \frac{32}{27} * \frac{C}{m}}\right)\right] = 6.9 \times \left[1 + \frac{27}{16} \times 0.128 \times \left(1 - \sqrt{1 + \frac{32}{27} \times 0.128.}\right)\right]$$

$$v_p = 6.9 \times \left[1 + 0.216 \times \left(1 - \sqrt{1.15}\right)\right] = 6.9 \times \left[1 + 0.216 \times \left(1 - \sqrt{1.15}\right)\right] = 6.792\,\text{km/s}$$

The flyer-plate velocity = 6.792 km/s = 6792 m/s

Example 7.11 Determining Collision Point Velocity and Commenting on Feasibility of EXW

A commercially-pure titanium (TA1) plate was explosively welded to a mild steel (Q235B) plate. The densities of the titanium plate and mild-steel plates' materials are 4.45 and 7.85 g/cm^3, respectively. The Vickers hardness numbers of the two plates' materials are 130 and 150 kgf/mm^2, respectively. (a) Determine the collision point velocity for the EXW operation.

(b) Comment on the successful feasibility of the operation.

Solution

(a) H_f = 130 kgf/mm^2 = 130 × 9.81 N/mm^2 = 1274 × 10^6 N/m^2, ρ_f = 4.45 g/cm^3 = 4450 kg/m^3

H_b = 150 kgf/mm^2 = 150 × 9.81 N/mm^2 = 1470 × 10^6 N/m^2, ρ_b = 7.85 g/cm^3 = 7850 kg/m^3

The average $R_e = \dfrac{8+13}{2} = 10.5$

By using Eq. 7.12,

$$v_c = \sqrt{\frac{2 * R_e * \left(H_f + H_b\right)}{\left(\rho_f + \rho_b\right)}} = \sqrt{\frac{2 \times 10.5 \times 10^6 \times (1274 + 1470)}{(4450 + 7850)}} = \sqrt{\frac{57624 \times 10^6}{12300}}$$
$$= 2.16 \times 10^3 \text{ m/s}$$

The collision point velocity = v_c = 2160 m/s

(b) Since the v_c value is well in the range of 1800–3000 m/s, the welding is feasible.

Example 7.12 Determining Collision Angle and Commenting on Feasibility of EXW

By using the data in Example 7.11, determine the collision angle for the EXW operation. Is the obtained β value conforms to successful feasibility of the welding operation?

Solution

(a) H_f = 1274 × 10^6 N/m^2, ρ_f = 4450 kg/m^3, H_b = 1470 × 10^6 N/m^2, ρ_b = 7850 kg/m^3
k_1 = 1.14 (for low hardness material—titanium), v_c = 2160 m/s.

$$H_v = \frac{1274 \times 10^6 + 1470 \times 10^6}{2} = 1372 \times 10^6 \text{ N/m}^2$$

$$\rho = \frac{4450 + 7850}{2} = 6150 \text{ kg/m}^3$$

By using Eq. 7.13,

$$\beta = \frac{k_1}{v_c} * \sqrt{\frac{H_v}{\rho}} = \frac{1.14}{2160} \times \sqrt{\frac{1373 \times 10^6}{6150}} = 527.77 \times 10^{-6} \times 0.4725 \times 10^3 = 0.24937 \text{ rad.}$$

$$\beta = \frac{180^o}{\pi} \times 0.24937 = 14.3°$$

The collision angle = β = 14.3°

(b) Since the β value is well in the range of 5—20°, the welding operation is feasible.

Example 7.13 Determining the Stand-off Distance in an Explosion Welding Operation

An explosion welding operation is performed using a 2-mm-thick flyer plate. Determine the stand-off distance.

Solution

t = 2 mm, k = 0.3

By using Eq. 7.14,

$$\text{Stand}-\text{off distance} = \text{k} * \text{t} = 0.3 \times 2 = 0.6\,\text{mm}$$

Example 7.14 Determining the Effective Length of a Weld in Roll Welding (RoW)

The total length of a weld, in a RoW operation, is 80 mm, and the weld size is 8 mm. Determine the effective length of the weld.

Solution

L_{tot} = 80 mm, s = 8 mm.

By using Eq. 7.15,

$$L_{eff} = L_{tot} - 2\text{s} = 80 - 2 \times 8 = 80 - 16 = 64\,\text{mm}$$

The effective length of the weld = 64 mm.

Questions and Problems

7.1. Encircle the most appropriate answers for each of the following statements.

(1) Which one of the following cold welding proceses requires the use of a detonator?

(a) Cold-pressure welding of sheets, (b) ultrasonic welding, (c) explosion welding, (d) RoW

(2) Which welding process is the most suitable for cold welding of cylindrical work-piece?

(a) Cold-pressure welding of sheets, (b) ultrasonic welding, (c) explosion welding, (d) RoW

(3) Which one of the following cold welding processes requires the use of a punch?

(a) Cold-pressure welding of sheets, (b) ultrasonic welding, (c) explosion welding, (d) RoW

(4) Which one of the following welding processes requires the use of a tranducer?

(a) Cold-pressure welding of sheets, (b) ultrasonic welding, (c) explosion welding, (d) RoW

(5) Which one of the following ranges of vibration-frequency is used in ultrasonic welding?

(a) 15–40 kHz, (b) 100–500 kHz, (c) 15–40 Hz, (d) 100–500 kHz.

(6) Which one of the following values is the approximate detonation velocity of TNT?

(a) 200 km/s, (b) 50 km/s, (c) 7 km/s, (d) 1.5 km/s.

(7) Which one of the following welding processes requires the creation of an *energy director*?

(a) Cold-pressure welding of sheets, (b) ultrasonic welding, (c) explosion welding, (d) RoW.

(8) Which one of the following welding processes is limited to welding of soft, ductile metals?

(a) Cold-pressure welding of sheets, (b) ultrasonic welding, (c) explosion welding, (d) RoW.

7.2. (a) Draw a labeled sketch showing cold-pressure welding of wires.
(b) Briefly describe cold-pressure welding of nanowires.

7.3. (a) List and elaborate the steps of operations in cold-pressure welding of sheets.
(b) What additional heat treatment is required in cold-pressure welding of sheets? And why?

7.4. (a) What is the working principle of ultrasonic welding (USW)?
(b) List the advantages and limitations of USW.
(c) List at least five components of an USW equipment.

7.5. Explain explosion welding (EXW) with the aid of sketches.

7.6. Explain briefly roll welding process with the aid of a sketch.

P7.7. Cold-pressure welding is performed to join annealed copper sheets with yield strength of 70 MPa. The contact diameter of the punch is 4 mm. Determine the: (a) required pressure, (b) contact area, and (c) punch force required for the welding operation.

P7.8. Cold-pressure welding is performed to join an annealed 1060 aluminum sheet (yield strength = 28 MPa) to an annealed brass sheet ($\sigma_{y,\,br}$ = 103 MPa). The contact dimensions of the punch are: 12 mm × 18 mm. Determine the required punch pressure. Take $k = 2.6$.

P7.9. Two poly-ether-ether-ketone (PEEK)-based composite sheets/adherents, each with a thickness of 1.9-mm, are joined by ultrasonic welding (USW); the welding stack is as follows: adherent-ED-adherent. The thickness of the energy director (ED) is 0.25 mm. The frequency of vibration, for the USW operation, is 20 kHz, and the peak-to-peak amplitude of vibration is 86 μm. The transverse storage modulus of ED, at its glass temperature T_g, is 0.97 GPa, while the transverse storage modulus of the adherent, at its T_g, is 7.61 GPa. The loss modulus of the *ED* material is 0.226 GPa. Determine *the: (a) cyclic strain on the ED, and (b)* viscoelastic heating rate in the USW operation.

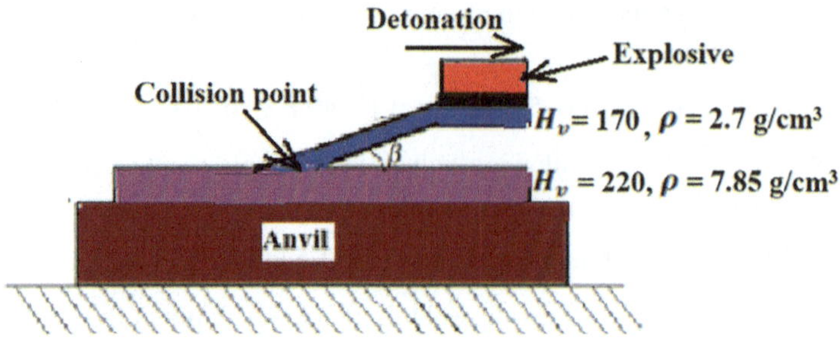

Fig. 7.9 Explosion welding data for P7.11

P7.10. An ultrasonic welding machine, with a frequency of 25 kHz, is operated for 0.3 seconds.
Determine the number of vibration cycles during the welding operation.

P7.11. A commercially-pure aluminum plate was explosively welded to a mild steel plate, as shown in Fig. 7.9. Determine the: (a) collision point velocity, and (b) collision angle. Take Reynolds number = R_e = 10.

P7.12. Two plastic components are welded using ultrasonic welding (USW). The weld area is 18 mm × 8 mm. The weldment was tested by a tensile test. The maximum force at break is 320 N. Determine the bonding strength of the ultrasonic weld.

References

Brito, C.B.G., Teuwen, J., Dransfeld, C.A., Villegas, I.F. (2025). Ultrasonic welding of thermoplastic composites: A comparison between polyetheretherketone and low-melt polyaryletherketone as resin in the adherends and energy directors. *Composites Part B: Engineering,* Vol. 296, 112264

Chen, W., Huang, F-l, Zeng, Z-h., *et al.* (2024). Microstructure and mechanical properties of heterogeneous-gradient composite plates with a sphere-array interlayer prepared using explosive welding. *Journal of Materials Research and Technology,* Vol. 30, pp. 6508-6519

Cowan, G.R., Bergmann, O.R., Holtzman, A.H. (1971). Mechanism of bond zone wave formation in explosion-clad metals, *Metallurgical Transactions*, Vol. 2, pp. 3145-3155.

Deribas, A.A., Zakharenko, I.D. (1974). Surface effects with oblique collisions between metallic plates. *Combustion, Explosion, and Shock Waves,* Vol. 10, pp. 358-367,

Émurlaeva, Y.Y., Bataev, I.A., Zhou, Q., *et al.* (2019). Welding Window: Comparison of Deribas' and Wittman's Approaches and SPH Simulation Results, *Metals*, Vol. 9, p, 1323

Grill, J. (2025). Cold Welding Explained: What is it? How It Works, *Weld Guru*. Online: https://weldguru.com/what-is-cold-welding/

El-Sobky, H. (1983) Mechanics of Explosive Welding, *In*: Explosive Welding, Forming and Compaction (Editor: T.Z. Blazynki), *Springer Dordrecht.*

Nguyen, V-T., Nhu, V.T.T.,Vo, X-T. (2024). Atomistic insights into cold welding process of high-entropy alloy nanowires, *materialstoday*, Vol. 41, 110755

Shan, S., Liu, Y., Zhang, J. *et al.* (2023). Explosion welding research on large-size ultra-thick copper-steel composites: a review. *Journal of Materials Research and Technology,* Vol. 24, pp. 4130-4142.

Salwan, S.K., Prasad, S., Deshpande, A.S. (1979). Explosive Welding Parameters. *Indian Welding Journal.* Jan. 1979, pp. 7–14.

Tateishi, N., North, T.H., Woodhams, R.T. (1992). Ultrasonic welding using tie-layer materials. Part I: analysis of process operation. Polymer Eng. Sci., Vol. 32.

Chapter 8
Hot Solid-State Welding Processes

Abstract This chapter presents both qualitative and quantitative discussions on all important hot solid-state welding processes, including friction welding (FRW), friction stir welding (FSW), diffusion welding/bonding, and inertia friction welding (IFW). These processes are explained with reference to their working principles, advantages, sketches and engineering analyses. The illustration of FRW includes a diagram showing the working components of FRW machines as well as a photo from industry. The engineering analysis of FRW includes the determinations of the rate of frictional heat generated, the torque, and the heat energy generated during the friction phase of the FRW operation; here, a new thermal model is presented based on the analysis of friction heat exchange on the faying surfaces. The mathematical modeling of FSW includes the determinations of the rates of heat generation and the feed. Diffusion bonding is explained by emphasizing its preference for joining of high-strength alloys in the aerospace and nuclear industries; here formulas for calculating the diffusion depth and the bonding force are included. The analysis of the IFW process covers calculations for the moment of inertia and the kinetic energy. This chapter includes 10 worked numerical examples, 7 diagrams, 10 formulas, and 6 MCQs; the answers to the MCQs and the selected problems are given at the end of the book.

Keywords Friction welding · Friction stir welding · Diffusion welding · Inertia friction welding · Probe

Abbreviations

CNC	Computer numerical control
FrW	Friction welding
FSW	Friction stir welding
IFW	Inertia friction welding
PLC	Programmable logic controller

Z. Huda, *Welding and Joining Engineering*,
https://doi.org/10.1007/978-3-032-20462-2_8

Nomenclature

$\dot{Q}_{FRW}$	Rate of heat energy generated during FrW
F	Axial force
μ	Coefficient of friction between the contacting surfaces
v	Peripheral speed of the rotating rod
T	Torque
r	Radius of the rotating rod
Q	Heat energy generated during the friction phase
t	Time elapsed during the friction phase
q_s	Rate of heat exchanged between unit area of free surfaces of the components and the environment
h_s	Coefficient of convective heat transfer
T_A	Ambient temperature or the initial temperature of the workpieces
T_s	Surface temperature of the components during heat exchange
x	Diffusion depth
D	Diffusion coefficient
t	Welding time
$\dot{Q}_{FSW}$	Local heat-generation rate during FSR
F_n	Normal force applied along the axis of the shouldered tool
R_s	Radius of the shouldered tool

8.1 Hot Solid-State Welding Processes—*Fundamentals*

Hot solid-state welding processes involve the coalescence of joining surfaces by the application of a combination of heat and pressure, but the temperature is less than the melting temperature of the base metal(s). Solid-state (SS) welding is advantageous over fusion welding, since neither filler metal is required in these processes, nor the undesirable heat-affected zone (HAZ) is formed. Bonding dissimilar metals can be achieved without problems of thermal expansions in SS welding (Akinlabi and Mahamood 2020). For successful atomic bonding in SS welding, the two surfaces must be very clean. Typical examples of hot solid-state welding processes include friction welding (FrW), friction stir welding (FSW), diffusion bonding/welding, and inertia friction welding (IFW). These processes are explained in the following sections.

8.2 Friction Welding

8.2.1 Overview of Friction Welding—Types and Advantages

Friction welding (*FrW*) is a hot solid-state welding process that produces a weld by generating heat under friction between workpieces in relative motion to one another. The friction-generated heat softens the material at the joint, allowing it to fuse under pressure, while impurities are pushed out (as flash), ensuring a strong bond. Friction welding can be widely used in industry owing to its precision of operation, high processing speed, and low cost. Friction welding has the advantages of the incomplete molten state of the welded parts at the weld interface region, which is particularly important for the welding of dissimilar materials, in which case a narrow heat-affected zone (HAZ) is favorable (Yilbas and Sahin 2014). Friction welding finds applications across a variety of industries, including automotive, aerospace, energy, military, and construction to create high-strength, reliable joints in critical components, such as vehicle chassis, aircraft parts, and pipeline connections.

There are three main methods of *FrW*: (a) rotary *FrW,* (b) radial *FrW,* and (c) linear *FrW.* In the rotary *FrW*, the rotation of one workpiece is controlled, while the other workpiece does not rotate. A linear friction welding machine controls the back-and-forth motion of one part while the other remains stationary. The rotary *FrW* is the most common method, and it is described in the following subsection.

8.2.2 Rotary Friction Welding

Rotary friction welding (rotary *FrW*) produces a weld by generating heat under friction by rapid rotation of one piece against a stationary piece with the application of a compressive force. Workpieces with a rotational symmetry (*e.g.* round rods, bars, tubes, etc.) are particularly suitable for friction welding. There are five phases in rotary *FrW*: (I) one workpiece rotates and the other is idle, (II) frictional contact phase, (III) friction phase, (IV) forge phase, and (V) holding phase (see Fig. 8.1).

Figure 8.1 illustrates the operational phases in a rotary FrW. Firstly, one of the components is rotated and the other component is brought close to it (Fig. 8.1a). The next is the frictional contact phase; here, with one work being rotated, the other component is pressed against it with a defined axial pressure (Fig. 8.1b). Then, the axial pressure is gradually increased, resulting in heat up and the formation of flash (friction phase: Fig. 8.1c). The next phase is the forge phase (Fig. 8.1d); here, the rotation is stopped, and the axial pressure is, once again, increased to complete the

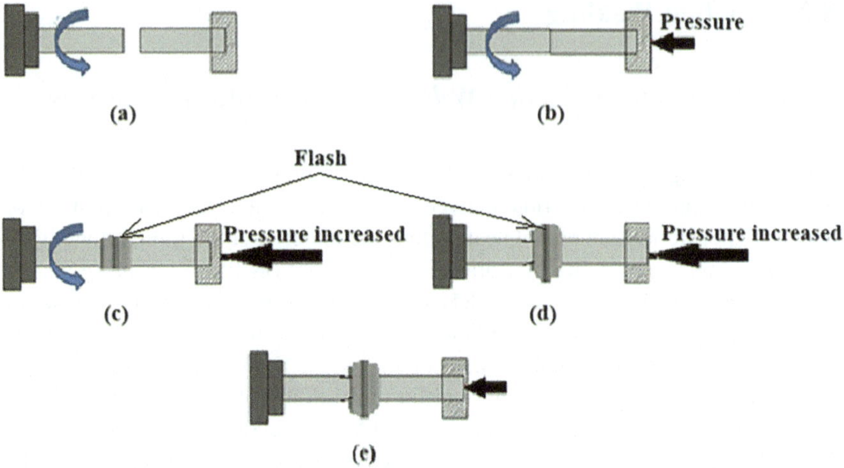

Fig. 8.1 Operational phases in rotary FRW; (**a**) one-work rotation, (**b**) frictional contact phase, (**c**) friction phase, (**d**) forge phase, (**e**) holding phase

formation of flash. Finally, both the components remain under pressure and the weldment is cooled (holding phase: Fig. 8.1e).

8.2.3 Friction Welding Equipment

A **friction welding machine** is a specialized device that uses rotational and linear motions to create heat through friction between two workpieces. The essential components of a friction-welding machine are illustrated in Fig. 8.2; these components include: a motor, a rotating chuck, a rotating workpiece, a non-rotating workpiece, and a hydraulic cylinder.

In modern industrial practice, the machine features advanced control systems, sensors, and monitoring tools to precisely adjust welding parameters and track the process, ensuring optimal results (see Fig. 8.3). The structural and operational components of a friction welding machine include (a) base and frame, (b) spindle and drive system, (c) clamps and workpiece fixturing system, and (d) a control and monitoring system. The base and frame form the machine's structure; they support all mechanical components, ensuring the machine's stability and rigidity. These parts are made of high-strength steel or cast iron to withstand the immense pressure and vibration during the welding process. The control and monitoring system, managed by a programmable logic controller (PLC) or computer numerical control (CNC) system, adjusts welding parameters such as rotation speed, pressure, and feed rate. It is equipped with sensors to measure force, temperature, and displacement, ensuring consistent and high-quality welds. A modern friction welding machine is ideal for high-volume production because it is efficient, highly repeatable, and requires minimal manual intervention.

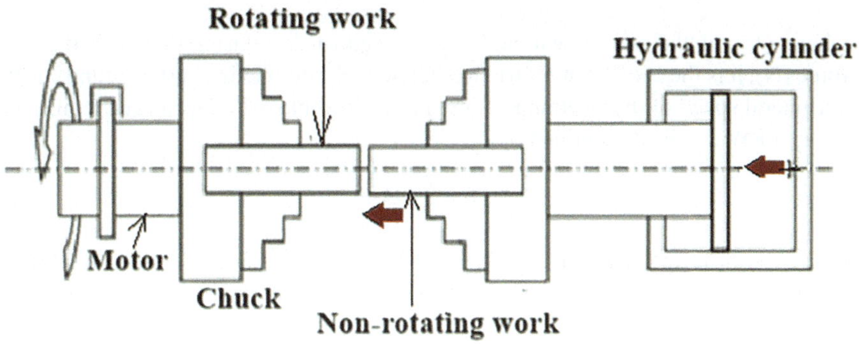

Fig. 8.2 Essential working components of a friction welding machine

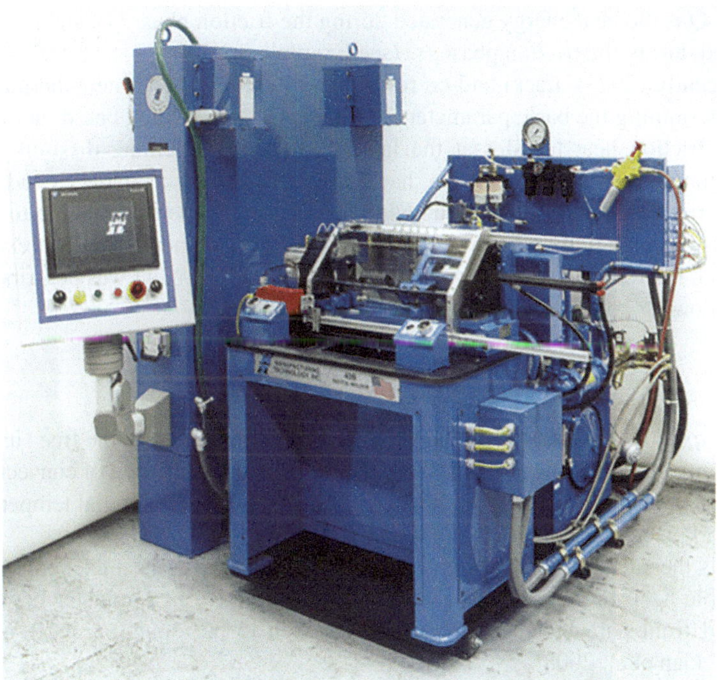

Fig. 8.3 The M40 B/V friction welding machine (Courtesy: *MTI Welding Tech. Ltd.*, UK)

8.2.4 Engineering Analysis of FRW

The rate of frictional heat generated during rotary friction welding can be expressed by:

$$\dot{Q}_{FrW} = \mu * \mathrm{F} * v \tag{8.1}$$

where $\dot{Q}_{FrW}$ is the rate of heat energy generated (in J/s) during FrW, F is the axial force (N), μ is the coefficient of friction between the contacting surfaces, and v is the peripheral speed of the rotating rod (m/s) (see Example 8.1). The torque required in rotary FRW can be determined by:

$$\mathrm{T} = \mu * \mathrm{F} * \mathrm{r} \tag{8.2}$$

where T is the torque (N•m) and r is the radius of the rotating rod (m) (see Example 8.2).

The heat energy generated during the friction phase of the FRW operation can be calculated by:

$$\mathrm{Q} = \dot{Q}_{FRW} * \mathrm{t} \tag{8.3}$$

where Q is the heat energy generated during the friction phase (J), and t is the time elapsed during the friction phase (s) (see Example 8.3).

Recently (2023), Lacki and co-researchers have reported a new thermal model for determining the basic parameters of rotary friction welding based on the analysis of friction heat transfer at the faying surfaces for joining dissimilar metals (titanium/aluminum alloys). They have proposed that the boundary conditions of the thermal model are the conductivity between components, machine tooling, as well as the radiation and convection within the environment. The heat exchanged between the free surfaces of components and the environment was described using the following convection equation (Lacki et al. 2023):

$$q_s = h_s * \left(T_s - T_A\right) \tag{8.4}$$

where q_s is the rate of heat exchanged between the unit area of the free surfaces of the components and the environment (W/m^2), h_s is the coefficient of convective heat transfer [in W/(m^2 • K)], T_A is the ambient temperature or the initial temperature of the workpieces (K), and.

T_s is the surface temperature of the components during heat exchange (K) (see Example 8.4). In general, the coefficient of convective heat transfer between the welded front part and the environment is taken to be around h_s = 20 W/(m^2·K) (Maalekian et al. 2008).

8.3 Friction Stir Welding

8.3.1 Working Principle—Advantages and Applications

Friction stir welding (*FSW*) is a hot solid-state welding process that involves feeding a rotating tool along the joint line between two workpieces thereby, generating frictional heat and stirring of the metal to form the weld seam. In the *FSW*, the

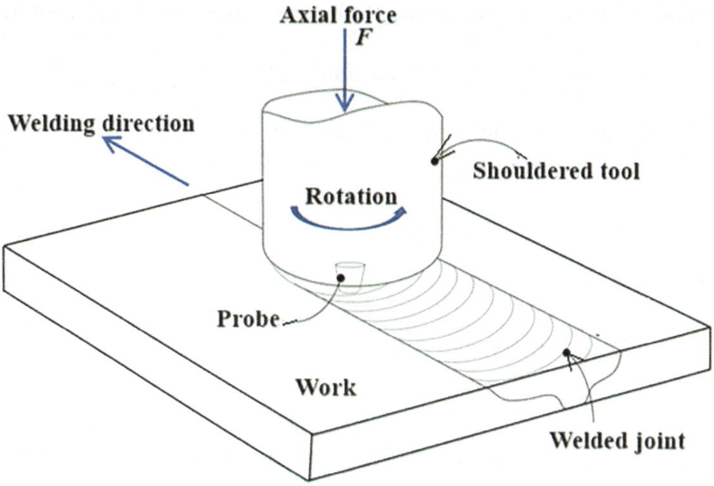

Fig. 8.4 Friction stir welding (FSW)—*schematic illustration*

rotating tool is stepped, consisting of a cylindrical shoulder and a smaller probe projecting beneath it (see Fig. 8.4). As the shouldered tool's probe or the work is fed in the direction of welding, frictional heat is generated between the probe and the material of the work pieces. Thus, the leading face of the probe forces plasticized material onto its back, while applying a substantial vertical force F to consolidate the weld metal. The welding of the material is facilitated by severe plastic deformation, resulting in good mechanical properties in the weld seam.

The FSW process finds application in the aerospace, automotive, railway, and shipbuilding industries, owing to its several advantages. The FSW process results in accurate part geometry and good mechanical properties in the weld. The FSW can join dissimilar materials and plastics. Additionally, this process can be easily automated for high volume production.

8.3.2 *Engineering Analysis of FSW*

The local heat generated during FSW can be determined by (Song and Kovacevic 2003):

$$\dot{Q}_{FSW} = 2 * \pi * \mu * F_n * R_s * \mathrm{N} \tag{8.5}$$

where $\dot{Q}_{FSW}$ is the local heat-generation rate during FSR (W), μ is the coefficient of friction, F_n is the normal force applied along the axis of the shouldered tool (N), R_s is the radius of the shouldered tool (m), and N is the rotational speed of the tool (rev/s) (see Examples 8.5 and 8.6).

The tool's feed rate or traverse speed (in mm/min) can be determined by:

$$\text{Tool feed rate} = N * f \tag{8.6}$$

where N is the tool's rotational speed (rpm), and *f* is the tool feed (mm/rev) (see Example 8.6).

8.4 Diffusion Welding

8.4.1 Working Principle

Diffusion welding (or diffusion bonding) is a hot *solid-state welding technique* that joins materials by applying heat and pressure, causing atomic diffusion to create a strong bond. In diffusion welding, a combination of heat and pressure is applied, usually in a controlled atmosphere, allowing sufficient time (about 1 h) for *diffusion* and coalescence to occur. The temperature in diffusion welding is in the range of 0.2–0.5 T_m. The diffusion bond is produced with loads below those that would cause macroscopic deformation to the part. Furthermore, a bonding aid can be used (such as an interface foil or coating) to facilitate the bonding. The primary mechanism of coalescence is solid-state diffusion, which involves *migration of atoms across the interface between contacting surfaces*. An electrical source can be used for resistance heating of the workpieces for diffusion welding, as shown in Fig. 8.5.

8.4.2 Advantages and Applications of Diffusion Welding

The following two conditions are necessary for successful diffusion welding: (a) mechanical intimacy of the faying surfaces, and (b) elimination of all contaminants and surface oxides. Diffusion welding results in excellent quality welds owing to less shrinkage and residual stresses as compared to other welding processes. This is why this process is preferred for the joining of high-strength alloys and refractory metals in the aerospace and nuclear industries. The diffusion welding process can be used to join both similar and dissimilar metals. However, for dissimilar metals, an interface filler layer of a different metal is often used between the two base metals to promote diffusion.

8.4.3 Mathematical Modeling of Diffusion Welding

The diffusion depth, in diffusion welding of two metals, depends on the welding time and the diffusion coefficient of one metal into the other. The diffusion depth can be determined by:

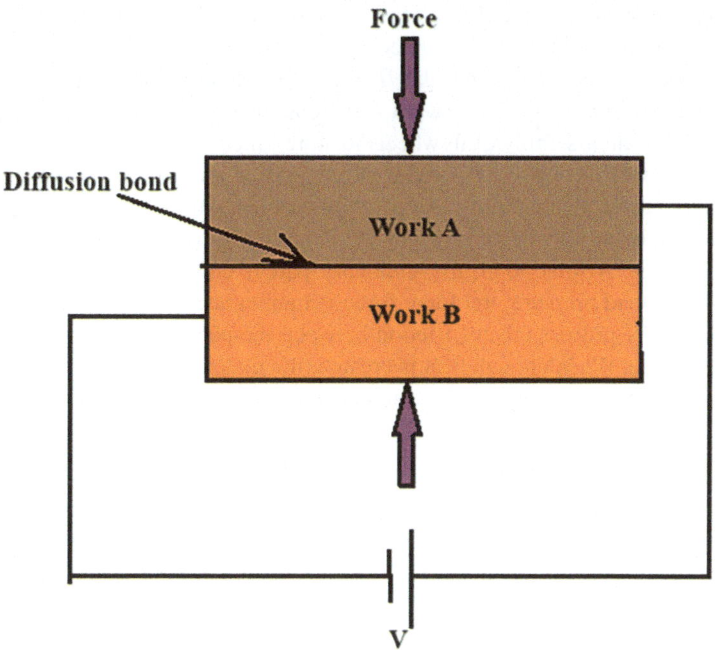

Fig. 8.5 Diffusion welding using electrical source for resistance heating

$$x = \sqrt{D \cdot t} \qquad (8.7)$$

where x is the diffusion depth (m), D is the diffusion coefficient (m^2/s), and t is the welding time (s) (see Example 8.7).

The applied pressure, in diffusion welding, lies in the range of 350–700 kgf/cm^2 (34.33–68.67 MPa). The bonding force can be calculated by:

$$\text{Bonding force} = p * (\text{contact area}) \qquad (8.8)$$

The significance of Eq. 8.8 is illustrated in Example 8.8.

8.5 Inertia Friction Welding

8.5.1 Working Principle

Inertia friction welding (*IFW*) (also known as *inertia welding*) uses kinetic energy, stored in a flywheel, with applied force to join parts together. In the IFW, one workpiece is spun rapidly while the other stays stationary; once the parts are pushed together, the friction at the surface creates intense heat. This heat softens the materials just enough to let them bond under pressure (see Fig. 8.6).

Like rotary friction welding, the *IFW* process is started by rotating one workpiece in a chuck while the other workpiece remains still in a clamp. In this process, the chuck has an assembly of flywheels to store kinetic energy. By accelerating the chuck and flywheels up to a desired rotational speed, the required kinetic energy is provided to bond the materials. Once the flywheels are up to speed, the motor driving the chuck and flywheels is disengaged and the flywheels coast, storing all the kinetic energy. At this point, the stationary part is engaged into the rotating part under forge load pressure; the forge force is maintained throughout the entire process. The IFW process creates friction between the parts and results in heat at the weld interface. When enough heat is created, the materials will begin to soften and will then begin to extrude and shorten, leading to joining of the materials.

8.5.2 Engineering Analysis of IFW

The moment of inertia of a flywheel (or a solid disc) can be determined by:

$$\mathrm{I} = \mathrm{m} * r^2 \tag{8.9}$$

where *I* is the moment of inertia (kg•m^2), *m* is the mass of flywheel (kg), and *r* is the radius of flywheel (m) (see Example 8.9). The initial kinetic energy stored in the flywheel can be determined by:

$$E_k = \frac{1}{2} * \mathrm{I} * \omega_o^2 \tag{8.10}$$

where E_k is the initial kinetic energy stored in the flywheel (J), and ω_o is the initial rotational speed of the flywheel (rad/s) (see Example 8.10).

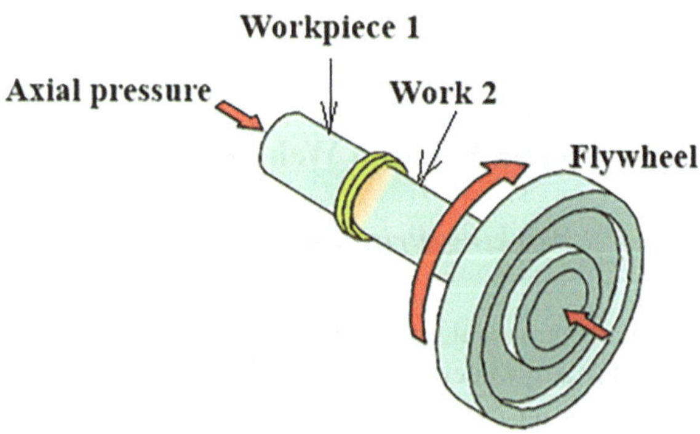

Fig. 8.6 Inertia friction welding—*working principle*

8.6 Worked Examples in Hot Solid-State Welding Processes

Example 8.1 Determining the Rate of Heat Energy Generated in FRW
A rotary FRW operation is performed to weld two cylindrical steel rods each of 25 mm in diameter. The rotational speed of the rotating workpiece is 1400 rpm, and the axial force applied to the other workpiece is 12 kN. The coefficient of friction between the steel surfaces is 0.3. Determine the rate of heat generated (in watts).

Solution

Rod radius = r = $\frac{D}{2} = \frac{25mm}{2}$ = 12.5 mm = 0.0125 m, F = 12,000 N, $\mu = 0.3$.

N = 1400 rev/min = $\frac{1400}{60}$ = 23.33 rev/s.

The peripheral speed of the rod = $v = 2 \bullet \pi \bullet r \bullet N = 2 \times \pi \times 0.0125 \times 23.33 = 1.83233$ m/s.

By using Eq. 8.1,

$$\dot{Q} = \mu * \text{F} * v = 0.3 \times 12000 \times 1.83233 = 6{,}596.4\,\text{J/s} = 6{,}596.4\,\text{W} = 6.6\,\text{kW}$$

The rate of heat generated during the friction-welding operation = $\dot{Q}$ = 6.6 kW.

Example 8.2 Determining the Torque Required in FRW
By using the data in Example 8.1, determine the torque required in the welding operation.

F = 12,000 N, μ = 0.3, r = 0.0125 m, T =?

By using Eq. 8.2,

$$\text{T} = \mu * \text{F} * r = 0.3 \times 12000 \times 0.0125 = 45\,\text{N} \bullet \text{m}$$

The torque required in the rotary FRW operation = 45 N•m

Example 8.3 Determining the Heat Energy Generated during the Friction Phase in FRW
By using the data in Example 8.1, determine the heat energy generated during the friction phase, if the friction phase lasts for 8 s.

Solution

$\dot{Q}$ = 6.6 kW = 6600 W, t = 8 s, Q =?

By using Eq. 8.3,

$$\text{Q} = \dot{Q} * \text{t} = 6600 \times 8 = 52{,}800\,\text{J} = 52.8\,\text{kJ}$$

The heat energy generated during the friction phase in friction welding = 52.8 kJ.

Example 8.4 Determining the Power Exchange b/w the Free Surfaces in FRW
Two dissimilar metals (alloys) are joined by rotary friction welding. The area of the free surfaces of the two components is 1.2 m^2. The ambient temperature is 20 °C,

and the surface temperature of the components during heat exchange is 450 °C. Determine the rate of heat energy exchange between the free surfaces of the components. Take the coefficient of convective heat transfer between the welded front part and the environment as 20 W/(m²·K).

Solution

$T_A = 20$ °C = 293 *K*, $T_s = 450$ °C = 723 K, $h_s = 20$ W/(m²·K), area = 1.2 m².

By using Eq. 8.4,

$$q_s = h_s * (T_s - T_A) = 20 \times (723 - 293) = 8600\,\mathrm{W/m^2}$$

Rate of heat energy exchange between the free surfaces = q_s* (area) = 8600 × 1.2 = 10,320 W.

The rate of heat energy exchange between the free surfaces = 10,320 W = 10.32 kW.

Example 8.5 Determining the Local Heat Energy Generated per Second in FSW

A FSW operation is performed to butt weld two metallic plates using a shouldered tool with a diameter of 50 mm. The rotational speed of the shouldered tool is 700 rpm, and the normal force applied along the tool axis is 20 kN. The coefficient of friction between the steel surfaces is 0.4. Determine the local heat energy generated per second (in watts).

Solution

$F_n = 20$ kN = 20,000 N, $D_s = 50$ mm, $R_s = \frac{D_s}{2} = \frac{50}{2} = 25$ mm = 0.025 m,

$N = 700$ rev/min = $\frac{700}{60} = 11.66$ rev/s, $\mu = 0.4$, $\dot{Q}_{FSW} = ?$

By using Eq. 8.5,

$$\dot{Q}_{FSW} = 2 * \pi * \mu * F_n * R_s * \mathrm{N} = 2 \times \pi \times 0.4 \times 20{,}000 \times 0.025 \times 11.66 = 14{,}652.38\,\mathrm{W}$$

The local heat energy generated per second = 14,652.38 W = 14.6524 kW.

Example 8.6 Determining the Feed Rate of the Shouldered Tool in FSW

By using the data in Example 8.5, determine the feed rate of the shouldered tool, if the tool's feed is 0.15 mm/rev.

Solution

$N = 700$ rev/min, f = 0.15 mm/rev.

By using Eq. 8.6,

$$\text{Tool feed rate} = \mathrm{N} * \mathrm{f} = 700 \times 0.15 = 105\,\mathrm{mm/min}$$

Example 8.7 Determining the Effective Diffusion Depth in Diffusion Welding

Two nickel plates have been diffusion welded in time duration of 1 hour. The diffusion coefficient of nickel, at 1000 °C, is 1.2×10^{-14} m²/s. Determine the effective diffusion depth.

Solution

$D = 1.2 \times 10^{-14}$ m²/s, t = 1 h = 3.600 s, x =?

By using Eq. 8.7,

$$x = \sqrt{D \bullet t} = \sqrt{1.2 \times 10^{-14} \times 3600} = \sqrt{4320 \times 10^{-14}} = 65.727 \times 10^{-7} \text{ m}$$

The diffusion depth = x = 6.5727×10^{-6} m = 6.57 μm

Example 8.8 Determining the Bonding Force in Diffusion Welding

The applied pressure, in a diffusion welding process, is 500 kgf/cm^2. Determine the bonding force, when the contact area is 8 cm^2.

Solution

p = 500 kgf/cm^2 = $500 \times \dfrac{9.81 N}{0.01^2\, m^2}$ = $500 \times \dfrac{9.81}{0.0001}$ N/m^2 = 500 × 98,100 Pa = 49,050,000 Pa.

Contact area = 8 cm^2 = $8 \times 0.01^2\ m^2$ = 0.0008 m^2.

By using Eq. 8.8,

$$\text{Bonding force} = \text{F} = \text{p} * (\text{contact area})$$
$$= 49050000\,\text{Pa} \times 0.0008\,\text{m}^2 = 39240\,\text{N} = 39.24\,\text{kN}.$$

Example 8.9 Determination of the Moment of Inertia of Flywheel in IFW

An *IFW* operation is performed to join two cylindrical steel rods; one of the rods is initially rotated using a flywheel. The flywheel's mass and radius are 80 kg and 40 cm, respectively. Determine the moment of inertia of the flywheel.

Solution

m = 80 kg, r = 40 cm = 0.4 m, I =?

By using Eq. 8.9,

$$\text{I} = \text{m} * r^2 = 80 \times 0.4^2 = 12.8\,\text{kg} \bullet m^2$$

The moment of inertia of the flywheel = I = 12.8 kg•m^2.

Example 8.10 Determination of the Kinetic Energy Stored in a Flywheel in IFW

Refer to the data in Example 8.9. Determine the initial kinetic energy stored in the flywheel, if the initial rotational speed of the flywheel is 180 rad/s.

Solution

I = 12.8 kg•m^2, ω_o = 180 rad/s.

By using Eq. 8.10,

$$E_k = \frac{1}{2} * \text{I} * \omega_o^2 = 0.5 \times 12.8 \times 180^2 = 207360\,\text{J} = 207\,\text{kJ}$$

The initial kinetic energy stored in the flywheel = 207 kJ

Questions and Problems

8.1.(MCQs). Encircle the most appropriate answers for each of the following statements.

(1) Which one of the following solid-state welding processes takes the longest time to complete?

(a) Friction welding, (b) FSW, (c) diffusion welding, (d) inertia welding

(2) Which one of the following phases of rotary FRW is referred to when one work being rotated, the other component is pressed against it with a defined axial pressure?

(a) Frictional contact phase, (b) friction phase, (c) forge phase, (d) holding phase.

(3) Which one of the following phases of rotary FRW is referred to when one component being rotated and the axial pressure is gradually increased resulting in the formation of flash?

(a) Frictional contact phase, (b) friction phase, (c) forge phase, (d) holding phase.

(4) Which one of the following phases of rotary FRW is involved when the rotation of one component is stopped and the axial pressure is increased with the other component?

(a) Frictional contact phase, (b) friction phase, (c) forge phase, (d) holding phase.

(5) Which one of the following solid-state welding processes involves the use of a rotating tool with a small probe?

(a) Friction welding, (b) FSW, (c) diffusion welding, (d) inertia welding.

(6) Which type of the following friction welding processes involves the back-and-forth motion of one component, while the other component remains stationary?

(a) Rotary FRW, (b) linear FRW, (c) radial FRW.

8.2. Draw diagrams showing various phases of friction welding process.

8.3. Explain friction stir welding (FSW) with the aid of a sketch.

8.4. (a) Why is reasonably long time allowed in diffusion welding?

(b) Highlight the advantages and applications of diffusion welding.

8.5. Explain, briefly, inertia welding process.

P8.6. Two dissimilar alloys are joined by rotary FRW. The area of the free surfaces of the two components is 1.4 m^2. The ambient temperature is 20 °C, and the surface temperature of the components during heat exchange is 500 °C. Determine the rate of heat energy exchange between the free surfaces of the components.

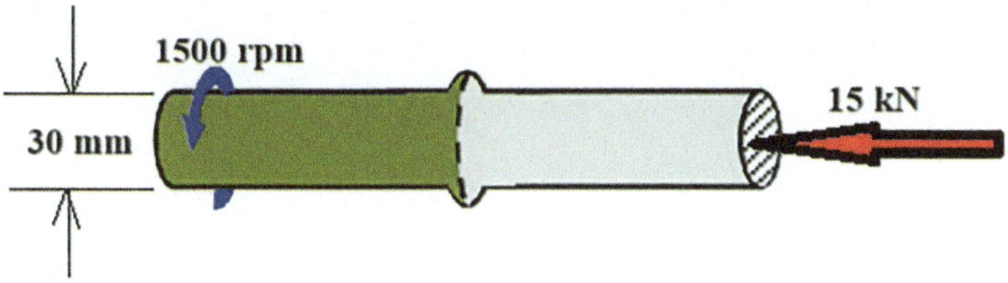

Fig. 8.7 The data for FRW's friction phase in P8.7

Take the coefficient of convective heat transfer between the welded front part and the environment as 20 W/(m^2·K).

P8.7. Refer to data for a FRW process, as shown in Fig. 8.6. Determine the: (a) rate of heat generated (in W), (b) torque in the welding operation, and (c) heat generated, if the friction phase lasts for 10 s. The coefficient of friction between the workpiece surface is 0.32.

P8.8. An FSW operation is performed to butt weld two metallic plates using a shouldered tool with a diameter of 45 mm. The rotational speed of the shouldered tool is 650 rev/min, and the normal force applied along the tool axis is 18 kN. The coefficient of friction between the steel surfaces is 0.4. Determine the local heat energy generated per second (in watts).

P8.9. Two nickel plates are diffusion welded in 55 min. The diffusion coefficient of nickel, at 1000 °C, is 1.2×10^{-14} m^2/s. Determine the effective diffusion depth.

P8.10. Refer to the data in P8.8. Determine the feed rate of the shouldered tool, if the tool's feed is 0.2 mm/rev.

P8.11. The applied pressure, in a diffusion welding process, is 550 kgf/cm^2. Determine the bonding force, when the contact area is 10 cm^2.

P8.12. An *IFW* operation is performed to join two cylindrical steel rods; one of the rods is initially rotated using a flywheel. The flywheel's mass and radius are 120 kg and 60 cm, respectively. Determine the moment of inertia of the flywheel (Fig. 8.7).

References

Akinlabi, E.T. and Mahamood, R.M. (2020). Solid-State Welding: Friction and Friction Stir Welding Processes, *Springer,* New York

Lacki, P. Adamus, J., Więckowski, W., Motyka, M. (2023). A new method of predicting the parameters of the rotational friction welding process based on the determination of the frictional heat transfer in Ti Grade 2/AA 5005 joints, *Materials*, Vol. *16*, Issue 13, p. 4787

Maalekian, M., Kozeschnik, E., Brantner, H.P., Cerjak, H. (2008). Comparative analysis of heat generation in friction welding of steel bars. *Acta Materialia,* Vol. 56, pp. 2843–2855.

Song, M. and Kovacevic, R. (2003). Numerical and experimental study of the heat transfer process in friction stir welding, *Proc. Instn Mech. Engrs*: *J. Engineering Manufacture,* Vol. 217, Part B, pp. 73–85

Yilbas, B.S. and Sahin, A.Z. (2014). Friction Welding—*Thermal and Metallurgical Characteristics, Springer,* New York.

Part IV
Brazing, Soldering, and Advanced Welding Processes

Chapter 9
Joining by Brazing and Soldering

Abstract This chapter covers both qualitative and quantitative aspects of brazing and soldering processes. Brazing, soldering, and welding processes have been distinguished with the aid of a table. Capillarity and surface tension have been explained owing to their role in brazing and soldering. Brazing of metals and ceramics is discussed with reference to its working principle, advantages, and applications. Brazing practice is described covering the operational steps in brazing, such as cleaning, joint clearance, brazing filler alloy (BFA), flux, jigging, heating, filling the joint area, and finishing; here, torch brazing, induction brazing, resistance brazing heating techniques are also explained. Engineering analysis of brazing includes braze joint design for flat components and tubular braze joints, and clearance in brazing of dissimilar metals. The soldering process is explained, highlighting its working principle and applications. Solder alloys are listed and explained with reference to phase diagrams and the factors in their selection. The steps in the operation of soldering are also described. The effects of contact angle and surface tension, interfacial energy on wetting in soldering are discussed and mathematically analyzed. This chapter includes 9 diagrams, 14 worked numerical examples, 2 tables, 7 exercise problems, and 5 multiple-choice questions (MCQs) with their answers given at the end of this book.

Keywords Braze joint design · Tubular braze joint · Torch brazing · Induction brazing · Resistance brazing · Brazing filler alloy (BFA) · Solder alloys · Wetting in soldering · Capillarity · Interfacial energy

Nomenclature

c_b	Joint clearance at the brazing temperature
c_o	Joint clearance at the ambient temperature
α_e	Coefficient of linear expansion for the metal with R_e
α_i	Coefficient of linear expansion for the metal with R_i
σ_{UTS}	Ultimate tensile strength of the weakest member

Z. Huda, *Welding and Joining Engineering*,
https://doi.org/10.1007/978-3-032-20462-2_9

τ_{BFA} Shear strength of the brazing filler alloy (BFA)
ΔT Temperature change
C Joint integrity factor (C = 0.8)
g Gravitational acceleration ($g = 9.81$ m s^{-2})
h Height the liquid rises
L_l Length of lap
r Radius of the capillary tube
R_e External pipe's internal radius
$R_{e(b)}$ External pipe's internal radius at the brazing temperature
R_i Internal pipe's external radius
$R_{i(b)}$ Internal pipe's external radius at the brazing temperature
t Thickness of the weakest member
γ Surface tension
θ Contact angle of the liquid on the tube
ρ Density of the liquid

9.1 Brazing, Soldering, and Welding: An Overview

It is mentioned in the introductory chapter that brazing and soldering are important joining processes. Brazing/soldering is a metal joining process in which a filler alloy is melted and distributed by capillary action between the faying surfaces of the metal parts (workpieces). These processes differ from welding in that here, heat is applied directly to the filler alloy (instead of base metal). In brazing, the filler alloy has a melting temperature above 450 °C but below the melting temperature of the base metal. In soldering, the filler alloy has melting point below 450 °C, which is usually the eutectic temperature in the corresponding phase diagram (see Chap. 2). The difference between brazing/soldering and welding is presented in Table 9.1. The brazing process is explained in Sect. 9.3, whereas soldering is discussed in Sect. 9.6.

Table 9.1 The difference between brazing/soldering and welding

Brazing/soldering	Welding
Heat is applied directly to the filler alloy	Heat is applied to the base metal
No pressure application is required	Pressure is applied in most solid-state welding processes
Capable to join the metals with poor weldability	Limited capability to do so
Capable to join dissimilar metals	Limited capability to do so
Recommended for low-strength requirements	Recommended for high-strength requirements
No thermal damage on the surface	Fusion welded parts have thermal damages
No heat affected zone (HAZ) is formed	HAZ is formed in fusion weldments

9.2 Capillarity and Surface Tension in Brazing/Soldering

The understanding of brazing and soldering processes requires a sound knowledge of capillarity and surface tension. *Capillarity* is the combined effect of cohesive and adhesive forces that causes a liquid to rise in thin tubes or other constricted spaces. By considering water as an example of a liquid rising in a thin glass tube, there exists an adhesive force inside the glass tube that draws water up the sides of the glass tube to form a meniscus. The cohesive force (the attraction of the water molecules to each other) then tries to minimize the distance between the molecules by pulling the bottom of the meniscus up against the gravitational force. This upward force per unit length is called *surface energy,* which is also defined as the *work per unit area done by the force that creates the new surface*. The surface tension of a liquid can be determined by measuring the height of the liquid rise in the capillary tube. The surface tension is related to the height by the following relationship:

$$\gamma = \frac{\rho g r}{2} \cdot \frac{h}{\cos\theta} \tag{9.1}$$

where γ is the surface tension (N•m^{-1} or J•m^{-2}), ρ is the density of the liquid (kg•m^{-3}), g is the acceleration due to gravity (g = 9.81 m•s^{-2}), r is the radius of the capillary tube (m), h is the height the liquid rises (m), and θ is the contact angle of the liquid on the tube (deg), which is a measure of the attraction of the liquid to the tube walls (see Example 9.1). The role of surface tension and capillary action in brazing/soldering is discussed in Sect. 9.8.

9.3 Brazing of Metals and Ceramics

9.3.1 Working Principle

Brazing is a metal-joining process in which a filler alloy is melted and flown (spread) by capillary action between the faying surfaces of the metal parts. In brazing, the filler metal has a melting temperature above 450 °C but below the melting temperature of the base metal (workpiece). The liquefied filler metal flows, by capillarity, between the parts to be joined and thereby creates a bond between them. Brazing operation can be performed either manually or using an automatic control system. In the case of manually controlled brazing, the heat source is a hand-held brazing torch (see Fig. 9.1).

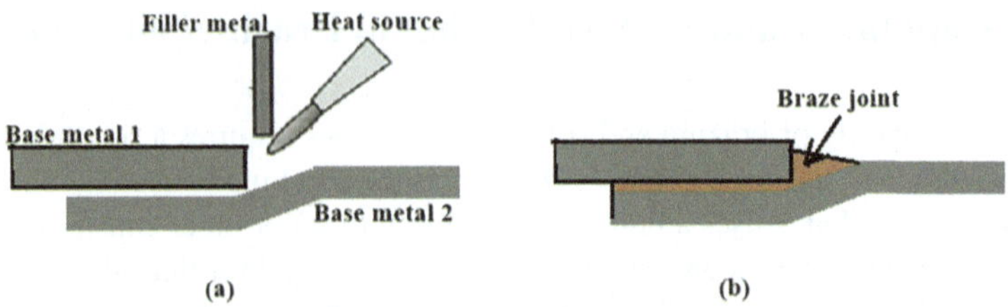

Fig. 9.1 Brazing of metal; (**a**) heating of filler metal, (**b**) brazed joint

9.3.2 Advantages and Applications of Brazing

Brazing has several advantages and applications. Brazing works well for joining dissimilar metals. The following are the metals that are used in brazing: aluminum, copper, gold, nickel, silver, stainless steel, and zinc-coated steel. Recently, Rozaimay et al. (2023) have reported brazing of porous nickel to copper and stainless steel for application in heat exchangers in industrial processes (especially in oil and gas production) (Rozaimay et al. 2023).

Brazing has diversified applications, such as refrigerators and ice machines, automobile manufacturing and repairing (*e.g.* vehicle air conditioning systems, radiator coils & fuel lines), electrical industry (*e.g.* fuses, motors, and packaging), HVAC systems (*e.g.* rooftop units for industrial cooling to residential heating and air conditioning systems) and fine jewelry. Besides metals, ceramics can also be effectively joined by brazing. In particular, the joining of the sapphire sensor chip to a sapphire disc, by brazing, is crucial in the fabrication of gas pressure sensors (Ning et al. 2003). Zaharinie and co-researchers (2022) have reported a reliable and economical method for direct brazing of sapphire using an active Ag-Cu-Ti filler alloy that can ensure effective brazing (Zaharinie et al. 2022).

9.4 Brazing Practice and Process Steps

9.4.1 Steps of Operations in Brazing

A successful brazing operation generally involves the following eight steps: (a) cleaning the workpieces, (b) joint clearance, (c) using the right brazing filler alloy (BFA), (d) using the right flux, (e) jigging the workpieces, (f) heating, (g) filling the joint area, and (h) finishing (see Fig. 9.2). These operational steps are explained in the following subsections.

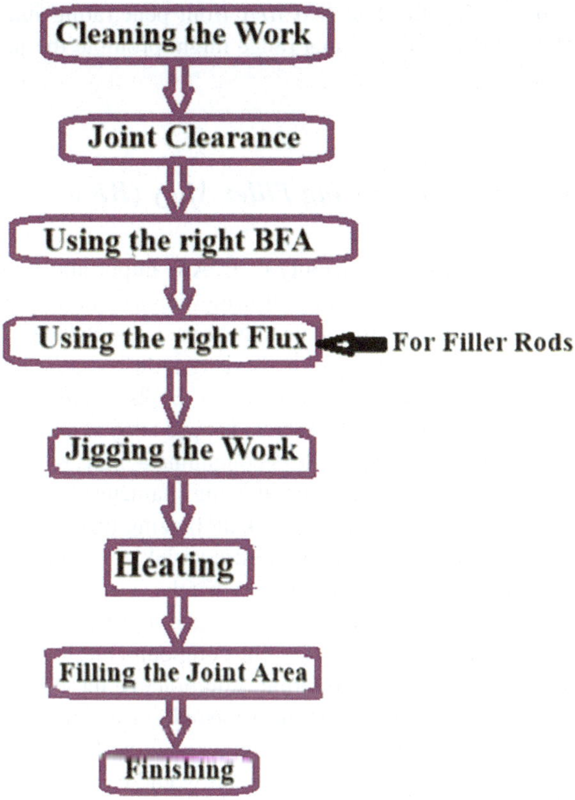

Fig. 9.2 Operational steps in brazing (BFA = brazing filler alloy)

9.4.2 Cleaning the Workpieces and Allowing the Joint Clearance

The presence of dirt, dust, and contaminants (*e.g.* oxides, paint, grease, oil, etc.) on the workpieces will affect the final result, giving rise to brittle joints and potentially to dangerous metal fumes. It is, therefore, important to thoroughly clean and degrease the workpieces to be joined by brazing.

A good braze design calls for a proper joint gap/clearance to ensure capillarity in brazing. The joint clearance is usually in the range of 0.2–1.5 mm, depending on the *BFA* used. The joint clearance should be determined at the brazing temperature, depending on the thermal expansion of the materials to be brazed (see Example 9.4). For example, when using a silver *BFA* to join a copper bushing in a steel sleeve, one must consider the coefficient of expansion of copper, which is higher than that of steel. Since the steel sleeve is the external element of the copper-steel assembly, the braze design calls for a higher tolerance compared with the one necessary in a steel-to-steel joint. Excessively wide joint clearances might decrease the capillarity force, thereby preventing the liquefied *BFA* from penetrating into the joint gap; however, an alloy with a wide melting range might promote the joint filling if the joint gap is wider.

9.4.3 Using the Right Brazing Filler Alloy (BFA)

The use of the right quality and quantity of BFA is important in brazing practice. The choice of a BFA depends on the following factors: the melting range of the BFA, its composition, the work materials to be joined, and the exposure to the operating environment during brazing. In the brazing industry, a *Safety Data Sheet* is provided that contains the features of a BFA. The BFAs suitable for joining copper tube are of two classes: (a) the BCuP series copper alloys containing phosphorus, and (b) the BAg series silver-copper alloys containing a high silver content. The two classes differ in their melting, fluxing and flowing characteristics and this should be considered in the selection of a BFA. The BCuP brazing filler alloys are commonly used in plumbing, HVAC refrigeration and fire sprinkler systems. The BAg series BFAs should be used when joining dissimilar metals.

The BCuP series BFAs are not recommended for brazing steels or nickel-containing alloys, since a brittle inter-metallic compound might develop, decreasing the mechanical properties of the braze joint. Additionally, these BFAs should not be exposed to atmospheres containing sulfur to avoid corrosion. The quantity of the BFA should be chosen reasonably, since a small amount of the BFA could result in an incomplete joint, while an excessive amount would cause a waste of resources. Huda et al. (2020) have reported a successful copper-stainless steel braze joint using a BFA with the following composition: Cu-7Ni-9.3Sn-6.3P (Huda et al. 2020). Besides metals, ceramics can also be joined by brazing; for example, joining of sapphire workpieces using an active Ag-Cu-Ti *BFA* has ensured effective sapphire-sapphire brazing (Zaharinie et al. 2022).

9.4.4 Using the Proper Flux for Brazing

A brazing flux provides the proper protection of the joint from oxides during a brazing operation. The selection of the right flux is based on several factors, including the type of base metal, the BFA, and the working temperature. It is important to select a flux with a minimum working temperature of 50 °C below the *solidus* temperature of the alloy. Brazing fluxes are available in powder, liquid, paste or gaseous form; the latter is called *Gasflux*, and it is used exclusively in torch brazing; these fluxes are classified according to European Standard DIN EN 1045 (European Standards 2008). Once the right flux has been chosen, it is important to use a sufficient amount of the flux in order to provide adequate protection to the overlapping areas during the whole brazing cycle. A low amount of flux would have a short protective action with the risk of incorporating oxides in the liquid braze.

9.4.5 Jigging the Workpieces

It is important to jig the workpieces for effective clamping during brazing, especially in manual brazing. In order to ensure a good braze-joint, the workpieces must remain steady when the *BFA* starts to solidify. However, moving the workpieces when the alloy is in the liquid phase, helps the outflow of the trapped gas. Brazing Jigs should be easy to load, have minimum contact with parts, have a low thermal mass, keep alignment, and be heat-resistant. The shape a brazing jig mainly depends on the geometry of workpieces and the BFA. Figure 9.3 illustrates the clamping of a sapphire-BFM sample system by use of a special jig tooling prior to brazing; here, both sapphire workpieces and the BFA are in disc shape.

9.4.6 Heating in Brazing

Heating techniques in brazing are broadly classified into: (1) localized heating, and (2) diffuse heating. There are three types of localized heating: (a) torch brazing, (b) induction brazing, and (c) resistance brazing. Diffuse heating can be performed either by furnace brazing, or dip brazing. In *torch brazing*, the heat is supplied by acetylene or other fuel gas (*e.g.* hydrogen, propane, etc.). The fuel gas is mixed with oxygen or air to form a flame that melts a BFA; the latter spreads to form a joint. Torch brazing involves the use of a flux, so a post-braze cleaning is often required. In *induction brazing*, heat is generated by a rapidly alternating current (AC), which is induced into the workpieces by an adjacent coil. High-frequency induction heating for brazing is clean and rapid, giving close control of temperature and location of heat. In *resistance brazing*, heat is created from resistance to an electrical current flowing through a circuit, which includes the workpieces. The process is most applicable to relatively simple joints in metals which have high electrical conductivity.

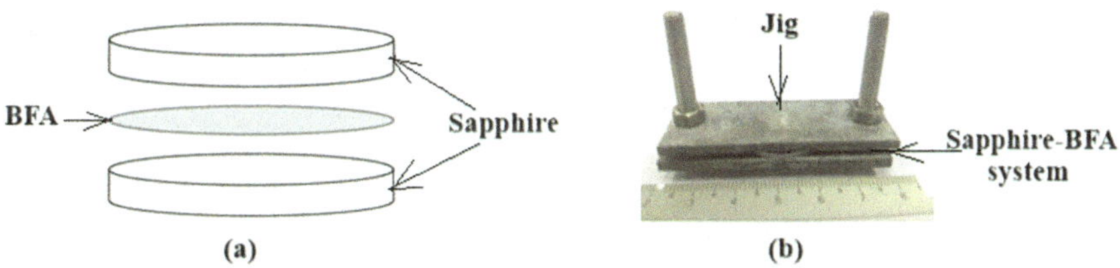

Fig. 9.3 (**a**) Pre-jigging sandwich arrangement of sapphire-BFA-sapphire, and (**b**) post-jigging arrangement of sapphire-BFA-sapphire system

Furnace brazing involves the use of a vacuum furnace or a protective atmosphere-controlled furnace (the use of vacuum negates the need for flux). Another advantage of furnace brazing is the ability to control accurately every stage of the heating and cooling cycles. Heating is either through the elements, or by gas firing. *Dip Brazing* involves the immersion of the entire assembly into a bath of molten *BFA* or molten flux. In both cases, the bath temperature is below the solidification point of the parent metal, but above the melting point of the *BFA*.

9.4.7 Filling the Joint Area

The BFA can be placed either in the joint area or applied when the brazing operation begins. In the latter case, the BFA is applied when the working temperature is reached. The use of a BFA rod necessitates protection by a flux. If the BFA and the flux are pre-placed, it is important to verify that the alloy has been completely melted, because the components and the jigging elements might absorb a significant portion of heat.

9.4.8 Finishing

Once the joint area has been filled, the brazed part should be finished by removing the flux residues; this can be done either using hot water or mechanically by brushing or sand blasting. It is necessary to remove flux residues to avoid corrosion.

9.5 Engineering Analysis of Brazing

9.5.1 Braze Joint Design for Flat Components

It is important to calculate an optimum lap length for a good braze-joint design. The length of lap for a flat braze joint (L_l) strongly depends on the thickness of the weakest member (t) and its tensile strength, joint integrity factor, and the shear strength of the BFA (see Fig. 9.4).

The lap length can be determined by:

$$L_l = \frac{\sigma_{UTS} * t}{C * \tau_{BFA}} \tag{9.2}$$

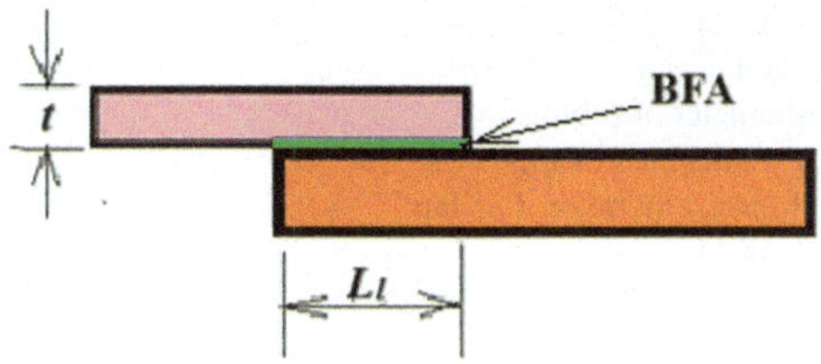

Fig. 9.4 Lap joint in brazing (BFA = brazing filler alloy)

where L_l is the length of lap (mm), σ_{UTS} is the tensile strength of the weakest member (MPa), t is the thickness of the weakest member (mm), C is the joint integrity factor (C = 0.8), and τ_{BFA} is the shear strength of the BFA (see Example 9.2).

9.5.2 *Design for Tubular Braze Joints: Clearance in Brazing Dissimilar Metals*

The design parameters for a tubular braze joint are illustrated in Fig. 9.5.

The lap length for a tubular braze joint can be determined by:

$$L_l = \frac{\sigma_{UTS} * t * (D - t)}{C * \tau_{BFA} * D} \tag{9.3}$$

where L_l is the lap length (mm), D is the outer diameter (OD) of the tube (mm), σ_{UTS} is the tensile strength of the weakest member (MPa), t is the thickness of the weakest member (mm), C is the joint integrity factor, and τ_{BFA} is the shear strength of the BFA (MPa) (see Example 9.3).

The clearance for a tubular braze joint, with two dissimilar metals, can be determined by analyzing the configuration, as illustrated in Fig. 9.6, which shows the cross-section of two co-axial pipes with different radii.

By reference to Fig. 9.6, the joint clearance at ambient temperature can be expressed as:

$$c_o = R_e - R_i \tag{9.4}$$

where c_o is the joint clearance at the ambient temperature, R_e is the external pipe's internal radius, R_i is the internal pipe's external radius (see Example 9.4). The calculation of the joint clearance at brazing temperature for brazing of dissimilar metals is critical, since the mismatch in thermal expansion coefficients can be a determining factor in the success or failure of a brazing setup (Stroppa et al. 2010). The clearance at brazing temperature for the configuration, (as shown in Fig. 9.6) can be computed by:

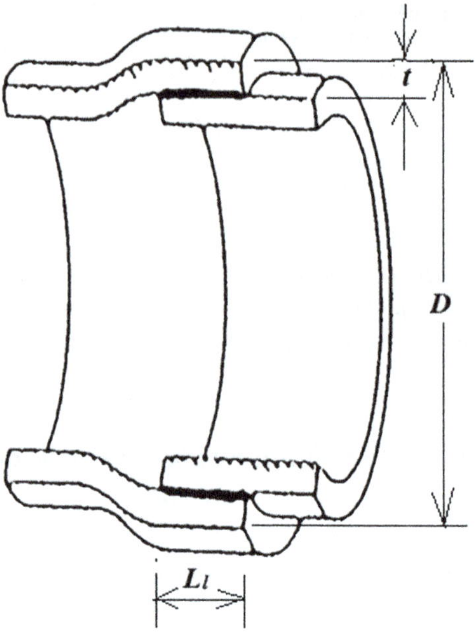

Fig. 9.5 A tubular braze joint design (D = outer diameter of tubing, t = thickness of the weakest member, L_l = lap length)

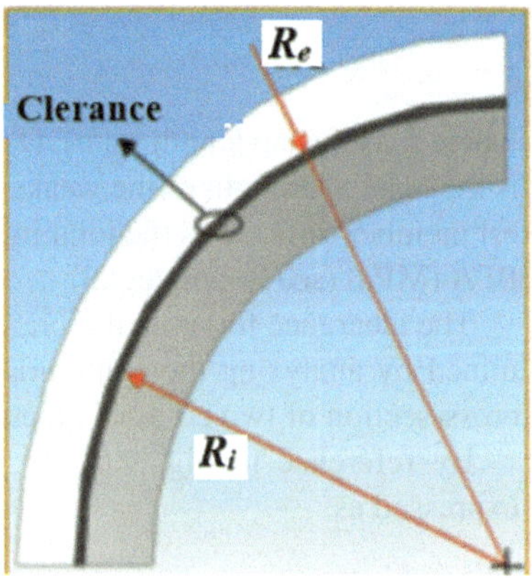

Fig. 9.6 Braze joint design for two co-axial pipes showing the joint clearnce (R_i = internal pipe's external radius, R_e = external pipe's internal radius)

$$c_b = R_{e(b)} - R_{i(b)} = \sqrt{\left[R_e^2 * \left(1 + 2\alpha_e * \Delta T\right)\right]} - \sqrt{\left[\mathrm{R}_\mathrm{i}^2 * \left(1 + 2\alpha_\mathrm{i} * \Delta T\right)\right]} \quad (9.5)$$

where c_b is the joint clearance at the brazing temperature (m), $R_{e(b)}$ is the external pipe's internal radius at the brazing temperature (m), $R_{i(b)}$ is the internal pipe's external radius at the brazing temperature (m), α_i is the coefficient of linear expansion for the metal with R_i (/°C), α_e is the coefficient of linear expansion for the metal with R_e (/°C), and ΔT is the temperature change (°C) (see Example 9.5).

9.5.3 Calculating the Volume and Mass of BFA

The volume of the filler alloy required in brazing can be determined by:

$$V_{BFA} = L_l * \mathrm{w} * \mathrm{c} \tag{9.6}$$

where V_{BFA} is the volume of the BFA, L_l is the lap length, w is the width of the braze joint, and c is the joint clearance (see Example 9.6). The mass of the BFA can be computed by multiplying the calculated volume by the density of the filler alloy.

9.5.4 Heat Required for Brazing

The heat energy required for brazing can be determined using the following formula:

$$\mathrm{Q} = \mathrm{m} * c_p * (T_b - T_o) \tag{9.7}$$

where Q is the heat energy (J), m is the mass of the work (kg), c_p is the specific heat at constant pressure (J/kg•°C), T_o is ambient temperature, and T_b is brazing temperature (see Example 9.7).

The rate of effective heat supply for brazing operation can be determined by:

$$\dot{Q}_{eff} = \eta * \dot{Q} \tag{9.8}$$

where $\dot{Q}$ is the rate of heat supply from the brazing torch (J/s) and η is the thermal efficiency of the torch (%) (see Example 9.8). The time taken in the brazing operation can be determined by:

$$\mathrm{t} = \frac{Q}{\dot{Q}_{eff}} \tag{9.9}$$

where t in the brazing time (s) (see Example 9.9).

9.5.5 Capillary Rise in Brazing

The capillary rise in brazing can be determined by modifying Eq. (9.1), as follows:

$$\mathrm{h} = \frac{2 * \gamma * cos\theta}{\rho * g * r} \tag{9.10}$$

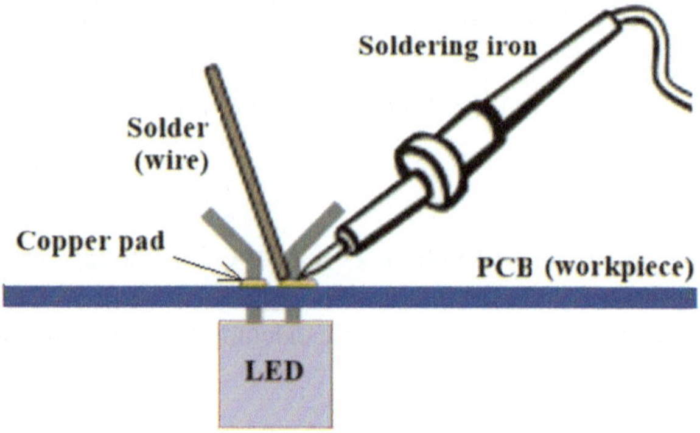

Fig. 9.7 Soldering of a PCB—*working principle* (PCB = printed circuit board, LED = light emitting diode)

where h is the height of capillary rise (m), γ is the surface tension (N/m), θ is the contact angle (deg), ρ is the density of the BFA (kg/m^3), g is the gravitational acceleration (g = 9.8 m/s^2), and r is the gap radius (= ½ of joint clearance) (see Example 9.10).

9.6 Soldering: Working Principle and Applications

Soldering is a metal joining process in which a filler alloy (called *solder*) with a melting point below 450 °C is melted and distributed between the faying surfaces of the base-metal parts. In soldering, the low-melting-point alloy (solder) is heated by a *soldering iron*, which is *a hand tool that* supplies heat to melt solder so that it can flow into the joint between two workpieces. Besides soldering iron, flux is often used to clean the metal surfaces, which ensures that the solder properly wets and adheres to them. Soldering finds extensive applications for creating connections in electronics, such as on printed circuit boards (PCBs); this process is also used to seal and assemble other metal components in jewelry and other industrial applications. Figure 9.7 illustrates the use of a soldering iron, solder, and copper pads in soldering of a PCB.

9.7 Solder Alloys and the Steps of Operation in Soldering

Prior to starting a soldering operation, it is important to select an appropriate solder alloy for the job. The choice of a specific *solder alloy* depends on its melting point, chemical reactivity, mechanical properties, toxicity, electrical conductivity, and other properties. Although the 60%Sn–40%Pb alloy is attractive for use as a solder owing to the lowest melting temperature (eutectic temperature) in the lead-tin phase diagram (see Chap. 2, Sect. 2.2), its use as a solder is prohibited since lead is toxic, hazardous and unsafe. This problem has been overcome by developing a safe lead-free eutectic solder alloy based on a bismuth-tin system: Sn–57%Bi (Huda 2020). *Alloys that melt between 180 and 190 °C* are the most commonly used solder alloys; these include alloys of gallium (Ga), bismuth (Bi), indium (In), tin (Sn), lead (Pb), and gold (Au). Some typical solder alloys include 66.5Ga-20.5In-13Sn, 58Bi-42Sn, 52In-48Sn, 63Sn-37Pb, 92.5Pb-5Sn-2.5Ag, 80Au-20Sn, and the like (Indium Corporation 2025).

The following are the steps involved in the soldering operation: (I) tinning the iron, (II) heating the connection, (III) applying solder, (IV) forming the joint, (V) removing the iron, and (VI) cooling the joint. Firstly, it is preferable to place a small amount of solder onto the hot tip of the soldering iron; this would help improve heat transfer to the components. Then, the hot soldering iron is touched to the two surfaces to be joined to ensure good contact with the flat edge of the tip of the soldering iron. Once both the workpiece components have been well heated, the solder is applied to the joint where the iron is touching. In the meantime, the solder melts and flows onto the heated parts. It is important to sufficiently heat the joint to ensure that the solder melts and forms a proper bond. Then, the solder wire is removed, followed by the removal of the soldering iron from the joint. Finally, the soldered joint is allowed to cool for a few seconds without moving it to ensure a strong, solid connection.

9.8 Engineering Analysis of Soldering

9.8.1 Effects of Contact Angle on Wetting

Surface tension and capillarity play an important role in soldering; this role can be emphasized by considering the liquid-drop model. Young treated the contact angle of a liquid as the result of the mechanical equilibrium of a drop resting on a plain surface of a solid metal under the action of three surface tensions (see Fig. 9.8).

The solid-liquid interfacial energy (γ_{sl}) can be related to the solid-vapor surface energy (γ_{sv}) and the liquid-vapor surface energy (γ_{lv}) by the following formula:

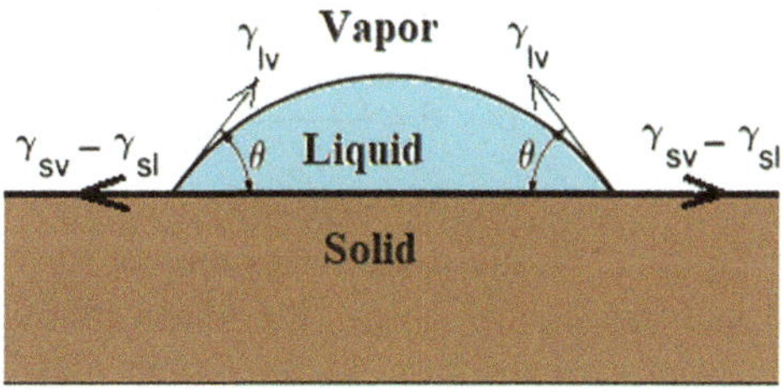

Fig. 9.8 Equilibrium of a liquid drop resting on a plain solid surface (θ = contact angle, γ_{lv} = the liquid-vapor surface energy, γ_{sl} = the solid-liquid interfacial energy, and γ_{sv} = the solid-vapor surface energy)

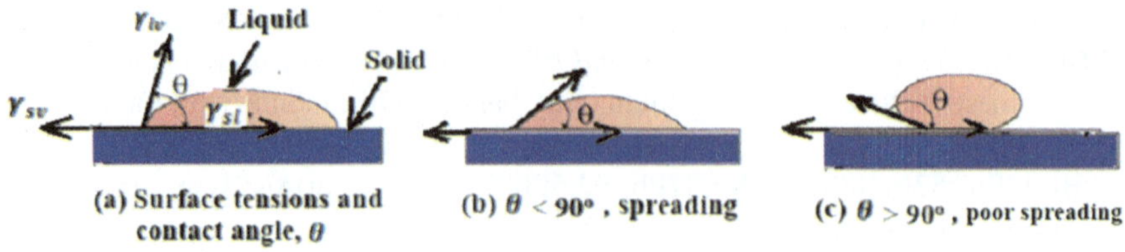

Fig. 9.9 The effects of contact angle on the spreading in soldering

Table 9.2 Interfacial energies and surface energies of metal-metal system (J/m^2)

γ_{sl}										
	Fe	Ni	Ti	W	Cu	Au	Ag	*Al*	γ_{lv}	
Ni	0.33	0.36	■	0.35					1.78	Ni
Cu	0.67	0.48	■	0.96	0.26				1.30	Cu
Au	0.42	0.44	■	0.98	0.08	0.20			1.17	Au
Ag	0.84	0.60	■	1.18	0.24	0.09	0.18		0.97	Ag
Al	■	■	■	■	■	■	0.07	0.15	0.91	*Al*
γ_{sv}	2.10	2.28	1.96	2.8	1.78	1.4	1.1	0.98		
	Fe	Ni	Ti	W	Cil	An	Ag	*Al*		

$$\gamma_{sl} = \gamma_{sv} - \gamma_{lv} * \cos\theta \tag{9.11}$$

where γ_{sl} is the surface tension between the solid metal and the liquid solder, γ_{sv} is the surface tension between the solid metal and vapor, γ_{lv} is the surface tension between the liquid solder and vapor, and θ is the contact angle (see Example 9.11). The assessment of the spreading power of a solder over a metallic surface involves the determinations of the contact angle between the solder and the metal surface, and the interfacial surface tensions. The effects of contact angle on the spreading of liquid solder are illustrated in Fig. 9.9, which indicates that the smaller the contact angle, the greater the spreading or wetting of the molten solder on the metal surface, thereby leading to successful soldering operation.

It is evident in Fig. 9.9 that the best spreading results are achieved when $\theta < 90°$. In order to assess the feasibility for a successful soldering operation, the terms in Eq. (9.11) may be re-arranged to obtain the contact angle, as follows:

$$\cos\theta = \frac{\gamma_{sv} - \gamma_{sl}}{\gamma_{lv}} \tag{9.12}$$

The significance of Eq. (9.12) is illustrated in Example 9.12.

The solid-liquid interfacial energies (γ_{sl}), the solid-vapor surface energies (γ_{sv}), and the liquid-vapor interfacial energies (γ_{lv}) for various metal-metal systems have been experimentally determined, and the data so obtained are presented in Table 9.2. It can be seen in Table 9.2 that solid-liquid interfacial energies (γ_{sl}) are generally significantly smaller than those of solid-vapor surface energies (γ_{sv}); this behavior is a necessary requirement for good wetting in brazing.

9.8.2 *Wetted Radius of a Solder Droplet*

The wetted area of a solder droplet approximates to the geometry of a spherical cap. The wetted radius of a solder droplet refers to the radius of the contact area on the substrate that the molten solder has spread to and bonded with. It is a measure of the extent of wetting, or how well the liquid solder spreads across the surface. The wetted radius is related to the contact angle (θ) and the radius of the solder droplet by the following formula:

$$r_w = r_d * \sin\,\theta \tag{9.13}$$

where r_w is the wetted radius of the droplet and r_d is the radius of droplet (see Example 9.13).

9.9 Worked Examples in Brazing and Soldering

Example 9.1 Computing the Surface Energy A capillarity experiment was performed using pure water in a clean glass tube with a diameter of 1.5×10^{-4} m. The contact angle of water on the tube is nearly zero. Determine the surface tension of water if the water rose to a height of 19.9 cm. The density of water is 1000 kg/m^3.

Solution

$$\mathrm{r} = \frac{Dia}{2} = \frac{1.5\times10^{-4}}{2} = 7.5\times10^{-5}\,\mathrm{m},\quad \theta = 0,$$
$$\mathrm{h} = 19.9\,\mathrm{cm} = 0.199\,\mathrm{m},\quad \rho = 1000\,\mathrm{kg\cdot m^{-3}}$$

By using Eq. (2.1),

$$\text{Surface tension} = \gamma = \frac{\rho\ g\ r}{2}\cdot\frac{h}{\cos\,\theta}$$
$$= \frac{1000\times9.81\times7.5\times10^{-5}}{2}\times\frac{0.199}{\cos 0} = 0.0732\,\mathrm{J\cdot m^{-2}}$$

Example 9.2 Determining the Lap Length for a Flat Braze Joint It is required to join by brazing a 1.2-mm-thick sheet of annealed Monel metal to a metal of equal or greater strength using a silver BFA with a shear strength of 172 MPa. The tensile strength of the Monel metal is 482 MPa. Determine the lap length for the brazing operation.

Solution
$\sigma_{UTS} = 482$ MPa, $t = 1.2$ mm, C = 0.8, $\tau_{BFA} = 172$ MPa, L_l =?
By using Eq. (9.2),

$$L_l = \frac{\sigma_{UTS} * t}{C * \tau_{BFA}} = \frac{482 \times 1.2}{0.8 \times 172} = 4.2\,\text{mm}$$

The lap length = 4.2 mm.

Example 9.3 Determining the Lap Length for a Tubular Braze Joint It is required to join by brazing a copper tubing (OD = 20 mm) to a 20-mm-ID steel tubing. The wall thickness of the copper tube is 1.5 mm. The tensile strength of annealed copper is 227 MPa, and the shear strength of the BFA is 172 MPa. Determine the lap length for the braze joint.

Solution
D = 20 mm, $\sigma_{UTS} = 227$ MPa, $t = 1.5$ mm, C = 0.8, $\tau_{BFA} = 172$ MPa.
By using Eq. (9.3),

$$L_l = \frac{\sigma_{UTS} * t * (D - t)}{C * \tau_{BFA} * D} = \frac{227 \times 1.5 \times (20 - 1.5)}{0.8 \times 172 \times 20} = 2.29 \cong 2.3 \text{ mm}$$

The lap length = 2.3 mm.

Example 9.4 Determining the Joint Clearance at the Ambient Temperature in Brazing Two dissimilar steel pipes (AISI-1020 and AISI-304) are required to be brazed using BCu-1 BFA. The 304 stainless-steel is the external pipe with 126-mm internal diameter, whereas the 1020 grade corresponds to the internal pipe with 125.80-mm external diameter. Determine the joint clearance at the ambient temperature.

Solution

$$R_e = \frac{126}{2} = 63 \text{ mm } 0.063 \text{ m}, R_i = \frac{125.8}{2} = 62.9 \text{ mm} = 0.0629 \text{ m}, c_o = ?$$

By using Eq. (9.4),

$$c_o = R_e - R_i = 0.063 - 0.0629 \text{ m} = 100 \times 10^{-6} \text{ m} = 100\,\mu m$$

The joint clearance at the ambient temperature = 100 μm.

Example 9.5 Determining the Joint Clearance at the Brazing Temperature Two dissimilar steel pipes (AISI-1020 and AISI-304) are required to be brazed using BCu-1 BFA. The 304 stainless-steel is the external pipe with 126-mm internal diameter, whereas the 1020 grade corresponds to the internal pipe with 125.80-mm external diameter. The coefficients of linear expansion of AISI-304 steel are AISI-1020 steel are 18.7 μm/(m•°C) and 13.9 μm/(m•°C), respectively. The ambient and the brazing temperatures are 20 °C and 1093 °C, respectively. Determine the joint clearance at the brazing temperature.

Solution

$R_e = 0.063$ m, $R_i = 0.0629$ m, $\alpha_e = 18.7$ μm/(m•°C), $\alpha_i = 13.9$ μm/(m•°C), $\Delta T = 1093 - 20 = 1070$ °C, c =?

By using Eq. (9.5),

$$c_b = \sqrt{\left[R_e^2 * \left(1 + 2\alpha_e * \Delta T\right)\right]} - \sqrt{\left[R_i^2 * \left(1 + 2\alpha_i * \Delta T\right)\right]}$$

$$c_b = \sqrt{\left[0.063^2 \times \left(1 + 2 \times 18.7 \times 10^{-6} \times 1070\right)\right]} - \sqrt{\left[0.0629^2 \times \left(1 + 2 \times 13.9 \times 10^{-6} \times 1070\right)\right]}$$

$$c_b = \sqrt{\left[0.003969 \times \left(1 + 0.040018\right)\right]} - \sqrt{\left[0.00395641 \times \left(1 + 0.029746\right)\right]}$$

$$c_b = 0.06424820186 - 0.06382865635 = 4.195455 \times 10^{-4} \text{ m} = 419.5\, \mu m$$

The joint clearance at the brazing temperature = 419.5 μm.

Example 9.6 Determining the Volume and Mass of a BFA A brazing operation is to be performed using a silver-based BFA with a density of 9.4 g/cm^3. The lap length is 6 cm. The joint clearance and width are 1.2 mm and 1 cm, respectively. Determine the required volume and mass of the BFA

Solution

$L_l = 6$ cm, w = 1 cm, c = 1.2 mm = 0.12 cm, $\rho = 9.4$ g/cm^3, V =?, m =?

By using Eq. (9.6),

$$\text{Volume of the BFA} = V_{BFA} = L_l * \text{w} * \text{c} = 6\,\text{cm} \times 1\,\text{cm} \times 0.12\,\text{cm} = 0.72\,\text{cm}^3$$

$$\text{Mass of the BFA} = \text{m} = \rho * \text{V} = 9.4 \times 0.72 = 6.77\,\text{g}.$$

Example 9.7 Determining the Heat Energy required for a Brazing Operation A brazing operation is to be performed at 620 oC on a 0.5-kg copper tube that is initially at 20 °C. The brazing torch delivers heat at the rate of 7600 J/s with 45% efficiency. Determine the heat energy required for the operation. The specific heat (c_p) of copper is 385 J/kg•°C.

Solution

$m = 0.5$ kg, $T_o = 20$ °C, $T_b = 620$ °C, $c_p = 385$ J/kg•°C, Q =?

By using Eq. (9.7),

$$\text{Q} = \text{m} * c_p * \left(T_b - T_o\right) = 0.5 \times 385 \times \left(620 - 20\right) = 115500\ \text{J} = 115.5\ \text{kJ}$$

The heat energy required for the brazing operation = Q = 115.5 kJ

Example 9.8 Determining the Rate of Effective Heat Supply from a Brazing Torch By using the data in Example 9.7, determine the rate of effective heat supply for the operation.

Solution
$\dot{Q}$ = 7600 J/s, $\eta = 45\% = 0.45$, $\dot{Q}_{eff}$ =?
By using Eq. (9.8),

$$\dot{Q}_{eff} = \eta * \dot{Q} = 0.45 \times 7600 = 3{,}420\,\text{J/s}$$

The rate of effective heat-energy supply = $\dot{Q}_{eff}$ = 3420 J/s

Example 9.9 Determining the Time for a Brazing Operation By using the data in Examples 9.7 and 9.8, determine the brazing time for the operation.

Solution
Q = 115,500 J, $\dot{Q}_{eff}$ = 3420 J/s, t =?
By using Eq. (9.9),

$$\text{t} = \frac{Q}{\dot{Q}_{eff}} = \frac{115500}{3420} = 33.7\,\text{s}$$

The brazing time = 33.7 s.

Example 9.10 Determining the Height of Capillary Rise in Brazing A brazing operation was performed with a joint clearance of 0.14 mm and contact angle of 40°. The density and surface tension of the molten BFA are 7350 kg/m^3 and 0.47 N/m, respectively. Determine the height of capillary rise.

Solution
Gap radius = r = ½*(joint clearance) = ½ × 0.14 = 0.07 mm = 0.07 × 10^{-3} m = 0.00007 m.
$\gamma = 0.47$ N/m, $\theta = 40°$, $\rho = 7350$ kg/m^3, g = g = 9.8 m/s^2, h =?
By using Eq. (9.10),

$$\text{Height of the capillary rise} = \text{h} = \frac{2 * \gamma * cos\theta}{\rho * g * r}$$

$$= \frac{2 \times 0.47 \times \cos(40)}{7350 \times 9.8 \times 0.00007} = \frac{0.72}{5.04} = 0.1428\ \text{m} = 14.28\ \text{cm}.$$

Example 9.11 Determining the Solid-Liquid Interfacial Energy in Soldering The solid-vapor and the liquid-vapor interfacial energies for a metal-metal system, in soldering, are 0.65 N/m and 0.55 N/m, respectively. The contact angle, for the soldering operation, has been measured to be 50°. Determine the solid-liquid interfacial energy for the operation.

Solution
$\gamma_{sv} = 0.65$ N/m, $\gamma_{lv} = 0.55$ N/m, $\theta = 50°$, $\gamma_{sl} =?$
By using Eq. (9.11),

$$\gamma_{sl} = \gamma_{sv} - \gamma_{lv} * \cos\ \theta = 0.65 - (0.55 \times \cos\ 50) = 0.65 - 0.3535 = 0.296$$

The solid-liquid interfacial energy ≅ 0.3 N/m

Example 9.12 Determining the Contact Angle to Assess the Feasibility of Soldering Operation The solid-vapor, the liquid-vapor, and the solid-liquid surface tensions for a metal-metal system are 0.63 N/m, 0.56 N/m, and 0.38 N/m, respectively. (a) Determine the contact angle for soldering. (b) Is the calculated contact angle feasible for successful soldering operation?

Solution
$\gamma_{sv} = 0.63$ N/m, $\gamma_{lv} = 0.56$ N/m, $\gamma_{sl} = 0.38$ N/m, $\theta =?$

(a) By using Eq. (9.12),

$$\cos\ \theta = \frac{\gamma_{sv} - \gamma_{sl}}{\gamma_{lv}} = \frac{0.63 - 0.38}{0.56} = 0.446$$

$$\text{Contact angle} = \theta = 63°$$

(b) Since the contact angle (63°) is less than 90°, the data is feasible for successful soldering.

Example 9.13 Determining the Wetted Radius of a Solder Droplet The radius of a solder droplet is 1.5 mm. Determine the wetted radius if the contact angle is 35°.

Solution
$r_d = 1.5$ mm, $\theta = 35°$, $r_w =?$
By using Eq. (9.13),

$$r_w = r_d * \sin\ \theta = 1.5 \times \sin 35° = 0.86\ \text{mm}$$

The wetted radius of the solder droplet = 0.86 mm.

Example 9.13 Verifying the Feasibility of Successful Soldering for Fe-Ag System Refer to the data in Table 9.2. (a) Determine the contact angle for soldering of steel workpiece using silver as solder. (b) Is the wetting good for successful soldering operation?

Solution
For the Fe-Ag system, γ_{sv}= 1.1 J/m^2, γ_{lv}= 0.97, γ_{sl}= 0.84

(a) By using Eq. (9.8),

$$\cos\,\theta = \frac{\gamma_{sv} - \gamma_{sl}}{\gamma_{lv}} = \frac{1.1 - 0.84}{0.97} = 0.268$$

$$\theta = 74^\circ < 90^\circ$$

(b) Since the contact angle, $\theta < 90^\circ$, wetting is good for successful soldering.

Example 9.14 Determining the Mass of Solder when the dimensions of the Joint Known A solder joint forms a 5-mm-high cylindrical ring. The outer and inner diameters are 3.0 and 2.6 mm, respectively. The density of solder is 7.5 g/cm^3. Determine the mass of the solder used.

Solution

OD = D = 3.0 mm, ID = d = 2.6 mm, l = 5 mm, ρ = 7.5 g/cm^3 = 7.5 × 10^3 kg/m^3,

$$V = \frac{\pi}{4} * l * \left(D^2 - d^2\right) = 0.785 \times 5 \times \left(3^2 - 2.6^2\right) = 3.926 \times 2.24$$

$$= 8.79\,\text{mm}^3 = 8.79 \times 10^{-9}\,\text{m}^3$$

$$\text{Mass} = \text{m} = \rho * \text{V} = 7.5 \times 10^3 \times 8.79 \times 10^{-9}$$

$$= 65.95 \times 10^{-6}\,\text{kg} = 0.0659\,\text{g} \cong 0.07\,\text{g}$$

Questions and Problems

9.1. (a) What makes a liquid rise in a thin tube?

(b) Draw a sketch of the Young's liquid-drop model, giving the meanings of the symbols in the sketch.

9.2. Differentiate between brazing, soldering, and welding with the aid of a Table.

9.3. (a) Explain the working principle of brazing with the aid of sketches.

(b) Highlight the advantages and applications of brazing.

9.4. (a) Draw the flow chart showing the operational steps in brazing.

(b) Why is it important to clean the workpieces and allow the joint clearance in brazing?

(c) Highlight the applications of the BCuP and BAg brazing filler alloys.

(d) Why are the BCuP series BFAs not recommended for brazing steels or nickel alloys?

9.5. Describe the various types of heating techniques in brazing.

9.6. (a) Explain the working principle of soldering with the aid of a sketch.

(b) Highlight the applications of soldering.

9.7. Draw sketches showing the effects of contact angle on the spreading in soldering.

9.8. What factors must be considered in selecting a solder alloy? List three solder alloys.

P9.9. It is required to braze a copper tubing (OD = 18 mm) to an 18-mm-ID steel tubing. The wall thickness of the copper tube is 1.6 mm. The tensile strength of annealed copper is 227 MPa, and the shear strength of the BFA is 172 MPa. Determine the lap length for the joint.

P9.10. It is required to braze a 1.4-mm-thick sheet of annealed Monel metal to a metal of equal or greater strength using silver BFA with shear strength of 172 MPa. The tensile strength of the Monel metal is 482 MPa. Determine the lap length for the brazing operation.

P9.11. Two dissimilar steel pipes (AISI-1020 and AISI-304) are required to be brazed using BCu-1 BFA. The 304 stainless-steel is the external pipe with 125-mm internal diameter, whereas the 1020 grade corresponds to the internal pipe with 124.70-mm external diameter. The coefficients of linear expansion of AISI-304 steel are AISI-1020 steel are 18.7 μm/(m•°C) and 13.9 μm/(m•°C), respectively. The ambient and the brazing temperatures are 20 °C and 1090 °C, respectively. Determine the joint clearance at the: (a) ambient temperature, (b) brazing temperature.

P9.12. The solid-vapor and the liquid-vapor interfacial energies for a metal-metal system, in soldering, are 0.62 N/m and 0.51 N/m, respectively. The contact angle, for the soldering operation, has been measured to be 55°. Determine the solid-liquid interfacial energy.

P9.13. A brazing operation is to be performed using a silver-based BFA with a density of 9.4 g/cm^3. The lap length is 5 cm. The joint clearance and width are 1.3 mm and 0.8 cm, respectively. Determine the required volume and mass of the BFA.

P9.14. The radius of a solder droplet is 1.8 mm. Determine the wetted radius, if the contact angle is 40°.

P9.15. The solid-vapor, the liquid-vapor, and the solid-liquid surface tensions for a metal-metal system are 0.60 N/m, 0.53 N/m, and 0.35 N/m, respectively. (a) Determine the contact angle for soldering. (b) Is the calculated contact angle feasible for the soldering operation?

9.17(MCQs). Encircle the most appropriate answers for each of the following statements

(1). Which one of the following processes is the best for the fabrication of gas pressure sensors?

(a) Welding, (b) brazing, (c) soldering, (d) adhesive joining.

(2). Which one of the following processes is the best for assembling printed circuit boards?

(a) Welding, (b) brazing, (c) soldering, (d) adhesive joining.

(3). Which one of the following joining processes is the best for high-strength requirements?

(a) Welding, (b) brazing, (c) soldering, (d) adhesive joining.

(4). Which one of the following brazing techniques involves the generation of heat by rapidly alternating current?

(a) Torch brazing, (b) induction brazing, (c) resistance brazing, (d) diffuse brazing.

(5). Which one of the following techniques involves the supply of heat by acetylene?

(a) Torch brazing, (b) induction brazing, (c) resistance brazing, (d) diffuse brazing.

References

European Standards (2008). DIN EN 1045 Fluxes for brazing - Classification and technical delivery conditions, ***EUROPEAN STANDARDS s.r.o,*** Plzen, Czech Republic

Huda, Z. (2020), Metallurgy for Phyiscists and Engineers, *CRC Press*, Boca Raton, FL, USA

Huda, Z., Zaharinie, T., Amir, Y. (2020). SEM-EDS analysis and microhardness variations in steel-copper braze joint, *Journal of Technological Sci. and Eng.* Vol. 1, RSE Press, USA

Indium Corporation (2025), Solder Alloys, Online: https://www.indium.com/products/alloys/solder-alloys/, Accessed on Sept. 16, 2025.

Ning, H., Geng, Z., Ma, J., Huang, F., Qian, Z., Han, Z. (2003). Joining of sapphire and hot pressed Al_2O_3 using Ag70.5Cu27.5Ti2 brazing filler metal, *Ceramic International*, Vol. 29, Issue 6, pp. 689-694.

Rozaimay, R., Zaharinie, T., Buys, Y.F., Huda, Z., Ganesan, P.B., Ariga, T. (2023). Investigation on microstructure and compressive strength of brazing porous nickel to copper and stainless steel, *Acta Metallurgica Slovaca*, Vol. 29, No. 3, pp. 148–154.

Stroppa, D.G., Hermenegildo, T., Unifried-Silgado, J., *et al.* (2010). Calculating joint clearance at brazing temperature, *Welding Journal*, Vol. 89, Issue 9, pp.32–35.

Zaharinie, T., Huda, Z., Ibrahim, S., Yusof, F., *et al.* (2022), Analysis on reaction layer formed during sapphire-sapphire brazing using Ag-Cu-Ti filler metal for gas-pressure sensors, *ACS Applied Electronic Materials*, Vol. 4, Issue 5, pp. 2405–2412

Chapter 10
Advanced Welding Processes

Abstract This chapter covers both qualitative and quantitative aspects of advanced welding processes. Firstly, an overview of commonly used advanced welding processes (magnetic pulse welding, friction stir welding, explosive welding, ultrasonic welding, laser beam welding, and electron beam welding) is given. Magnetic pulse welding (MPW) is discussed with reference to its working principle, applications, equipment set up, and mathematical modeling. Electron beam welding (EBW) is explained by taking into account its working principle, applications, and in-depth engineering analysis. Finally, laser beam welding (LBW) is described, including its working principle, advantages, applications, and engineering analysis; here a very recent advancement in LBW is also mentioned. This chapter contains 14 worked numerical examples, 4 diagrams, 9 exercise problems, and 8 multiple-choice questions (MCQs) with their answers given at the end of the book.

Keywords Magnetic pulse welding (MPW) · Electromagnetic force · Inductive coil · Lorentz force · Magnetic impact welding · Impact weld · Impact velocity · Impact angle · Electron beam welding (EBW) · Cathode filament · Deflection coil · Depth of penetration · Laser beam welding (LBW) · Convex lens · Laser beam welding in vacuum (LaVa)

Abbreviations

EBW	Electron beam welding
LaVa	Laser beam welding in vacuum
LBW	Laser beam welding
MPW	Magnetic pulse welding
PD	Power density

Z. Huda, *Welding and Joining Engineering*,
https://doi.org/10.1007/978-3-032-20462-2_10

Nomenclature

E_k	Kinetic energy of an electron in the beam (J)
P_i	Magnetic pressure (Pa)
v_i	Impact velocity (m/s)
v_w	Welding speed
ρ_f	Density of the flyer material
B	Magnetic flux density (Tesla)
C	Capacitance (F)
d	Depth of penetration
e	Charge on an electron
E	Energy (J)
f	Circuit (discharge pulse) frequency (Hz)
k	Material-dependent constant (k = 0.5 for steel)
L	Total inductance (H)
m	Mass of an electron (kg)
P	Power
V	Charge voltage (volts)
v	Speed of electron (m/s)
γ	Impact angle
δ	Skin depth (m)
ρ	Density of the work material
ρ	Electrical resistivity of the material in contact with the coil (Ω•m)

10.1 Advanced Welding Processes: An Overview

Most of the commonly used welding processes have been discussed in the preceding chapters (Chaps. 4–8). However, there are technical shortcomings in most of these welding processes; for example, GMAW and other arc welding processes are unable to weld over a wide area and are just used for linear and point welds. In addition, it is hard to focus them on very small regions. In order to overcome these technical problems, the following six advanced welding processes are in use in industry: (a) magnetic pulse welding (MPW), friction stir welding (FSW), explosive welding (EXW), ultrasonic welding (USW), laser beam welding (LBW) and electron beam welding (EBW) (Wiley, 2026). This chapter explains only three welding processes (the MPW, the LBW, and the EBW) because the remaining processes (the FSW, the EXW, and the USW) have been explained in the previous chapters.

10.2 Magnetic Pulse Welding

10.2.1 Working Principle and Applications

Magnetic pulse welding (MPW) (also known as magnetic impact welding or electromagnetic pulse welding) is an advanced solid-state welding process that uses electromagnetic pressure to accelerate one workpiece resulting in an impact against another workpiece. The MPW creates a strong metallic bond similar to the bond created by explosion welding. This process enables us to produce multi-metallic hybrid materials by joining dissimilar metals in their solid state (Lapovok et al. 2012). The MPW technology involves a significantly high-speed collision and a high strain-rate plastic deformation (Sapanathan et al. 2016).

The MPW has been successfully applied to various similar and dissimilar metal combinations. In particular, the materials with high conductivity, such as aluminum and copper, are the easiest to weld by MPW. For example, aluminum has been successfully welded to steel and stainless steel. In another example, copper has been successfully welded to steel and stainless steel. This process has been used to join fuel pipes, fuel filters, exhaust system components, power cables, and for the construction of automotive body parts.

10.2.2 Equipment Set Up in MPW

The MPW equipment setup consists of a pulse generator, a coil and an optional field shaper; the generator contains a transformer which transforms a low-voltage power supply into a high voltage charge that is stored in a capacitor bank (see Fig. 10.1). This configuration is used to weld a core-clad combination (workpieces). The generator set, connected to an inductive coil through a control switch, supplies a high discharge current, which flows through the coil, thereby generating a magnetic field.

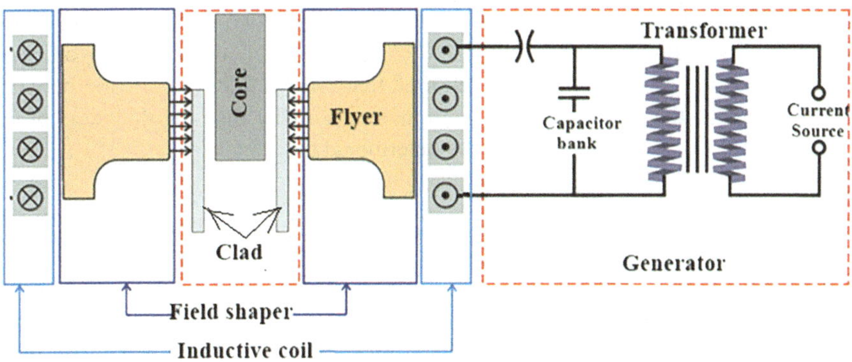

Fig. 10.1 Magnetic pulse welding (MPW)—*working set up*

The generated magnetic field creates a significantly large electromagnetic (Lorentz) force within the flyer tube. Thus, the flyer tube collides onto the fixed inner rod to produce an impact weld (see Fig. 10.1).

10.2.3 Mathematical Modeling of MPW

The energy stored in the capacitor bank can be determined from the size (capacitance) of the bank and the charge voltage using the following formula:

$$E = \frac{1}{2} * C * V^2 \tag{10.1}$$

where E is the energy stored (J), C is the capacitance (F), and V is the charge voltage (volts, V) (see Example 10.1). A significant amount of energy, usually between 5 and 200 kJ is stored in capacitors charged to a high voltage that may be in the range of 3000–30,000 V.

The circuit frequency is related to the inductance and the capacitance by:

$$f = \frac{1}{2\pi} * \sqrt{\frac{1}{L * C}} \tag{10.2}$$

where f is the circuit (discharge pulse) frequency (Hz), L is the circuit's total inductance (H), and C is the capacitance of the generator.(F) (see Example 10.2). The circuit's frequency depends on the parameters of the electromagnetic circuit; it lies in the range of 10–200 kHz, but usual operational frequencies are in between 10–20 kHz.

The magnetic field creates a significantly large electromagnetic force within the flyer workpiece, since it is closest to the inductive coil (see Fig. 10.1). The depth to which the field penetrates is called "skin depth" (δ); without this "skin depth" it would be impossible to carry out the welding operation. Recently (2025), Cartagena and co-researchers have investigated the MPW of two dissimilar materials: AA6082-T6 aluminum alloy with HC420LA cold-rolled micro-alloyed steel (Cartagena et al. 2025). They have reported that for an effective collision between both materials, the generated skin depth must be smaller than the thickness of the flyer workpiece. The skin depth can be determined by:

$$\delta = \sqrt{\frac{\rho}{\pi * f * \mu}} \tag{10.3}$$

where δ is the *skin depth* (m), ρ is the electrical resistivity of the material in contact with the coil (Ω•m), f is the pulse discharge frequency (Hz), and μ is the flyer magnetic permeability (H/m). In the case of non-ferrous flyer materials, the magnetic permeability is the same as that of the magnetic permeability in free space ($\mu = \mu_o = 4\pi \times 10^{-7}$ H/m) (see Example 10.3).

It is learnt in the preceding section that the generated magnetic field around the inductive coils creates a large electromagnetic force within the flyer tube. This force per unit area is referred to as the *magnetic pressure* or the *impact pressure*, which can be determined by:

$$P_i = \frac{B^2}{2^* \mu} \tag{10.4}$$

where P_i is the magnetic pressure (Pa) and B is the magnetic flux density (Tesla) (see Example 10.4). The impact velocity in the MPW can be determined by:

$$v_i = \frac{gap}{t} * \sqrt{\frac{P_i}{\rho_f}} \tag{10.5}$$

where v_i is the impact velocity (m/s), gap is the initial stand-off distance between the two parts to be joined (m), P_i is the impact pressure (Pa), and t and ρ_f are the thickness (m) and the density of the flyer (kg/m^3), respectively (see Example 10.5).

The impact angle is the angle that the two materials form at the point of impact (see Fig. 10.2). At the point of impact, a shock wave is created in the two materials in the radial direction. This shock wave creates the jet of air that cleans the surface to be welded.

The impact angle (γ) can be determined by reference to Fig. 10.2, as follows.

$$\tan \gamma = \frac{Gap}{Cantiver - Overlap} \tag{10.6}$$

The significance of Eq. (10.6) is illustrated in Example 10.6.

10.3 Electron Beam Welding

10.3.1 Working Principle and Equipment

Electron beam welding (EBW) uses a stream of highly-focused high-energy electrons to melt and fuse joint surfaces of a workpiece. A conventional *EBM* machine consists essentially of a device for producing a focused beam of electrons (an electron beam column) mounted on, or in, an evacuated chamber that contains devices for holding and moving a workpiece, as shown in Fig. 10.3. By applying a high voltage, electrons are ejected from the cathode; these electrons are accelerated (by use of the anode) in the electron beam column to form a long, fine beam moving at

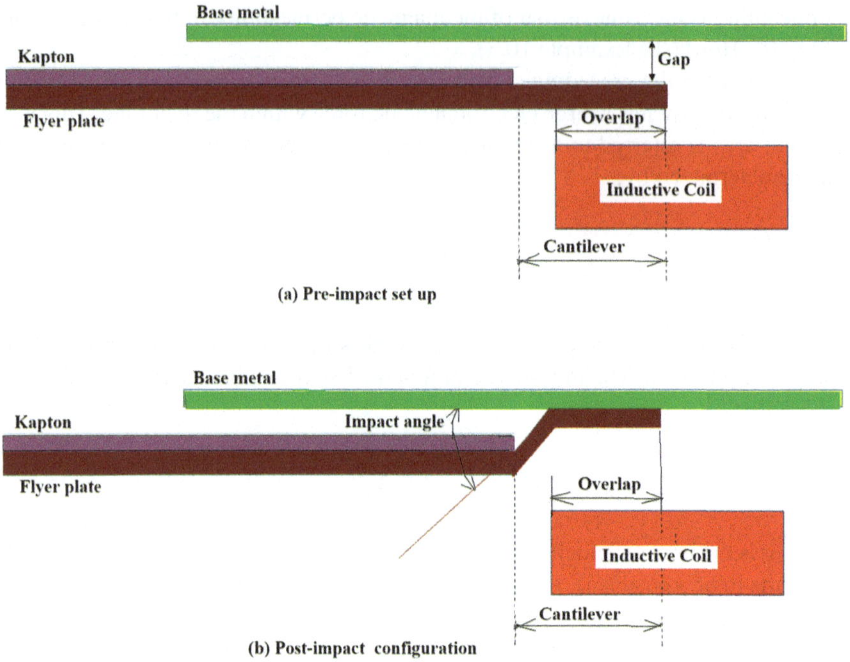

Fig. 10.2 The impact angle, the gap, the overlap, and the cantilever in the MPW

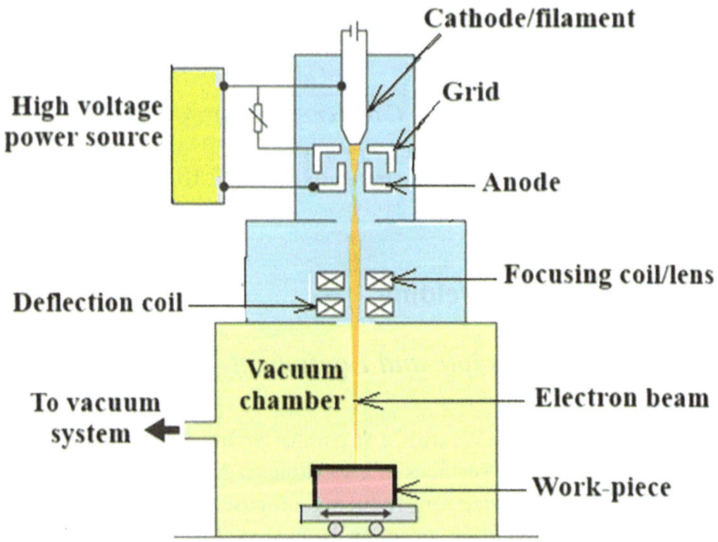

Fig. 10.3 Electron beam welding—*essential equipment components*

a very high velocity. The magnetic coils/lenses then focus the beam to produce an intense concentration of kinetic energy that can penetrate deeply into the work. The deflection coils, situated below the focusing lens, move the electron beam in complex patterns for fusion zone improvement when supplied by appropriate voltages and waveforms. The electron beam travels through the abutting surfaces, and the joint surfaces are then progressively fused, thereby creating a weld. The high heat intensity of the electron beam results in a very narrow fusion zone with minimal distortion in the work.

10.3.2 Engineering Analysis of EBW

In the *EBM* process, the electrical energy of the electron gun is first transferred to the electron beam, and then the high kinetic energy of the electron beam is translated into the heat energy of the workpiece. The main weld parameters, all of which are readily adjustable, are accelerating voltage (V), beam current (I), beam focus, and transverse speed. In order to determine the energy of an electron in electron beam welding (EBW), one is typically interested in the kinetic energy gained by each electron, as it is accelerated by a high-voltage electric field. The kinetic energy of an electron accelerated through a potential difference V is given by:

$$\mathrm{E} = \mathrm{e} * \mathrm{V} \tag{10.7}$$

where E is the energy (J), e is the charge on an electron ($e = 1.6 \times 10^{-19}$ Coul), and V is the potential difference between the filament and the anode (volts) (see Example 10.7). The electron starts from rest (from the electron gun) so the kinetic energy gained by electrons is:

$$E_k = \frac{1}{2} * \mathrm{m} * v^2 \tag{10.8}$$

where E_k is the kinetic energy of an electron in the beam (J), m is the mass of an electron (kg), and v is the speed of electron (m/s). Each electron gains kinetic energy equal to the amount of energy transferred from the electron gun. By combining Eqs. (10.7) and (10.8), we obtain:

$$\frac{1}{2} * \mathrm{m} * v^2 = e * V \tag{10.9}$$

or

$$v = \sqrt{\frac{2eV}{m}} \tag{10.10}$$

The significance of Eqs. (10.7)–(10.10) is illustrated in Example 10.8. The power of an electron beam can be compute by:

$$P = V * I \tag{10.11}$$

where P is the power (W), and I is the current in the cathode (filament) (A) (see Example 10.9).

The depth of penetration of the electron beam in the work, in EBW, can be calculated by:

$$d = 2.6 \times 10^{-17} * \left(\frac{V^2}{\rho} \right) \tag{10.12}$$

where d is the depth of penetration (mm), and ρ is the density of the work material (kg/mm^3) (see Example 10.10). The power density, in EBW, is determined by using the following formula:

$$\mathrm{PD} = \frac{f_1^* V^* I}{A} \tag{10.13}$$

where PD is the power density (W/mm^2), f_l is the heat transfer factor (in the range of 0.8–0.95) (AWS 2007), V is the accelerating voltage (volts), I is the beam current (A), and A is the work surface area on which the electron beam is focused (mm^2) (see Example 10.11).

10.4 Laser Beam Welding

10.4.1 Working Principle, Advantages and Applications

Laser beam welding (LBW) is an advanced fusion welding process that uses a concentrated heat source in the form of a laser to melt the materials, which fuse together as they cool down. The *LBW* works on the principle that when the electrons of an atom are excited by receiving some energy, and then after an instant they return to their ground states, they emit a photon of light. The concentration of this emitted photon is increased by the excited emission of radiation, resulting in a high-energy, focused laser beam. The laser beam heats and melts the edges of the workpiece, thereby forming a joint (see Fig. 10.4).

The *LBW* is a versatile process since it can weld thin materials at rapid welding speeds while running narrow and deep welds for thicker materials. A notable advantage of LBW is its capability to produce welds with very high penetration depth—even higher than EBM. Recently, Krichel et al. (2022) have experimentally proved the achievement of 115-mm-deep penetration in single-layer welds using LaVa

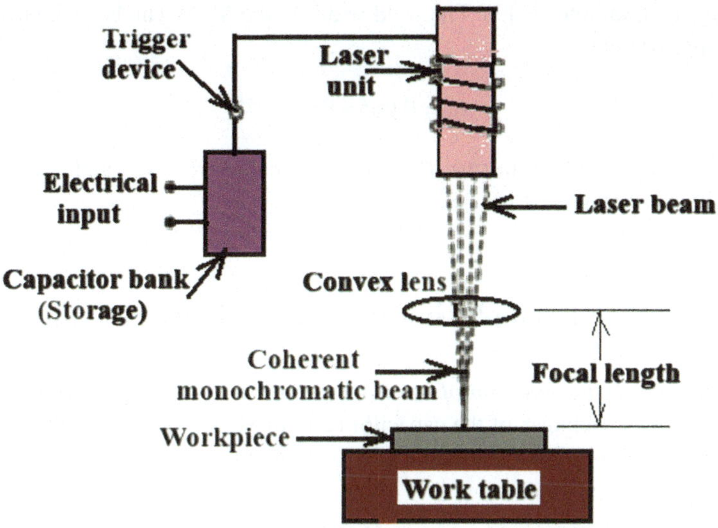

Fig. 10.4 Laser beam welding—*working principle*

(laser beam welding in vacuum) process (Krichel et al. 2022). LBW is known for its accuracy, speed, and ability to work on small, delicate components, thereby making it ideal for industries, such as electronics, batteries, automotive, aerospace, and the like.

10.4.2 Engineering Analysis of LBW

The optical spot size, in LBW, can be determined by (Amada 2016):

$$\text{Focused spot size} = 1.22 \times \frac{\lambda * M^2 * F}{D} \tag{10.14}$$

where λ is the laser-beam's wave length (m), *M* is laser-beam quality factor, *F* is the focal length of focusing optic/lens (m), and *D* is the laser-beam diameter at focus optic (m). For penetration welding, the typical spot size is in the range of 0.25–0.75 mm (see *Example 10.12*).

The penetration depth can be approximated by:

$$\mathrm{h} = \mathrm{k} * \sqrt{\frac{P}{v_w}} \tag{10.15}$$

where *h* is the penetration depth (mm), *k* is the material-dependent constant (k = 0.5 for steel), P is the laser power (W), and v_w is the welding speed (or travel speed) (mm/s) (see Example 10.13). The weld width in the MPW can be estimated by the following formula:

$$w = 0.005 * P^{0.6} * v^{-0.4} \tag{10.16}$$

where w is the weld width (m), P is the electrical power (W), and v is the speed (mm/s) (see Example 10.14).

10.5 Worked Examples in Advanced Welding Processes

Example 10.1 Determination of the Energy Stored in the Capacitor Bank in the MPW The capacitance of the capacitor bank in MPW set up is 2 mF. The charge voltage is 10,000 V. Determine the energy stored in the capacitor bank.

Solution

C = 2 mF = 2 × 10^{-3} F, V = 10,000 V, E =?

By using Eq. (10.1),

$$\mathrm{E} = \frac{1}{2} * \mathrm{C} * V^2 = 0.5 \times 2 \times 10^{-3} \times 10{,}000^2 = 10^{-3} \times 10^8 = 10^5 \text{ J}$$

The energy stored in the capacitor bank = 100 kJ.

Example 10.2 Determining the Circuit's Total Inductance in the MPW The discharge pulse frequency during a MPW operation is 20 kHz. The capacitance of the capacitor bank is 1.5 mF. Determine the circuit's total inductance.

Solution

f = 20 kHz = 20 × 10^3 Hz, C = 1.5 mF = 1.5 × 10^{-3} F, L =?

By using the modified form of Eq. (10.2),

$$\sqrt{\frac{1}{L * C}} = 2\pi * \mathrm{f}$$

$$\frac{1}{L * 0.0015} = (2 * \pi \times 20000)^2 = 1.58 \times 10^{10}$$

$$\mathrm{L} = \frac{1}{0.0015 \times 1.58 \times 10^{10}} = 4.22 \times 10^{-8} \text{ H}$$

The circuit's total inductance = 0.0422 mH.

Example 10.3 Determining the Skin Depth in the MPW A 1.5-mm-thick flyer plate made of AA6082-T6 aluminum alloy is used to collide with an alloy- steel plate to perform the MPW using a discharge pulse frequency of 20 kHz. The electrical resistivity of the flyer plate is 4 × 10^{-8} Ω•m. Determine the skin depth in the welding operation.

Solution

ρ = 4 × 10^{-8} Ω•m, f = 20,000 Hz, $\mu = 4 \times \pi \times 10^{-7}$ H/m, δ =?

By using Eq. (10.3),

$$\delta = \sqrt{\frac{\rho}{\pi * f * \mu}} = \sqrt{\frac{4\times10^{-8}}{\pi\times20000\times4\times\pi\times10^{-7}}} = \sqrt{\frac{10^{-1}}{197392}} = 7.12\times10^{-4}\,\text{m} = 0.71\,\text{mm}$$

The skin depth = 0.71 mm.

Example 10.4 Determining the Magnetic Pressure in a Magnetic Pulse Welding Operation A high-amperage current flows through an inductive coil during a magnetic pulse welding operation, thereby inducing a magnetic flux density of 1.8 Tesla. The flyer plate is made of an aluminum alloy. Determine the magnetic pressure in the welding operation.

Solution

Since the flyer plate is made of a non-ferrous alloy, $\mu = \mu_o = 4\pi \times 10^{-7}$ H/m, B = 1.8 Tesla.

By using Eq. (10.4),

$$P_i = \frac{B^2}{2 * \mu} = \frac{1.8^2}{2\times4\times\pi\times10^{-7}} = \frac{3.24}{25.1327}\times10^7 = 0.1289\times10^7 = 1.289\,\text{MPa}$$

The magnetic pressure = 1.289 MPa.

Example 10.5 Determining the Impact Velocity in MPW A 1.5-mm-thick flyer plate made of AA6082-T6 aluminum alloy is used to collide with a cold-rolled alloy steel plate to perform MPW. The density of the flyer material is 2.7 g/cm^3. The *gap* between the two plates is 1.3 mm. The impact pressure is 3500 MPa, and the impact angle is 20°.

Determine the impact velocity.

Solution

t = 1.5 mm = 0.0015 m, P_i = 3500 MPa = 35 × 10^8 N/m^2), ρ_f = 2.7 g/cm^3 = 2700 kg/m^3, gap = 1.3 mm = 0.0013 m, v_i =?

By using Eq. (10.5),

$$v_i = \frac{gap}{t} * \sqrt{\frac{P_i}{\rho_f}} = \frac{0.0013}{0.0015}\times\sqrt{\frac{35\times10^8}{2700}}$$
$$= 0.8667\times0.11385\times10^4 = 986.7 \simeq 1000\,\text{m/s}$$

The impact velocity = 1000 m/s.

Example 10.6 Determination of the Impact Angle in an MPW Operation The *gap* in a magnetic-pulse welding operation is 2 mm. The overlap and the cantilever are 6.5 mm and 9.5 mm, respectively. Determine the impact angle.

Solution
By using Eq. (10.6),

$$\tan\gamma = \frac{Gap}{Cantiver - Overlap} = \frac{2}{9.5-6.5} = 0.667$$
$$\gamma = \tan^{-1} 0.667 = 33.68°$$

The impact angle = 33.68 ≅ 34°.

Example 10.7 Determining the Energy gained by each Electron in EBW The voltage between cathode and anode of an electron gun in an EBW process is 80 kV. Determine the energy gained by an electron.

Solution
V = 80,000 V, e = 1.6×10^{-19} Coul., E =?
By using Eq. (10.7),

$$\mathrm{E} = \mathrm{e} * \mathrm{V} = 1.6\times10^{-19} \times 8\times10^{4} = 1.28\times10^{-14}\ \mathrm{J}$$

The energy gained by an electron = 1.28×10^{-14} J

Example 10.8 Determining the Speed of an Electron in the beam in EBW By using the data in *Example 10.7*, determine the speed of the electron in the EBM process.

Solution
V = 80,000 volts, e = 1.6×10^{-19} Coul., mass of an electron = m = 9.11×10^{-31} kg
By using Eq. (6.10),

$$v = \sqrt{\frac{2\ e\ V}{m}} = \sqrt{\frac{2\times1.6\times10^{-19}\times80000}{9.11\times10^{-31}}} = \sqrt{28101\times10^{12}} = 167.633\times10^{6}\ \mathrm{m/s}$$

The speed of the electron = 167.633×10^{6} m/s.

Example 10.9 Determining the Power of the Electron Beam in EBW By using the data in Example 10.8, determine the power of the electron beam, if the current applied to the cathode is 11 mA.

Solution
V = 80,000 volts, I = 11×10^{-3} A = 0.011 A, P =?
By using Eq. (10.11),

$$\mathrm{P} = \mathrm{I} * \mathrm{V} = 0.011\times80{,}000 = 880\ \mathrm{W}$$

The power of the electron beam = 880 W.

Example 10.10 Determining the Depth of Penetration in EBW By using the data in Example 10.9, determine the depth of penetration of electron beam in the work-piece in the EBW process, if the work material is steel with a density of 8000 kg/m^3.

Solution

V = 80,000 volts, ρ = 8000 kg/m^3 = 8 × 10^{-6} kg/mm^3, d =?

By using Eq. (10.12),

$$d = 2.6\times 10^{-17} * \left(\frac{V^2}{\rho}\right) = 2.6\times 10^{-17}\times\left(\frac{80000^2}{8\times 10^{-6}}\right)$$
$$= 2.6\times 10^{-17}\times 8\times 10^{14} = 0.0208\ \text{mm}$$

The depth of penetration of electron beam in the work = 0.0208 mm.

Example 10.11 Determining the Power Density in EBW An EBW operation is performed with the electron beam focused on a circular area that is 0.30 mm in diameter. The voltage and the beam current are 50 kV and 55 mA, respectively. The heat transfer factor is 0.85. Determine the average power density in the area.

Solution

Work surface diameter = D = 0.30 mm, V = 50,000 volts, I = 0.055 A, f_l = 0.85, PD =?

$$\text{Area} = \text{A} = \frac{\pi}{4}D^2 = \frac{\pi}{4}0.3^2 = 0.07065\,\text{mm}^2$$

By using Eq. (10.13),

$$\text{PD} = \frac{f_1 * V * I}{A} = \frac{0.85\times 50{,}000\times 0.055}{0.0706} = 33109\ \text{W}/\text{mm}^2$$

The average power density = 33,109 W/mm^2 = 33.109 kW/mm^2.

Example 10.12 Determination of the Focused Spot Size in LBW A laser beam of wave length 1 μm is used to perform a laser-beam welding operation. The laser-beam quality factor is 1.01, and the diameter of the laser beam at focus optics is 0.4 mm. The focal length of the focusing lens is 160 mm. Determine the focused spot size.

Solution

λ = 1 μm = 10^{-6} m, M = 1.02, F = 160 mm = 0.16 m, D = 0.4 mm = 0.4 × 10^{-3} m.

By using Eq. (10.14),

$$\text{Focused spot size} = 1.22 \times \frac{\lambda * M^2 * F}{D} = 1.22 \times \frac{10^{-6} \times 1.02^2 \times 0.16}{0.4 \times 10^{-3}} = 0.5 \times 10^{-3}\,\text{m}$$

Focused spot size = 0.5 mm.

Example 10.13 Determining the Heat Input and Penetration Depth in LBW A LBW operation is performed with a 2-kW-power laser at a travel speed of 15 mm/s. The work material is steel. Determine the: (a) heat input, and (b) penetration depth in the weld.

Solution
P = 2000 W = 2000 J/s, v_w = 15 mm/s, H_i =?, h =?

(a) By using Eq. (2.11),

$$H_i = \frac{P}{v_w} = \frac{2000\ J/s}{15\ mm/s} = 133.33\ \text{J/mm}$$

(b) By using Eq. (10.15),

$$\text{h} = \text{k} * \sqrt{\frac{P}{v_w}} = 0.5 \times \sqrt{133.33} = 5.77\,\text{mm}$$

The heat input is 133.33 J/mm, and the penetration depth = 5.77 mm.

Example 10.14 Determination of the Weld Width in LBW The power of the electrical source, in a laser-beam welding operation, is 1200 W. Determine the weld width, if the speed is 6 mm/s.

Solution
P = 1200 W, v = 6 mm/s, w =?
By using Eq. (10.16),

$$w = 0.005 * P^{0.6} * v^{-0.4} = 0.005 \times 1200^{0.6} \times 6^{-0.4} = 0.005 \times 70.389 \times 0.488 = 0.17\ \text{mm}$$

The weld width = 0.17 mm.

Questions and Problems

10.1. MCQs. Encircle the most Appropriate Answers for the Following Statements

(1) Which one of the following processes is the most suitable for welding dissimilar metals?

(a) SMAW, (b) MPW, (c) EBW, (d) LBW.

(2) Which force is resulted by generating a magnetic field around an inductive coil?

(a) Gravitational force, (b) upward force, (c) rotary force, (d) Lorentz force.

(3) What is the range of voltage to which capacitors are charged in the MPW?

(a) 3–30 kV, (b) 40–100 kV, (c) 200–500 kV, (d) 1–5 MV.

(4) The initial stand-off distance between the two workpieces in the MPW, is called ....

(a) Cantilever, (b) overlap, (c) gap, (d) clearance.

(5) Which of the following welding processes can result in the deepest-penetration weld?

(a) MPW, (b) LBW, (c) EBW, (d) thermit welding.

(6) Which of the following welding processes involves the use of a cathode filament?

(a) MPW, (b) LBW, (c) EBW, (d) thermit welding.

(7) Which one of the following welding processes involves the use of optical lens?

(a) MPW, (b) LBW, (c) EBW, (d) thermit welding.

(8) Which one of the following welding processes involves the use of electromagnetic lens?

(a) MPW, (b) LBW, (c) EBW, (d) thermit welding.

10.2. (a) Explain the working principle of magnetic pulse welding (MPW).
(b) What are the advantages and applications of the MPW?
10.3. Draw a sketch showing the working set up of the MPW process.
10.4. Illustrate electron beam welding (*EBW*) with the aid of a sketch.
10.5. What are the functions of the following coils in the *EBW* equipment:

(a) focusing coils, and (b) deflection coils?

10.6. An electron of mass m with a charge of e is accelerated through a voltage V. Derive an expression for the speed gained by the electron.
10.7. Explain the working principle of laser beam welding (*LBW*) with the aid of a sketch.
10.8. Highlight the advantages and applications of the *LBW*.
P10.9. The discharge pulse frequency during a MPW operation is 18 kHz. The capacitance of the capacitor bank is 1.6 mF. Determine the circuit's total inductance.
P10.10. The capacitance of the capacitor bank in MPW set up is 2.5 mF. The charge voltage is 8000 V. Determine the energy stored in the capacitor bank.
P10.11. A 1.8-mm-thick flyer plate made of AA6082-T6 aluminum alloy is used to collide with an alloy-steel plate to perform the MPW using a discharge pulse

frequency of 17 kHz. The electrical resistivity of the flyer plate is 4×10^{-8} Ω•m. Determine the skin depth in the welding operation.

P10.12. A high-amperage current flows through an inductive coil during an MPW operation, thereby inducing a magnetic flux density of 2 Tesla. The flyer plate is made of an aluminum alloy. Determine the magnetic pressure in the welding operation.

P10.13. A 1.8-mm-thick flyer plate made of AA6082-T6 aluminum alloy is used to collide with a cold-rolled alloy steel plate to perform MPW. The density of the flyer material is 2.7 g/cm^3. The *gap* between the two plates is 1.5 mm. The impact pressure is 3300 MPa, and the impact angle is 22°. Determine the impact velocity.

P10.14. The voltage between cathode and anode of an electron gun in an *EBW* process is 90 kV. Determine the energy gained by an electron.

P10.15. By using the data in P10.14, determine the: (a) speed of the electron in the EBM process, and (b) power of the electron beam, if the current applied to the cathode is 10 mA.

P10.16. By using the data in P10.14–P10.15, determine the depth of penetration of electron beam in the work-piece in the EBW process, if the work material is steel with a density of 8000 kg/m^3.

P10.17. An LBW operation is performed with a 3-kW-power laser at a travel speed of 13 mm/s. The work material is steel. Determine the heat input, and the penetration depth in the weld.

References

Amada, Inc. (2016). Laser Welding Fundamentals, *Amada Miyachi America Inc.*, CA, USA. https://www.amadaweldtech.eu/sites/default/files/documents/whitepapers/Laser%20Welding%20Fundamentals%202016.pdf

AWS (2007). Welding Handbook (9th Ed.), Vol. 1, Welding Science and Technology. *American Welding Society,* Miami, Florida, USA.

Cartagena, M.A.R., Plaza, E.I., Larruscain, A.T., *et al.* (2025), Magnetic Pulse Welding of Dissimilar Materials: Weldability Window for AA6082-T6/HC420LA Stacks, *Metals—Welding Experiemnts and Simulation, 15*(6), 619; https://doi.org/10.3390/met15060619

Lapovok, R., Ng, H. P., Tomus, D., Estrin, Y. (2012). Bimetallic copper–aluminum tube by severe plastic deformation, *Scripta Materialia*, Vol. 66, pp. 1081-1084

Krichel, T., Olschok, S., Reisgen, U. (2022). Comparison of penetration depth in bead-on-plate welds of thick-walled steel sheets with laser beam welding in vacuum and electron beam welding, *Vacuum*, Vol. 203, 111314

Sapanathan, T., Raoelison, R.N., Buiron, N., Rachik, M. (2016). Magnetic Pulse Welding: An Innovative Joining Technology for Similar and Dissimilar Metal Pairs, *In*: Joining Technologies (Ed. M. Ishak). *IntechOpen Limited*, London, UK.

Wiley, E. (2026). 6 Advanced Welding Processes and Their Applications Explained, Wiley Metal Fabricating Online: https://www.wileymetal.com/6-advanced-welding-processes-and-their-applications-explained/

Part V
Welding Design and Quality Assurance

Chapter 11
Welding Design

Abstract The design of a correct weld joint is critical to the successful fabrication of metallic products. This chapter develops the skill of designing a safe weld joint by covering both qualitative and quantitative aspects of welding design. This is why this chapter can be considered as the heart of the book. Weld joints have been introduced with the aid of sketches as per the AWS standards.

Welding symbols, as per the *ASME* code items (the arrow, the reference line, the tail, groove welds, complete joint preparation, and fillet welds) have been explained with the aid of diagrams. Welding codes, developed by the AWS, have been summarized; the AWS D1.1 codes are presented as a table. The discussion on weld sizing covers both fillet weld design and butt weld design; here, mathematical models for calculating the throat area, shear stress, and tensile stress are presented with the aid of sketches. Finally, the factor of safety (FoS) in the weld design is elaborated with reference to the yield-strength and the tensile-strength data for various electrodes. This chapter contains 25 worked numerical examples/design problems, 19 diagrams, 2 tables, 20 formulas/equations, and 5 MCQs with their answers given at the end of the book.

Keywords Weld joint · CJP · Welding symbols · Welding codes · AWS · ASME · Fillet weld · Butt weld · Throat area · Factor of safety (FoS)

Abbreviations

AISC	American Iron and Steel Construction
ASME	American Society of Mechanical Engineers
AWS	American Welding Society
CJP	Complete joint preparation
FoS	Factor of Safety
WPS	Welding procedure specification

Z. Huda, *Welding and Joining Engineering*,
https://doi.org/10.1007/978-3-032-20462-2_11

Nomenclature

A_{dpfw}	Throat area of double parallel fillet weld
A_{pfw}	Throat area of parallel fillet weld
W_{UTS}	Weld ultimate tensile strength, MPa
W_{ys}	Weld yield strength, MPa
$W_{\tau s}$	Weld shear strength, MPa
a	Throat thickness
F	Force or load, N
h	Leg size, mm
l	Weld length, mm
σ	Tensile stress, MPa
τ	Shear stress, MPa

11.1 Weld Joint and Design: An Overview

The design of a correct weld joint is critical to the successful fabrication of metallic parts and structures. A poor weld design or substandard welding practice would result in a poor quality product, posing a threat to society owing to its premature failure. A number of historical disasters have resulted due to faulty design/practice in welding of pressure vessels, boilers, ships, and the like (see Chap. 14). It is, therefore, very important that a welding engineer ensures the welded part or fabricated structure has high reliability and quality. To achieve the safe design of welded products, a number of engineering technologies must be taken into consideration; these include: structural and mechanical engineering, materials engineering, weldability and performance, corrosion science, physics and chemistry of gases, and health and safety aspects (Hicks 2001).

A welding joint is the configuration of two metal plates that will be welded together. According to the AWS, American Welding Society, there are basically five types of welding joints: (1) butt joint, (2) lap joint, (3) corner joint, (4) T-joint, and (5) edge joint (see Fig. 11.1).

A butt joint, or square groove joint, is formed by placing the ends of two parts together. It is the simplest and the commonest type of welding joint. It is generally used for flanges, valves, fittings, and the like. A lap joint is formed when two pieces of metal are stacked on top of each other. It is usually used for sheet metal welding. A corner joint meets with the corner either in a closed or an open manner, creating an 'L' shape. It is used to make frames, boxes, and other similar items. A tee (T) joint is made by intersecting two parts at a right angle and one part lies at the center of the other. An edge joint is formed by welding the edges of two parts. Besides these five basic types of joints, another important weld joint is *fillet weld*. It has been shown that the SMAW fillet welds have the greatest thermal efficiency (see Chap. 12, Table 12.1).

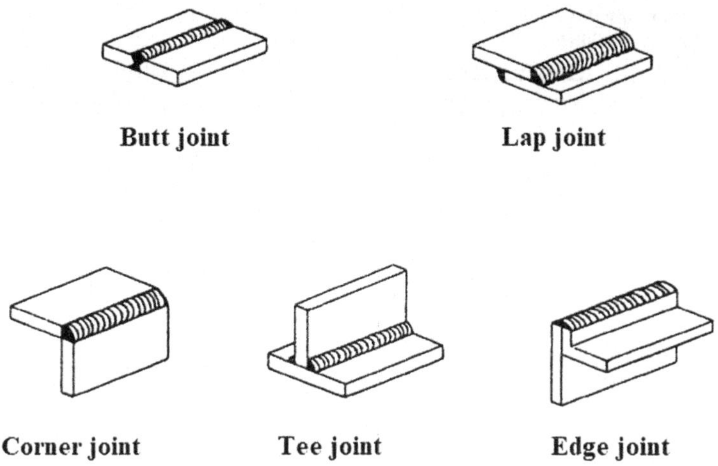

Fig. 11.1 The five basic welding joints

A weld designer as well as a welder should have a sound knowledge of the following weld-design terms: weld joints, welding symbols, welding codes, weld sizing, and the like. These aspects of welding design are discussed in the following sections.

11.2 Welding Symbols

A welding symbol is a graphic representation of the elements needed to fully specify weld requirements. The welding symbols provide a considerable advantage in developing a shop standard for use by the welder. The following are the welding symbols associated with the *ASME* code items: (a) the arrow, (b) the reference line, (c) the tail, (d) groove welds, (e) complete joint preparation (CJP), and (f) fillet welds. These welding symbols are elaborated, as follows.

The Arrow. The arrow is an essential part of every welding symbol; it must point to the joint to be welded. The side of the joint to which *the arrow* points, is referred to as the "arrow side" of the joint, and the opposite side of the joint is called the "other side" (see Fig. 11.2). Sometimes, a multiple-arrow welding symbol is used on a drawing; this welding-symbol system involves the use of more than one arrow in a single welding symbol if each joint to which an arrow is pointing is to be welded in exactly the same manner. A multiple-arrow welding symbol can be very helpful, particularly around closed corners.

The Reference Line. *The reference line* is the horizontal line that makes up the main body of the welding symbol. It is the anchor to which all the other welding symbols are tied.

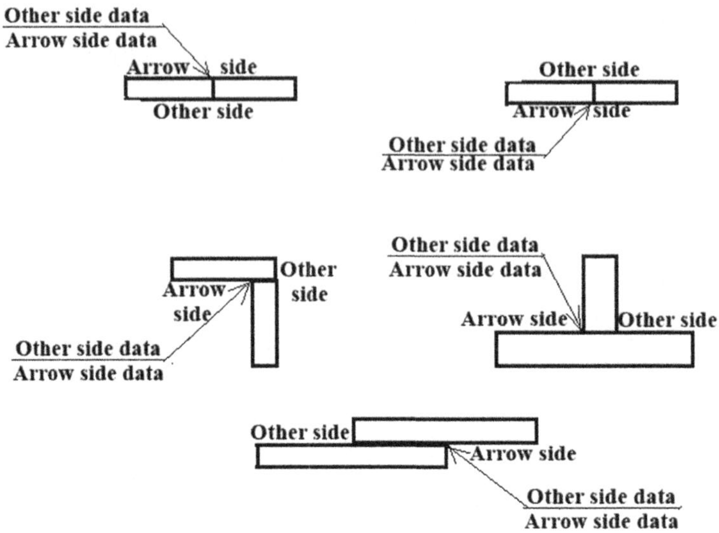

Fig. 11.2 *The arrow* welding symbol for various weld joints

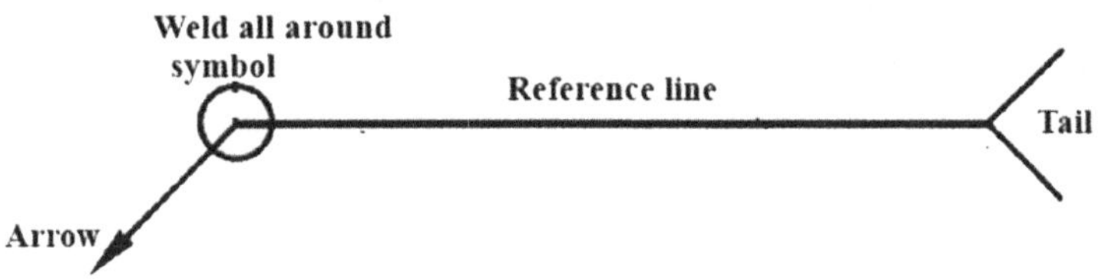

Fig. 11.3 Welding symbols: the arrow, the reference line, the tail, etc.

It is evident in Fig. 11.2 that the "arrow side data" is placed below the reference line, whereas the "other side data" is placed above the reference line.

The Tail. *The tail* welding symbol is connected at the end of the reference line opposite the arrow; it is drawn as a greater-than (>) or less-than (<) symbol, as shown in Fig. 11.3. Reference to the approved welding procedure specification (WPS) is an example of information appropriate in the tail of a welding symbol.

Groove Welds. A groove weld (*e.g.* V-groove, U-groove, etc.) symbol may be added to a welding symbol below the reference line to specify a weld only on the "arrow side" of the joint (single weld); the symbol may be added above the reference line to specify a weld only on the "other side" of the joint (also a single weld) (see Fig. 11.4).

Complete Joint Preparation (CJP). The general practice for specifying the CJP is to use an arrow, a reference line, and add the CJP to the tail of the symbol; this symbol specifies the CJP with no detail as to how the final condition is to be achieved. Such a symbol is used when there is uncertainty as to what specific equipment will be available when the work is to go through the shop. Later, when

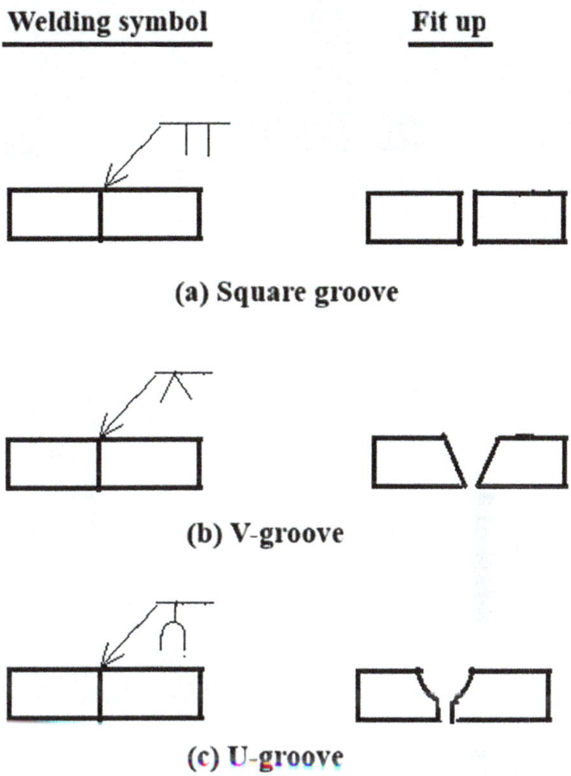

Fig. 11.4 Groove weld symbols in butt joints; (**a**) square groove, (**b**) V-groove, and (**c**) U-groove

equipment availability is known, the shop should submit assembly drawings to the engineering department for final approval, with welding symbols containing all pertinent details.

Fillet Welds. The fillet weld symbol is a right triangle placed on the reference line with the perpendicular leg always on the left. Fillet welds are used extensively in industries, including boiler and pressure vessel manufacturers. The dimension, specifying the size of a fillet weld, is placed to the left of the fillet weld symbol; it appears on the same side of the reference line (see Fig. 11.5). A complete welding symbol for a fillet weld includes the fillet weld symbol and its size, length, pitch, contour, method of countering, and the like (see Examples 11.1 and 11.2).

The welding symbol for a double fillet weld is illustrated in Fig. 11.6 (see Example 11.3).

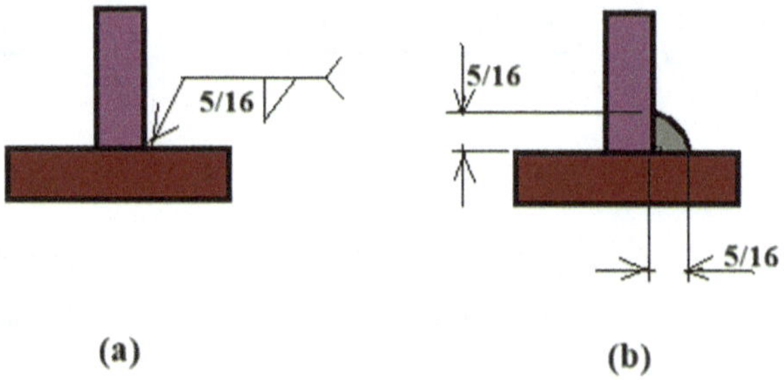

Fig. 11.5 Fillet weld symbol (**a**), and weldment with fillet weld (**b**)

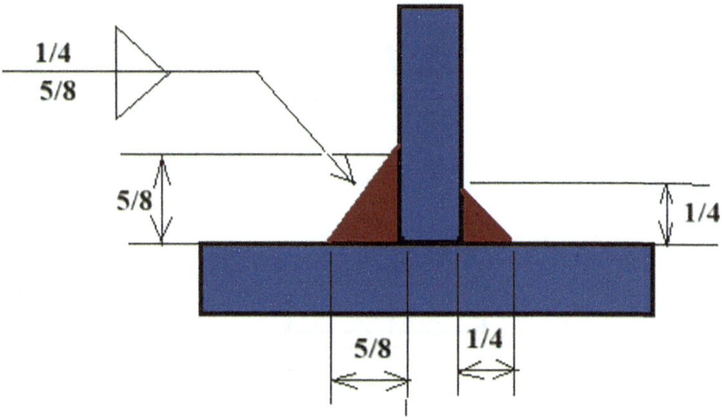

Fig. 11.6 The welding symbol and the dimensions for a double fillet weld

11.3 Welding Codes

Welding codes are developed and compiled, in the United States, by the AWS, American Welding Society. The AWS has authored more than 350 codes/standards for welding practices and procedures (AWS 2025). These welding codes are partially arranged in the following sections: *Structural and Sheet Metal* (AWS D1.1 to AWS D8.9), *Reference and Business* (AWS: A1.1 to A-3), *Safety and Health* (AWS: Z-49.1 to C4.2), *Qualification* (AWS: B2.1 to G1.6), *Training* (AWS: CM to GWF), *Certification* (AWS: QC1 to CM), *Testing and Inspection* (AWS: B1.10 to G1.10), *Operation* (D16.2), *Thermal Spraying* (AWS: C2.16 to C2.25), *Plastics* (AWS: B2.4 to G1.10), *Food Processing System* (AWS: D18.1 to D18.3), *Welding and Cutting Processes* (AWS: C1.1 to D10.13), Standard Welding Procedure Specifications: *Brazing and Soldering* (AWS: C3.2 to D10.13), *Pipe and Tubing* (AWS: D10.4 to A5.02), *Filler Material Specifications* (AWS: A5.1 to

Table 11.1 AWS codes for structural and sheet metal

No.	Code	Materials/method
1.	AWS D1.1/D1.1M	Structural welding code—steel
2.	AWS D1.1-SWJ-WC	Welded joint details for structural applications wall chart
3.	AWS D1.1-BI	The official book of D1.1 interpretations
4.	AWS D1.2/D1.2M	Structural welding code—aluminum
5.	AWS D1.3/D1.3M	Structural welding code—sheet steel
6.	AWS D1.4/D1.4M	Structural welding code—reinforcing steel
7.	AWS D1.5/D1.5M	Bridge welding code
8.	AWS D1.6/D1.6M	Structural welding code—stainless steel
9.	AWS D1.7/D1.7M	Guide for strengthening and repairing existing structures
10.	AWS D1.8/D1.8M	Structural welding code—seismic supplement
11.	AWS D1.9/D1.9M	Structural welding code—titanium
12.	AWS D9.1/D9.1M	Sheet metal welding code
13.	AWS D8.9M	Recommended practices for test methods for evaluating the resistance spot Welding behavior of automotive sheet steel materials

A5.34), *Automotive* (AWS D8.1 to D14.3), *Aerospace* (AWS: D17.1 to D17.3), *Marine* (AWS: D3.5 to D3.7), *Machinery* (AWS: D14.1 to D14.8), *Railroad* (AWS: D15.1 to D15.2), and *Resistance Welding* (AWS: RWMA Bulletin # 5 to D17.2). The welding codes for *Structural and Sheet Metal* are presented in Table 11.1. The AWS D1.1 is one of the most-referenced codes in the world (see Table 11.1).

11.4 Weld Sizing

11.4.1 Design of Fillet Welds

The design of a fabricated structure must be based on the avoidance of stress concentration sites; this objective can be achieved by providing fillet welds, which have roughly a right-angled triangular geometry. The simple design philosophy of fillet welds states that the distributed or demand stress on the critical throat plane created due to external forces in fillet welds shall be equal to or smaller than the strength criterion of the weld material (Picón and Cañas 2009).

Leg Length and Throat Thickness. There are two important dimensions of a fillet weld: (1) leg length (h) and (2) throat thickness (a) (see Fig. 11.7). The *leg*

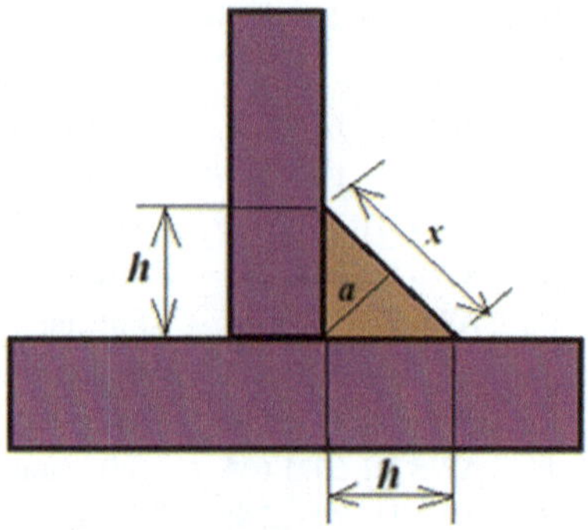

Fig. 11.7 The dimensions in a fillet weld with equal leg lengths; h = leg length, a = throat thickness

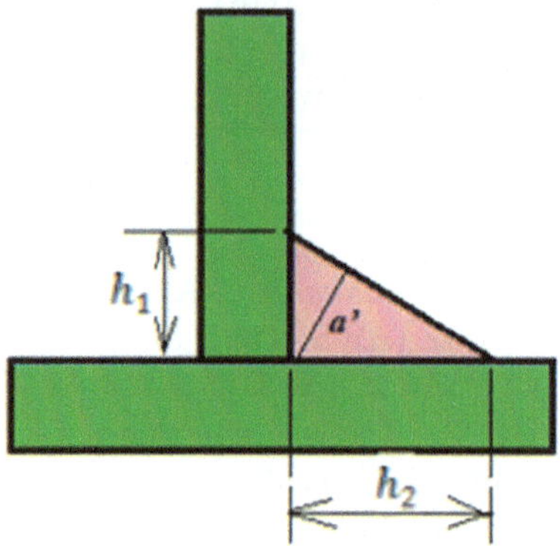

Fig. 11.8 Fillet weld with unequal leg lengths

length is the distance from the root of the weld to its toe. A fillet weld joint has two legs: one in the horizontal direction, and another in the vertical plane. The *throat thickness* is the shortest thickness from the root of the weld to its face; it determines the strength of the weld.

The leg size, h, can be easily measured (see Fig. 11.7). By using trigonometry, the throat thickness (a) can be determined by using Pythagoras theorem, as follows:

$$x^2 = h^2 + h^2 \tag{11.1}$$

$$(2a)^2 = 2 * h^2 \tag{11.2}$$

$$a = 0.707 * \mathrm{h} \tag{11.3}$$

It must be noted that Eq. (11.3) holds good when both the leg lengths, in the fillet weld, are equal (see Example 11.4). Figure 11.8 illustrates a fillet weld with unequal leg lengths: h_1 and h_2. In the case of a fillet weld with unequal lengths, the throat thickness (a') can be determined by:

$$a' = \frac{h_1 * h_2}{\sqrt{(h_1^2 + h_2^2)}} \tag{11.4}$$

where a' is the throat thickness of the fillet welt with leg lengths h_1 and h_2 (see Example 11.5).

Throat Area. The throat area (A_{fw}) of a fillet weld can be calculated, with reference to Fig. 11.7, by using the following formula:

$$A_{fw} = \frac{1}{2} * \text{base} * \text{height} = \frac{1}{2} * h * h = \frac{1}{2} * h^2 \tag{11.5}$$

The significance of Eq. (11.5) is illustrated in Example 11.6. Fillet welds are designed according to the American Iron and Steel Construction code: AISC 360—Chapter J2. *When a load acts parallel to the length of the weld,* the weld is called *parallel fillet weld.* A parallel fillet weld fails by *shear* under the action of shear forces acting parallel to the weld length, as shown in Fig. 11.9.

The throat area of a parallel fillet weld (A_{pfw}) can be determined by:

$$A_{pfw} = 0.707 * h * l \tag{11.6}$$

where h is the leg length, and l is the length of the weld (see Example 11.7). Accordingly, the throat area for a double parallel fillet weld (Fig. 11.10) can be expressed as:

$$A_{dpfw} = 2 \times 0.707 * h * l = 1.414 * h * l \tag{11.7}$$

The terms in Eq. (11.7) have their usual meaning (see Example 11.8).

Geometric Limitations. The geometric limitations for the leg size of a fillet weld are specified by the AISC code and are based on the thickness of the thinner part

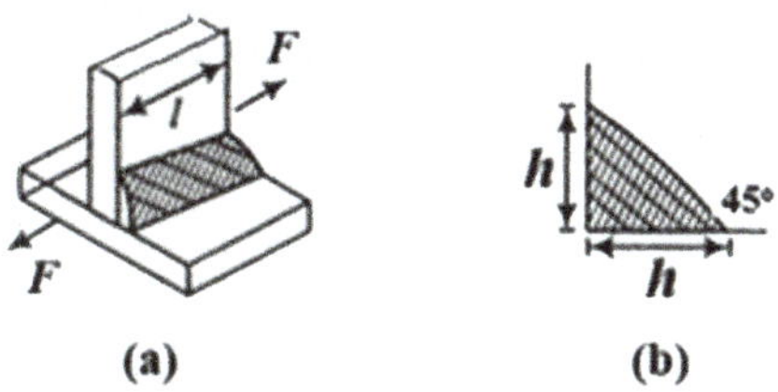

Fig. 11.9 The shear forces acting on a weldment (**a**), thefillet weld (**b**)

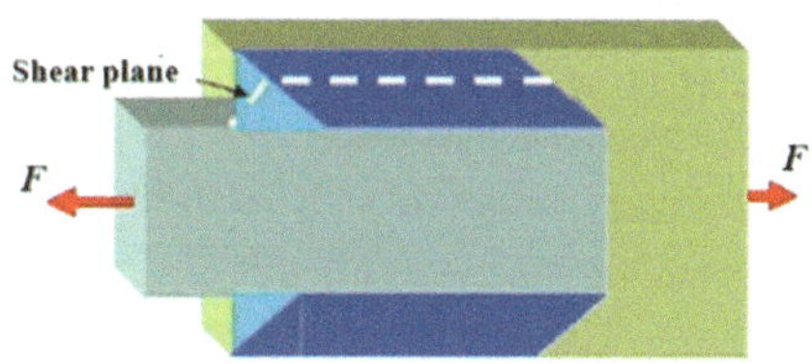

Fig. 11.10 A weldment with a double parallel fillet weld

joined. For a 6-mm (or ¼ in.) thick plate, the minimum weld size is 3 mm (or 1/8 in.), whereas the maximum weld size is the thickness of the plate itself, *i.e.* 6 mm. The maximum size of a fillet weld along an edge of a plate less than 6 mm thick is also the thickness of the plate itself.

The fillet-weld leg size can be approximately related to heat input by (Funderburk 1999):

$$\mathrm{h} = \sqrt{\frac{H_i}{500}} \tag{11.8}$$

where h is the fillet-weld leg size (mm) and H_i is the heat input (J/mm) (see Example 11.9).

Shear Stress. Once the throat area has been calculated by using Eq. (11.6), the shear stress acting on the parallel fillet weld can be computed by using the *AISC* code as follows:

$$\tau = \frac{F}{A_s} = \frac{F}{0.707 * \mathrm{h} * l} \tag{11.9}$$

where τ is the shear stress (MPa), F is the shear force acting parallel to the weld length (N), A_s is the shear-plane area or the throat area of the parallel fillet weld (mm^2), h is the leg length (mm), and l is the weld length (mm) (see Example 11.10). The shear stress can be related to the allowable shear stress by:

$$\tau = \frac{F}{0.707 * \mathrm{h} * l} \leq \tau_{all} \tag{11.10}$$

where τ is the design shear stress and τ_{all} is the allowable shear stress (see Example 11.11).

Transverse Fillet Welds. When load acts perpendicular to the fillet weldment, tensile stress is induced and the corresponding weld is called a *transverse fillet weld* (see Fig. 11.11). The tensile stress is related to the load and weld length by:

$$\sigma = \frac{F}{0.707^{*} h^{*} l} \tag{11.11}$$

where σ is the tensile stress (MPa), and F is the tensile force (N) (see Examples 11.12 and 11.13).

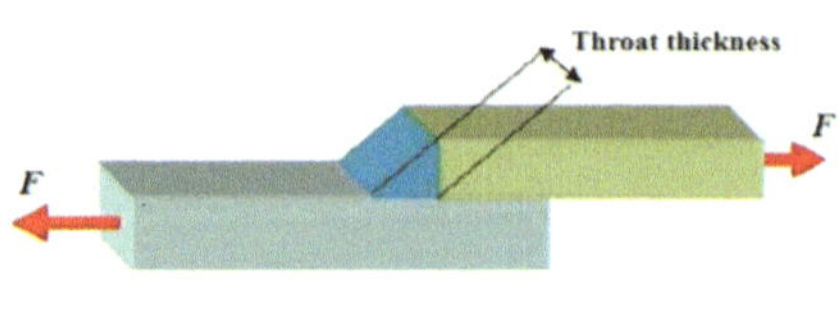

Fig. 11.11 A weldment with a transverse fillet weld

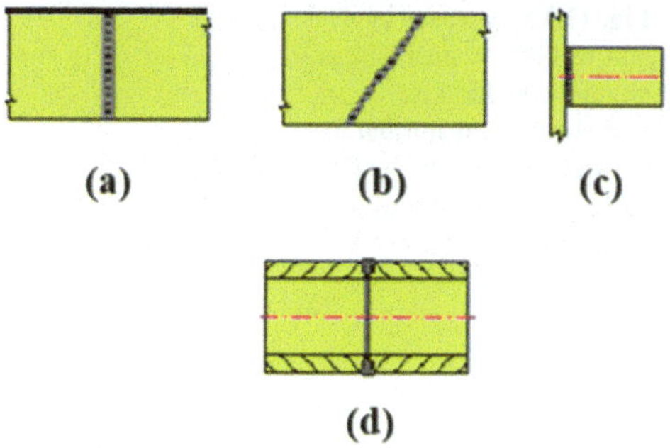

Fig. 11.12 Various butt weld orientations; (**a**) normal, (**b**) angled, (**c**) end tube, (**d**) circumferential tube

11.4.2 Design of Butt Welds

It is important that in a butt joint, the surfaces of the workpieces being joined are on the same plane and weld metal remains within the planes of the surfaces (TWI 2025).

Throat Area. The throat area of a butt-weld joint strongly depends on the orientation of the weld (see Fig. 11.12). The throat area of a butt weld oriented normal to the applied load (Fig. 11.12a) is expressed as:

$$A_1 = s * l \tag{11.12}$$

where A_1 is the throat area of the weld (mm^2), s is the thickness of thinner joined part (mm), and l is the weld length (mm) (see Example 11.14).

The throat area of a butt weld angled relative to applied load (Fig. 11.12b) can be determined by:

$$A_2 = \frac{s * l}{\cos\delta} \tag{11.13}$$

where A_2 is the throat area of the butt weld, and δ is the weld angle (deg) (see Example 11.15).

The throat area of the butt end-tube weld (Fig. 11.12c) can be calculated by:

$$A_3 = s * l \tag{11.14}$$

where A_3 is the throat area of the butt end-tube weld (see Example 11.16). The throat area of a circumferential tube butt weld (Fig. 11.12d) can be determined by:

$$A_4 = \pi s(D - s) \tag{11.15}$$

where A_4 is the throat area of the circumferential tube butt weld, and D is the tube outer diameter (see Example 11.17).

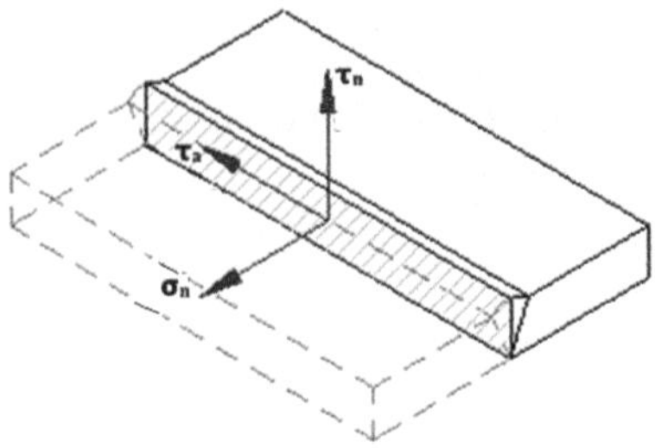

Fig. 11.13 Stresses in a butt weld (τ_a = shear stress along the weld axis, τ_n = shear stress normal to the weld axis, σ_n = tensile stress normal to the weld axis, and)

Stresses in Butt Welds. The stresses in butt welds are determined based on whether they are in tension or shear (see Fig. 11.13). The *tensile stress* in a butt weld can be calculated using *AISC* code as follows:

$$\sigma = \frac{F}{A} \tag{11.16}$$

where σ is the tensile stress acting normal to the weld axis (MPa), *F* is the tensile force (N), and *A* is the throat area of the weld (see Example 11.18).

The shear stress acting on a butt weld can be determined using *AISC* code as follows:

$$\tau = \frac{F_s}{A} \tag{11.17}$$

where τ is the shear stress (MPa) and F_s is the shear force (N) (see Example 11.19).

11.5 The Factor of Safety: Design Problems (Solved)

In arc welding, the weld shear strength ($W_{\tau s}$) is related to the yield strength of the electrode by:

$$W_{\tau s} = 0.6 * W_{ys} \tag{11.18}$$

where W_{ys} is the weld yield strength or the yield strength of the electrode. Table 11.2 presents the yield strength and tensile strengths of various electrodes in accordance with their AISC welding codes. The significance of Eq. (11.18) and Table 11.2 is illustrated in Example 11.20.

Once the weld shear strength ($W_{\tau s}$) has been calculated by using Eq. (11.18), the safety in the weld can be ensured by using the following formula:

Table 11.2 Welding electrode strengths using AISC welding codes

Welding electrode code	E60xx	E70xx	E80xx	E100xx	E110xx
Yield strength (W_{ys}), MPa	345	390	470	600	670
Ultimate tensile strength (W_{uts}), MPa	427	482	550	698	760

$$\tau = \frac{W_{\tau s}}{(FoS)} \quad (11.19)$$

where τ is the design shear stress and FoS is the factor of safety in the weld design.

For a safe design, the tensile stress in a weld is related to the ultimate tensile strength of the weld or the electrode by:

$$\sigma = \frac{W_{UTS}}{FoS} \quad (11.20)$$

where σ is the design tensile stress and W_{UTS} the ultimate tensile strength of the weld or the electrode. The significance of Eqs. (11.19) and (11.20) is illustrated in Examples 11.21–11.25.

11.6 Worked Numerical Examples in Welding Design

Example 11.1 Using Welding Symbols to Show the Dimensions in a Weldment Refer to the welding symbol in Fig. 11.14. Re-draw Fig. 11.14 showing the missing dimension(s).

Solution

The re-drawn Fig. 11.14 with the dimension(s) is shown as Fig. 11.15.

Example 11.2 Using Welding Symbols to Write Welding Information Refer to Fig. 11.16. Write the values/information for the following welding-design parameters: (a) weld type, (b) weld face contour, (c) method of contouring, (d) fillet weld size, (e) length of weld, (f) pitch of weld.

Solution

(a) Fillet weld, (b) weld face contour: convex, (c) G = grinding, (d) fillet weld size = 5/16″, (e) length of weld = 4″, (f) pitch of weld = 6″.

Example 11.3 Drawing a Sketch with Dimensions using Welding Symbols Figure 11.17 shows the welding symbol for a weld. What is the type of weld? Draw a sketch of the weldment showing the dimensions of the weld.

Solution

The type of the weld is double fillet weld. Figure 11.18 shows the weldment with the dimensions.

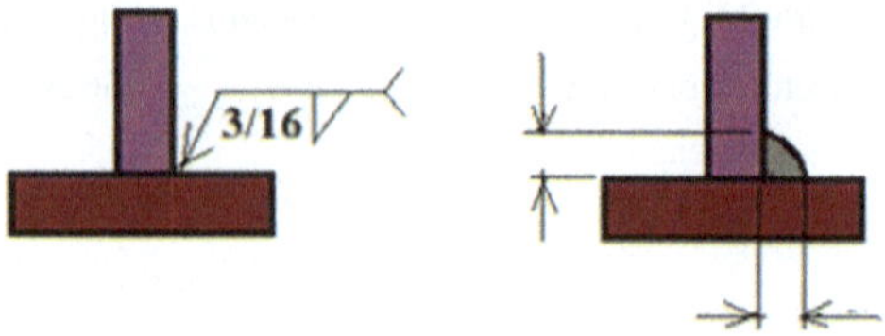

Fig. 11.14 Figure for Example 11.1

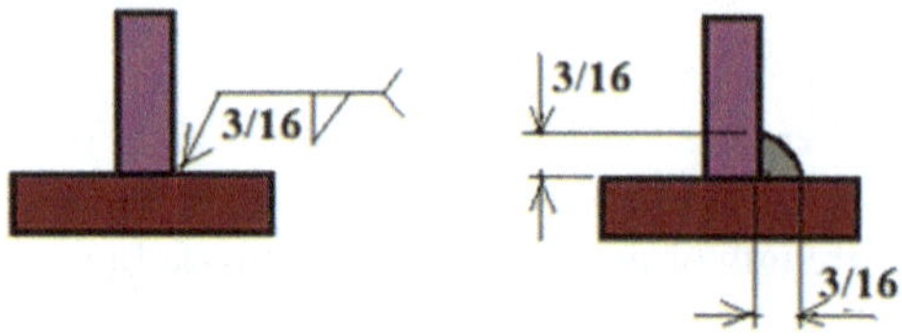

Fig. 11.15 Redrawn figure for Example 11.1

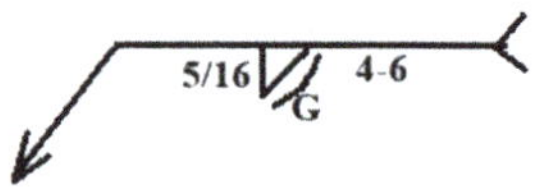

Fig. 11.16 Welding symbol for Example 11.2

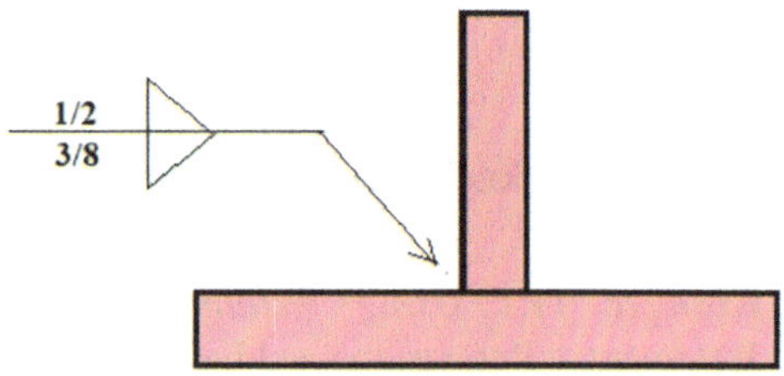

Fig. 11.17 The welding symbol for Example 11.3

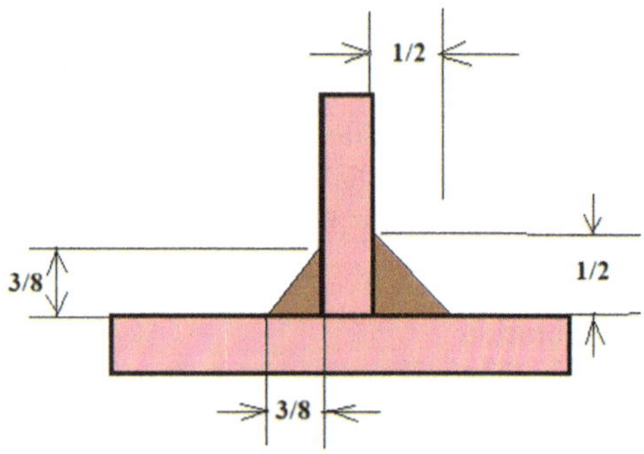

Fig. 11.18 The weldment with the dimensions of the double fillet welds

Example 11.4 Determination of the Throat Thickness of a Fillet Weld with Equal Legs A fillet weld has equal leg lengths. Determine the throat thickness if the leg length is 6 mm.

Solution

By using Eq. (11.3),

$$a = 0.707 * \mathrm{h} = 0.707 \times 6 = 4.24$$

The throat thickness = 4.24 mm.

Example 11.5 Determination of the Throat Thickness of a Fillet Weld with Unequal Legs A fillet weld has leg lengths of 9 mm and 7 mm. Determine the throat thickness of the weld.

Solution

h_1= 9 mm, h_2= 7 mm, a' =? By using Eq. (11.4),

$$a' = \frac{h_1 * h_2}{\sqrt{\left(h_1^2 + h_2^2\right)}} = \frac{9 \times 7}{\sqrt{\left(9^2 + 7^2\right)}} = \frac{63}{11.4} = 5.53$$

The throat thickness = 5.53 mm.

Example 11.6 Determination of the Throat Area of a Fillet Weld By using the data in Example 11.4, determine the throat area of the fillet weld.

Solution

By using Eq. (11.5),

$$A_{fw} = \frac{1}{2} * h^2 = \frac{1}{2} \times 6^2 = 18$$

The throat area of the fillet weld = 18 $(mm)^2$.

Example 11.7 Determination of the Throat Area of a Parallel Fillet Weld Two steel plates, each of 6-mm thickness, are welded with a parallel fillet weld. The weld length is 55 mm. Determine the throat area of the weld. Take the minimum recommended leg length.

Solution

For a 6-mm thick plate, the minimum weld leg, h = 3 mm, l = 55 mm, A_{pfw} =?
By using Eq. (11.6),

$$A_{pfw} = 0.707 * \mathrm{h} * l = 0.707 \times 3 \times 55 = 116.65\ (mm)^2$$

The throat area of the parallel fillet weld = A_{pfw}= 116.65 $(mm)^2$.

Example 11.8 Determination of the Throat Area of a Double Parallel Fillet Weld A double parallel fillet weld has a leg length of 6 mm and a weld length of 58 mm. Determine the throat area of the weld.

Solution

By using Eq. (11.7),

$$A_{dpfw} = 1.414 * \mathrm{h} * l = 1.414 \times 6 \times 58 = 492$$

The throat area of the double parallel fillet weld = 492 $(mm)^2$

Example 11.9 Determination of the Fillet-weld Leg Size when the Heat Input in Known The heat input in an arc welding operation is 5 kJ/mm. Determine the fillet-weld leg size.

Solution

H_i = 5 kJ = 5000 J, h =? By using Eq. (11.8),

$$\mathrm{h} = \sqrt{\frac{H_i}{500}} = \sqrt{\frac{5{,}000}{500}} = \sqrt{10} = 3.16 \text{ mm}$$

The weld leg size = 3.16 mm.

Example 11.10 Determination of the Shear Stress in a Parallel Fillet Weld By using the data in Example 11.7, determine the shear stress acting on the parallel fillet weld if the shear force is 60 kN.

Solution

A_{pfw}= 116.65 $(mm)^2$, F = 60,000 N, σ =? By using Eq. (11.9),

$$\tau = \frac{F}{A} = \frac{60000}{116.65} = 514.36\,\text{MPa}$$

The shear stress = τ = 514.36 MPa.

Example 11.11 Designing the Fillet Weld by Determining its Leg Size A shear force of 58 kN is applied on the weldment with a parallel fillet weld (weld length = 50 mm). The allowable shear stress is 144.6 MPa. Design the fillet weld by determining its leg size.

Solution

F = 58,000 N, l = 50 mm, $\tau_{all} \leq$ 144.6 MP, h =? By using Eq. (11.10),

$$\tau = \frac{F}{0.707 * \mathrm{h} * l} \leq 144.6\text{MPa}$$

$$\frac{58000}{0.707 * \mathrm{h} * 50} \leq 144.6$$

$$\mathrm{h} \geq \frac{1640.7}{144.6} \geq 11.3$$

Hence, the design of the fillet weld calls for a leg length of at least 11.3 mm.

Example 11.12 Determination of the Tensile Stress in a Transverse Fillet Weld

A tensile force of 35 kN acts on a weldment with a transverse fillet weld of leg length 8 mm and weld length of 60 mm. Determine the tensile stress induced in the weld.

Solution

F = 35,000 N, h = 8 mm, l = 60 mm, σ =?

By using Eq. (11.11),

$$\text{Tensile stress induced} = \sigma = \frac{F}{0.707 * h * l} = \frac{35000}{0.707 * 8 * 60} = 103.13\,\text{MPa}$$

Example 11.13 Selecting the Most Economical Welding Electrode for a Weld Design By using the data in Example 11.12, select the most economical electrode for the weld design.

Solution

In order to select an electrode for a welding design, the induced tensile stress must be less than the ultimate tensile strength of the electrode. By using Table 11.2, the **E60xx electrode**, with σ_{uts} = 427 MPa, is the most economical one.

Example 11.14 Determining Throat Area in a Butt Weld Oriented Normal to Applied Load Two metal plates each of thickness 3 mm are butt welded to form a 6-cm long welded joint such that the butt weld is normal to the applied load. Determine the throat area.

Solution

s = 3 mm, l = 6 cm = 60 mm, A_1 =? By using Eq. (11.12),

$$A_1 = s * l = 3 \times 60 = 180\ \text{mm}^2$$

The throat area of the but weld = 180 mm^2

Example 11.15 Determining Throat Area in a Butt Weld Angled to Applied Load Two metal plates are welded such that the butt joint is angled at 35° relative to the applied load. The thinner part is 6-mm-thick, and the weld length is 45 mm. Determine the throat area.

Solution

s = 6 mm, l = 45 mm, δ = 35°, A_2 =?

By using Eq. (11.13),

$$A_2 = \frac{s * l}{\cos\delta} = \frac{6 \times 45}{\cos 35} = 329.6\left(mm\right)^2$$

The throat area of the angled butt-weld joint = 329.6 $(mm)^2$

Example 11.16 Determining Throat Area in a Butt Weld with an End Tube A metal tube is welded to a metal plate, as shown in Fig. 11.12c; the weld length is

20 mm. The thickness of the thinner part (plate) is 5 mm. Determine the throat area of the butt weld.

Solution
By using Eq. (11.14),

$$\text{Throat area} = A_3 = s * l = 5 \times 20 = 100\ (mm)^2$$

Example 11.17 Determining Throat Area in a Butt Weld with a Circumferential Tube A weldment is produced, as shown in Fig. 11.12d. The tube outer diameter is 15 mm. The thickness of the tube is 6 mm. Determine the throat area.

Solution
D = 15 mm, s = 6 mm. By using Eq. (11.15),

$$\text{Throat area} = A_4 = \pi s(D - s) = \pi \times 6 \times (15 - 6) = 170\ (mm)^2$$

Example 11.18 Determining the Tensile Stress in a Butt Weld A tensile force of 70 kN acts normal to a butt-weld axis in a weldment (Fig. 11.12a). The weld length is 60 mm and the thickness of the thinner plate is 5 mm. Determine the tensile stress.

Solution
F = 70,000 N, l = 60 mm, s = 5 mm.

$$\text{Throat area} = A_3 = s^* l = 5 \times 60 = 300\,\text{mm2}$$

By using Eq. (11.16),

$$\sigma = \frac{F}{A} = \frac{70000}{300} = 233.33\ \text{N}/\text{mm}^2$$

The tensile stress = σ = 233.33 MPa.

Example 11.19 Determining the Shear Stress in a Butt Weld A shear force of 20 kN acts on a butt weld with a weld length of 52 mm. The thickness of the thinner plate is 6 mm. Determine the shear stress induced in the weld.

Solution
F_s = 20 kN = 20,000 N, s = 6 mm, l = 52 mm, τ =? By using Eq. (11.17),

$$\tau = \frac{F_s}{s * l} = \frac{20000}{6 \times 52} = 64.1\,\text{MPa}$$

The shear stress = τ = 64.1 MPa

Example 11.20 Determining the Shear Strength of a Weld when the Electrode is Identified With reference to the data in Table 11.2, determine the shear strength of the fillet weld made by using E80xx electrode in an arc welding operation.

Solution

The yield strength of E80xx electrode = W_{ys} = 470 MPa (from Table 11.2).
By using Eq. (11.18),

$$W_{\tau s} = 0.6 * W_{ys} = 0.6 \times 470 = 282\,\text{MPa}$$

The weld shear strength = $W_{\tau s}$ = 282 MPa

Example 11.21 Determining the Design Shear Stress in a Fillet Weld when the FoS is Known By using the data in Example 11.20, determine the design shear stress in a fillet weld if the factor of safety is 2.5.

Solution

$W_{\tau s}$ = 282 MPa (from Example 11.19), FoS = 2.5, τ =? By using Eq. (11.19),

$$\tau = \frac{W_{\tau s}}{(FoS)} = \frac{282}{2.5} = 112.8\,\text{MPa}$$

The design shear stress = τ = 112.8 MPa.

Example 11.22 Determining the Design (Maximum) Force for a Double Parallel Fillet Weld An arc welding operation was performed using E80xx electrode to join two plates together by double parallel fillet welds, as shown in Fig. 11.19. Determine the design (maximum) force that can be sustained by the weld with a factor of safety of 4.

Solution

W_{ys} = 470 MPa (from Table 11.2), h = 5 mm, l = 35 mm, FoS = 4, F =? By using Eq. (11.18)

$$W_{\tau s} = 0.6 * W_{ys} = 0.6 \times 470 = 282\,\text{MPa}$$

For a safe weld, we must use Eq. (11.19), as follows:

$$\tau = \frac{W_{\tau s}}{(FoS)} = \frac{282}{4} = 70.5\,\text{MPa}$$

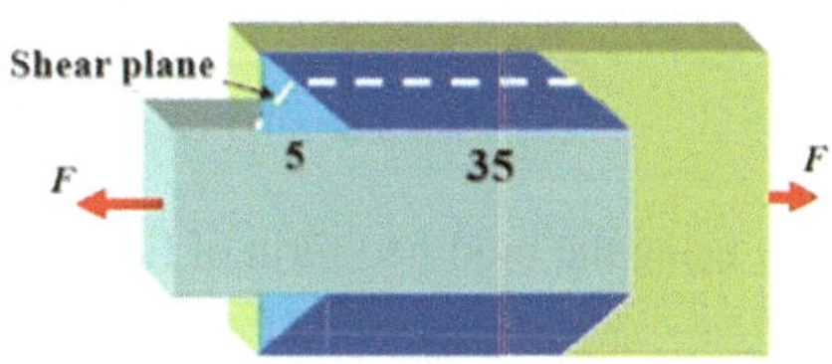

Fig. 11.19 Figure for Example 11.21

For the double parallel fillet weld (Fig. 11.19), we must use Eq. (11.7) as follows:

$$A_{dpfw} = 1.414 * h * l = 1.414 \times 5 \times 35 = 247.45\,\text{mm}^2$$

By using Eq. (11.9),

$$70.5 = \frac{F}{A_s} = \frac{F}{247.45}$$
$$\text{F} = 70.5 \times 247.45 = 17,445\text{N}$$

The design load = 17.445 kN.

Example 11.23 Determining the Design Tensile Load for a Butt Weld—E70xx Electrode An arc welding operation was performed to butt weld two steel plates (each of thickness = 7 mm) using an E70xx electrode. The weld length is 100 mm. Determine the design (maximum) tensile load that can be safely applied to the joint if the butt weld is oriented normal to the applied load. The factor of safety is 3.

Solution

$W_{UTS} = 482$ MPa (from Table 11.2), butt weld: $l = 100$ mm, s = 7 mm, FoS = 3, $F = ?$

By using Eq. (11.20),

$$\sigma = \frac{w_{UTS}}{FoS} = \frac{482}{3} = 160.66\ \text{MPa}$$
$$\text{Throat area} = \text{A} = \text{s} * l = 7 \times 100 = 700\ \text{mm}^2$$

By using the modified form of Eq. (11.16),

$$\text{F} = \sigma * \text{A} = 160.66 \times 700 = 112.466\ \text{N}$$

The design (maximum allowable) load = 112.4 kN.

Example 11.24 Determining the Design Tensile Load for a Butt Weld—E60xx Electrode By using the data in Example 11.23, determine the design (maximum) tensile load that can be safely applied to the joint if E60xx electrode is used instead of E70xx.

Solution

$W_{UTS} = 427$ MPa (from Table 11.2), butt weld: $l = 100$ mm, s = 7 mm, FoS = 3, $F = ?$

$$\sigma = \frac{W_{UTS}}{FoS} = \frac{427}{3} = 142.33\ \text{MPa}$$
$$\text{Throat area} = \text{A} = \text{s}^* l = 7 \times 100 = 700\ \text{mm}^2$$

By using the modified form of Eq. (11.16),

$$F = \sigma^* A = 142.33 \times 700 = 99.633.3 \ \text{N}$$

The design (maximum allowable) load = 99.6 kN.

Example 11.25 Determining the Factor of Safety for Two Electrodes for their Comparison It is required to butt weld two steel coupons (100 × 40 × 4 mm each) longitudinally. The design tensile load of the weld joint is 90 kN. Determine the weld-joint safety factor for the use of the following electrodes: (a) E60xx, and (b) E70xx. (c) Comment on the results.

Solution

F = 90,000 N, butt weld: l = 100 mm, s = 4 mm.

$$\text{Throat area} = \text{A} = \text{s}^* l = 4 \times 100 = 400\,\text{mm}^2$$

$$\sigma = \frac{F}{A} = \frac{90000}{400} = 225\text{MPa}$$

(a) For E60xx electrode, W_{UTS} = 427 MPa.

$$\sigma = \frac{W_{UTS}}{FoS} = \frac{427}{FoS} = 225$$

$$\text{FoS} = \frac{427}{225} = 1.89$$

(b) For E70xx electrode, W_{UTS} = 482 MPa.

$$\sigma = \frac{W_{UTS}}{FoS} = \frac{482}{FoS} = 225$$

$$\text{FoS} = \frac{482}{225} = 2.14$$

(c) The safety factor for E70xx electrode is greater due to its higher tensile strength.

Questions and Problems

11.1. Highlight the importance of welding design in societal context.

11.2. Draw sketches for the AWS recommended weld joints.

11.3. Define the following welding symbols with the aid of diagrams: (a) the arrow, (b) the reference line, (c) the tail, (d) groove welds, (e) the CJP, and (f) fillet welds.

11.4. Draw diagrams showing the following terms in a fillet weld: (a) leg with equal length, (b) throat thickness, (c) leg with unequal length, (d) parallel fillet weld, (e) double parallel fillet weld, and (e) transverse fillet weld.

11.5. Draw sketches for various butt weld joints.

11.6 (MCQs). Encircle the most appropriate answers for the following statements.

(1) Which AWS code is the most referenced one in the world?

(a) AWS D1.1, (b) AWS A1.1, (c) AWS B2.1, (d) B1.10.

(2) Which AWS code is referred for *Structural and Sheet Metal*?

(a) AWS D1.1, (b) AWS A1.1, (c) AWS B2.1, (d) B1.10.

(3) Which welding joint is formed when two metallic pieces are stacked on top of each other?

(a) Butt joint, (b) lap joint, (c) T-joint, (d) corner joint.

(4) Which welding joint is used to make frames, boxes, and other similar items?

(a) Butt joint, (b) lap joint, (c) T-joint, (d) corner joint.

(5) Which weld joint is generally used for flanges, valves, fittings, and the like?

(a) Butt joint, (b) lap joint, (c) T-joint, (d) corner joint.

P11.7. A fillet weld has legs of 9 mm and 7 mm. Determine the throat thickness of the weld.

P11.8. The heat input in an arc welding operation is 4 kJ/mm. Determine the fillet-weld leg size.

P11.9. Two steel plates, each of 6-mm thickness, are welded with a parallel fillet weld. The weld length is 55 mm. Determine the shear stress induced in the weld if the shear force is 58 kN.

P11.10. A metal tube is welded to a metal plate (Fig. 11.12c). The weld length is 25 mm. The thickness of the thinner part is 5 mm. Determine the throat area of the butt weld.

P11.11. A fillet weld is made using E90xx electrode. Determine the shear stress induced in the weld if the safety factor is 3.5. Hint: refer to the data in Table 11.2.

P11.12. Two steel plates (each of thickness = 6 mm) are butt welded using an E60xx electrode.

The weld length is 80 mm. Determine the design tensile load that can be safely applied if the butt weld is oriented normal to the applied load. The factor of safety is 2.5.

P11.13. A shear force of 60 kN is applied on the weldment with a parallel fillet weld (weld length = 45 mm). The allowable shear stress is 150 MPa. Design the fillet weld by determining its leg size.

P11.14. Two metal plates are welded such that the butt joint is angled at 30° relative to the applied load. The thinner part is 5-mm-thick, and the weld length is 55 mm. Determine the throat area of the weld.

References

AWS (2025). AWS: Codes and Standards, Publisher: American Welding Society, USA. Online: https://www.aws.org/standards-and-publications/codes-and-standards/#numberOfResults=5. Accessed on Nov. 6, 2025.

Funderburk, S.R. (1999). Key Concepts in Welding Engineering. *Welding Innovation*, Vol. XVI, No. 1, 1-4.

Hicks, J. (2001). Welded Design – theory and practice, *Woodhead Publishing Co.*, Sawston, UK

Picón, R. and Cañas, J. (2009). On strength criteria of fillet welds. *International Journal of Mechanical Sciences*, 51(8), 609-618. https://doi.org/10.1016/j.ijmecsci.2009.06.003

TWI (2025). What is a Butt Weld? *The Welding Institute (TWI) Ltd.*, Cambridge, UK. Online: https://www.twi-global.com/technical-knowledge/faqs/what-is-a-butt-weld

Chapter 12
Residual Stresses and Distortion: *Prevention*

Abstract This chapter first introduces readers to residual/thermal stresses and the various types of distortions resulting from fusion welding. Then engineering analyses of longitudinal shrinkage, transverse shrinkage, and angular distortion are presented with reference to butt joint and T-joint in welding. Residual stress and strain have been discussed and analyzed, including the distribution of residual stresses in a butt-joint weldment. Residual distortions in welded structures are also discussed. A workable mathematical expression for calculating the heat input and the residual strain in a simply-supported (welded) beam has been derived, with reference to butt joint, SMAW fillet weld, and GMAW fillet weld. Finally, the techniques to reduce residual distortion are presented. This chapter contains 14 worked numerical examples, 20 mathematical models/formulae, 12 diagrams, 6 exercise problems, and 6 MCQs with their answers given at the end of the book.

Keywords Residual stress · Residual strain · Angular distortion · Longitudinal shrinkage · Transverse shrinkage · Fillet weld · Weldment

Nomenclature

δl	Change in length (shrinkage)
l_o	Original weld length
α	Co-efficient of thermal linear expansion
T_0	Original temperature
T	Temperature after cooling
δ_L	Longitudinal shrinkage
I	Electrode current
L	Weld length
t	Plate thickness
δ_T	Transverse shrinkage
k	Empirical factor (for steel $k = 0.1$)
A_w	Cross-sectional area of weld
w_{av}	Average width of the weld
$\delta_{T(butt)}$	Transverse shrinkage in butt welding

Z. Huda, *Welding and Joining Engineering*,
https://doi.org/10.1007/978-3-032-20462-2_12

d	Root opening
ε	Linear strain
ΔT	Difference in temperature
ε_b	Residual strain in the base metal
ε_w	Residual strain in the weld metal
σ_b	Residual stress in the base metal
σ_w	Residual stress in the weld metal
E	Modulus of elasticity
σ_r	Residual stress in the welded pipe section
ΔL	Circumferential opening
Δz	Axial misalignment
Δr	Radial misalignment
D	Original outer diameter of welded pipe
h	Wall thickness of welded pipe

12.1 Thermal Stresses and Distortion

12.1.1 Thermal Stresses and Shrinkage: Causes of Distortion

The stresses that would exist in a body on removal of external loads are called *residual stresses*. The residual stresses that result from non-uniform temperature changes may be termed as *thermal stresses*. When a fusion weld is laid, the weld metal is in molten state and therefore hot. As the molten metal cools, it shrinks which develops stresses on the weld and the base metal near the weld. Initially the shrinkage stresses are small since the metal is hot and weak. As the metal solidifies the weld imposes increasing shrinkage stresses on the weldment. The shrinkage in a fusion weldment occurs in three stages: (a) liquid contraction, (b) solidification shrinkage, and (c) solid contraction (Khan 2007). The shrinkage stresses may result in one or more of the following phenomena: (a) transverse shrinkage, (b) longitudinal shrinkage, and (c) distortion (Fig. 12.1). Distortion is also caused by

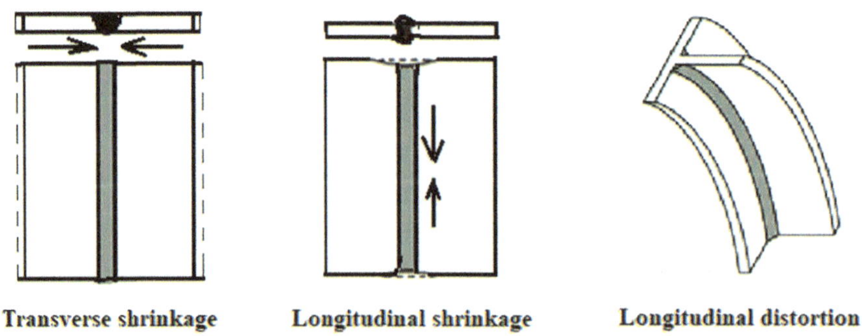

Fig. 12.1 Shrinkage and distortion due to thermal stresses in welding

localized heating of the base metals which may result in local expansion and contractions during the welding operation.

It is evident in Fig. 12.1 that transverse shrinkage occurs when the contracting weld metal tends to pull the plates towards the center line of the joint; this results in a reduction of the dimension of the weldment in the transverse direction. Longitudinal shrinkage involves reduction in the overall length of the weld joint; this shrinkage occurs due to the thermal stresses that are developed when the weld metal tends to contract and its contraction is constrained.

12.1.2 Engineering Analysis for Linear Shrinkage

The shrinkage/contraction of weld metal is volumetric; it can be estimated along the length and across it. The linear shrinkage can be estimated by:

$$\delta l = l_o \cdot \alpha \cdot (\mathrm{T} - \mathrm{T}_0) \tag{12.1}$$

where δl is the change in length (shrinkage), l_o is the original weld length, α is the co-efficient of thermal linear expansion, $(C^o)^{-1}$; T_o is the original temperature, °C; and T is the temperature after cooling, °C (see Example 12.1). It may be noted that, in practice, the actual shrinkage may be considerably lower than the calculated one due to the constraint provided by the adjoining cold plates (see Fig. 12.1). As a rule of thumb, the following allowances should be made to cover shrinkage, when welding steels, at the assembly stage: (a) fillet weld—0.8 mm per 3 m of weld, and (b) butt weld—3 mm per 3 m of weld (TWI-global 2023).

The longitudinal shrinkage, in arc welding, mainly depends on three factors (a) weld length, (b) plate thickness, and (c) welding current. The longitudinal shrinkage, in arc welding, can be calculated by the following formula:

$$\delta_L = 3.15 \times 10^{-5} \left(\frac{I * L}{t} \right) \tag{12.2}$$

where δ_L is the longitudinal shrinkage, mm; I is the electrode current, A; L is the weld length, mm; and t is the plate thickness, mm (see Example 12.2).

12.1.3 Engineering Analysis for Transverse Shrinkage

In transverse shrinkage, the whole weld-joint area is in transverse tension. The amount of transverse shrinkage depends on the total volume of weld metal. Transverse shrinkage, in a T-joint, can be related to the weld cross-sectional area and the plate thickness by (Khan 2007):

$$\text{Transverse shrinkage for a T-joint} = \delta_T = k\left(\frac{A_w}{t}\right) \tag{12.3}$$

where δ_T is the transverse shrinkage, mm; k is an empirical factor (for steel $k = 0.1$); A_w is the cross-sectional area of weld, mm^2; and t is the plate thickness, mm (see Fig. 12.2a). With reference to Fig. 12.2(a), the weld cross-sectional area in a T-joint can be expressed as:

$$A_w = \left(\tfrac{1}{2} * a * a\right) + \left(\tfrac{1}{2} * a * a\right) = a^2 \tag{12.4}$$

The significance of Eqs. 12.3–12.4 is illustrated in Example 12.3. The expression for the transverse shrinkage for a V-butt joint can be derived by reference to Fig. 12.2(b), as follows.

By using Eq. 12.3 for the V-butt joint in Fig. 12.2(b),

$$\text{Transverse shrinkage in a V butt joint} = \delta_T = k\left(\frac{A_w}{t}\right) = A_w\left(\frac{k}{t}\right) = \left(\tfrac{1}{2}\,\mathrm{w} * \mathrm{t}\right)\left(\frac{k}{t}\right)$$

or

$$\delta_T = \mathrm{k}\left(\tfrac{1}{2}\,\mathrm{w}\right) = \mathrm{k}\, w_{av} \tag{12.5}$$

where w_{av} is the average width of the weld (see Example 12.4).

A better approach, in determining the transverse shrinkage in a butt-joint weld, is to take into consideration the *root opening*—the gap between the plates to be joined to assure 100% penetration in groove-type welds (see Fig. 12.3). Thus, the transverse shrinkage can be more accurately calculated by the following formula:

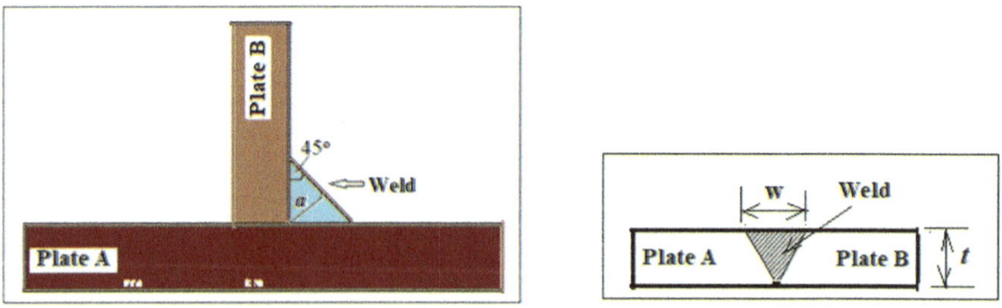

(a) (b)

Fig. 12.2 Weld cross-sectional parameters in a T-joint (**a**), and in a V-butt joint (**b**)

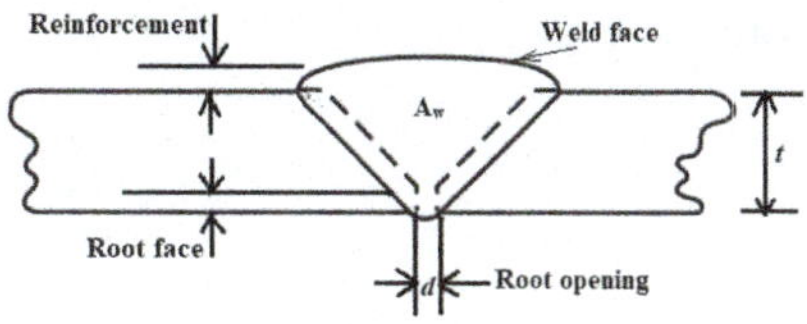

Fig. 12.3 Parameters for transverse shrinkage in a butt-joint weld

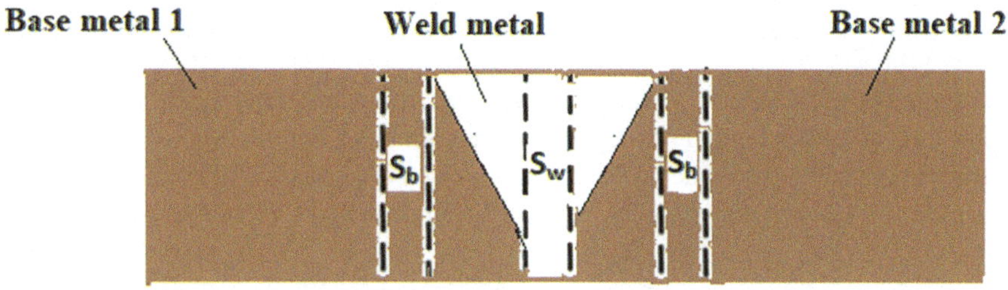

Fig. 12.4 Weldment showing the residual stresses. (S_w = residual stress in the weld metal, S_b = residual stress in the base metals)

$$\delta_{T(butt)} = 5.08 * \left(\frac{A_w}{t}\right) + 1.27 * d \tag{12.6}$$

where $\delta_{T(butt)}$ is the transverse shrinkage in butt welding, m; A_w is the cross-sectional area of the weld, m^2; t is the plate thickness, m; and d is the root opening, m (see Example 12.5).

12.2 Residual Stresses and Strains: Engineering Analysis

We learnt in the preceding section that *residual stresses* are the stresses that would exist in a body on removal of external loads. The residual stresses in welding are actually thermal stresses since the former result from non-uniform temperature changes. When fusion welding has been completed, thermal stresses are left behind. These thermal (residual) stresses are caused by the solidifying weld metal contracting more than the adjacent (colder) base metal that constraints it. In order to estimate the magnitude of residual stresses, we may divide the weldment into three elements (strips) of equal size and unit depth, as shown in Fig. 12.4.

As the weld metal solidifies, the central strip will tend to contract by an amount equal to linear shrinkage, which can be expressed by re-writing Eq. 12.1 as:

$$\delta l = l_o \cdot \alpha \cdot (\Delta T) \tag{12.7}$$

or

$$\frac{\delta l}{l_o} = \frac{l_o}{l_o} \cdot \alpha \cdot (\Delta T)$$

or

$$\varepsilon = \alpha \cdot \Delta T \tag{12.8}$$

where ε is the linear strain, and ΔT is the difference in temperature. In the meantime, the adjacent (colder) base metal constraints the contraction of weld *i.e.* the strain imposed by the adjacent base metal (ε_b), on either side of the weld metal, will restrain it by an amount ε_w. Accordingly, the strains can be related as follows:

$$\varepsilon_b = \varepsilon_w - (-\alpha \cdot \Delta T) = \varepsilon_w + \alpha \cdot \Delta T \tag{12.9}$$

where ε_b is the residual strain in the base metal, and ε_w is the residual strain in the weld metal.

By using Hook's law ($\sigma = E \bullet \varepsilon$), Eq. 12.9 transforms to:

$$\frac{\sigma_b}{E} = \frac{\sigma_w}{E} + \alpha \cdot \Delta T$$

or

$$\sigma_b = \sigma_w + E \cdot \alpha \cdot \Delta T \tag{12.10}$$

where σ_b is the residual stress in the base metal, σ_w is the residual stress in the weld metal, E is the modulus of elasticity, and α is the coefficient of linear thermal expansion (see Example 12.6).

For equilibrium, the force exerted by the strip in the weld must be equal to opposing force exerted by the two strips (on either side) in the base metal *i.e.* $F_w = -F_b - F_b = -2F_b$ or

$$\sigma_w \cdot A_w = -2\sigma_b \cdot A_b \tag{12.11}$$

Since the lengths and width of all strips are equal, $A_w = A_b$, Eq. 12.11 simplifies to:

$$\sigma_w = -2\sigma_b \tag{12.12}$$

It means that the residual stress in the weld metal is twice that of the stress in the base metal; this stress behavior is shown in Fig. 12.5. The significance of Eqs. 12.10–12.12 is illustrated in Examples 12.7 and 12.8.

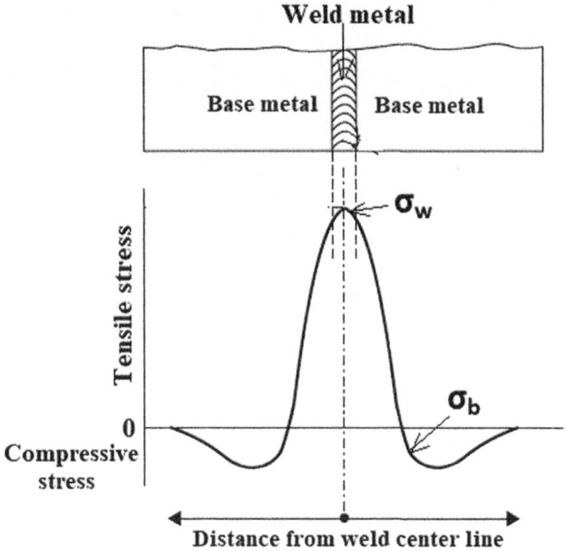

Fig. 12.5 Distribution of residual stresses in a butt-joint weldment (σ_w = residual stress in the weld metal, and σ_b = residual stress in the base metal)

12.3 Residual Welding Distortions in Beams

12.3.1 Importance of Residual Welding Distortion in Welded Structures

The major requirements of modern welded structures are the safety, fitness for production and economy. In the optimum design, the safety and production requirements are fulfilled by design and fabrication constraints. Fabrication aspects are very important for welded structures, since the distortions caused by shrinkage of welds should be minimized (Farkas and Jármai 2006).

12.3.2 Residual Strain Due to Shrinkage of Longitudinal Welds in Simply-Supported Beams: Engineering Analysis

The residual strain at the center of gravity (ε_G) for a welded steel simply-supported beam (Fig. 12.6) can be calculated by:

$$\varepsilon_G = 0.844 \times 10^{-3} \frac{H_i}{A} \quad (12.13)$$

where H_i is the heat input, J/mm; and A is the beam cross-sectional area, (mm)^2.

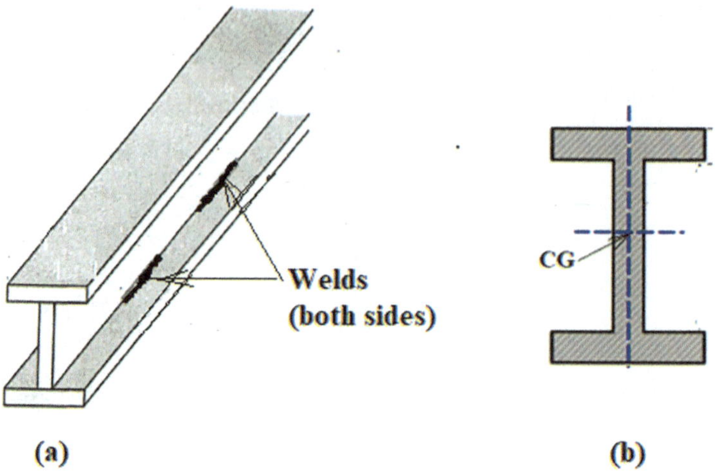

Fig. 12.6 Welded simply-supported beam (**a**), cross-sectional view (**b**)

The heat input H_i, in arc welding, can be determined by using Eq. 4.1. However, in order to express Eq. 4.1 in a workable form, we may proceed with engineering analysis as follow. The coefficient of penetration, in arc welding, is defined as the mass penetrated per unit amperage per unit time, *i.e.*

$$\alpha_N = \frac{m}{I_W \cdot t} \tag{12.14}$$

where α_N is the coefficient of penetration, kg/A•s ($\alpha_N = 2.44 \times 10^{-6}$ kg/A•s for steel) (Farkas and Jármai 1997; Farkas and Jármai 1998); m is the mass penetrated, kg; I_w is the welding current, A; and t is the time duration current is passed, s (see Example 12.9). Equation 12.14 can be re-written as:

$$\mathrm{t} = \frac{m}{\alpha_N \cdot I_W} \tag{12.15}$$

The length of weld can be expressed in terms of density by:

$$\mathrm{L} = \frac{m}{\rho \cdot A_w} \tag{12.16}$$

where L is the weld length, mm; m is the mass penetrated into the weld, kg; ρ is the density of weld metal, kg/(mm)^3; and A_w is the cross-sectional area of the weld, (mm)^2. Now, the expressions for time t and the length L from Eqs. 12.15–12.16 can be substituted in the numerical definition of welding speed, as follows:

$$v_w = \frac{L}{t} = \frac{\alpha_N \cdot I_w}{\rho \cdot A_w} \tag{12.17}$$

where v_w is the welding speed, mm/s. By combining Eqs. 12.1 and 12.17, we obtain:

$$H_i = \eta_o \frac{U \cdot I_w}{v_w} = \eta_o \frac{U \cdot \rho \cdot A_w}{\alpha_N} \tag{12.18}$$

The recommended thermal efficiency η_o values for various weld joints are listed in Table 12.1.

The cross-sectional area of a fillet weld mainly depends on its leg length, **z** (see Fig. 12.7). The fillet weld cross-sectional area (A_w) can be determined by the following formula:

$$A_w = \frac{z^2}{2} \tag{12.19}$$

The significance of Eq. 12.19 and Table 12.1 are illustrated in Examples 12.10–12.12. Once the H_i has been calculated by using Eq. 12.18, the residual strain at the center of gravity, for a steel simply-supported beam, can be calculated using Eq. 12.13 (see Example 12.13).

Table 12.1 Thermal efficiency η_o values for various weld joints

Weld joint	SMAW butt welds	SMAW fillet welds	GMAW fillet welds
η_o	0.7	0.91	0.68

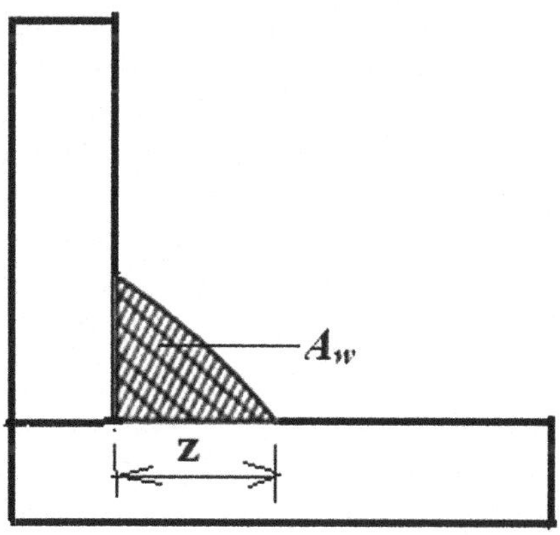

Fig. 12.7 Cross-sectional view of a fillet weld (z = leg length, A_w = weld cross-sectional area)

12.4 Reducing Residual Stresses and Distortion in Welding

Residual stresses can cause distortion and cracking in the material, potentially resulting in the failure of the welded structure. Post-weld heat treatment (PWHT) helps to reduce stress by heating the material to a reasonably high temperature for a certain period of time, allowing the metal to relax and release the stress. The PWHT is often required for pressure vessel and piping components for relieving residual stresses and/or improving weldment properties.

Weld distortions can be reduced by several techniques, which include (Kou 2003): (a) reducing the volume of the weld metal; (b) intermittent welding, rather than continuous welds, will minimize the volume of weld metal (see Fig. 12.6), (c) minimizing the joint preparation angle and the root pass, (d) the use of advanced fusion welding (*e.g.* EBW, LBW, etc.), (e) balancing welding by using a double-V joint in preference to a single-V joint, (f) welding alternately on either side of the double-V joint, (g) placing welds about the neutral axis (see Fig. 12.8), (f) placing another weld on the opposite side of the neutral axis, and the like.

12.5 Recent Advances in Reducing Residual Stress in Welded Pipes and Pressure Vessels

With increasing demand for welded pipes in the pipeline industry for oil and gas transportation, the issue of residual stresses in welded pipes is receiving increasing attention. There are two main types of high-grade large-diameter welded pipes: (a) longitudinal submerged arc welded (LSAW) pipes with high equipment and operating costs, and (b) helical submerged arc welded (HSAW) pipes with lower equipment and operating costs. Recently (2024), Xiong and co-researchers have conducted ring splitting tests on HSAW pipes and LSAW pipes; the analysis of results for HSAW pipe showed that relative displacement generally occurs in three directions: circumferential, axial, and radial, and the degree of deformation varies greatly. On the other hand, for SAWL pipe segment, only circumferential opening generally occurs (Xiong et al. 2024). Based on the mechanical analysis of the macro-stress distribution in the deformed condition of the welded pipes, and according to the superposition principle of elasticity, the following prediction model for

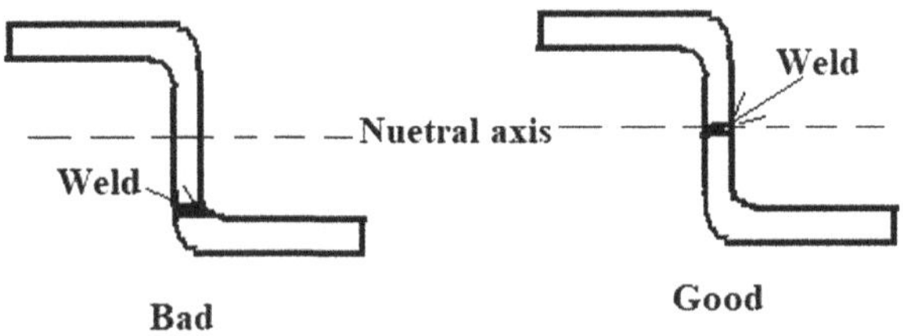

Fig. 12.8 Placing a weld about the neutral axis

the residual stress of high-grade, large-diameter HSAW pipes is presented (Xiong et al. 2024; Qingren et al. 2009):

$$\sigma_r = \sqrt{\left(\frac{Eh\Delta L}{\pi \cdot D^2 + D \cdot \Delta L} + \frac{2Eh\ \Delta r}{\pi \cdot D^2}\right)^2 + 4\left(\frac{Eh\ \Delta z}{1.3 * \pi * D^2}\right)^2} \tag{12.20}$$

or

$$\sigma_r = (E * h) * \sqrt{\left(\frac{\Delta L}{\pi \cdot D^2 + D \cdot \Delta L} + \frac{2\ \Delta r}{\pi \cdot D^2}\right)^2 + 4\left(\frac{\Delta z}{1.3 * \pi * D^2}\right)^2} \tag{12.21}$$

where σ_r is the residual stress in the welded pipe section, MPa; ΔL is the circumferential opening, mm; Δz is the axial misalignment, mm; Δr is the radial misalignment, mm; D is the original outer diameter of welded pipe, mm; h is the wall thickness of welded pipe, mm; E is the elastic modulus, MPa (see Example 12.14).

Besides welded pipes, the complex residual stress is also the main cause of stress-corrosion cracking (SCC) accidents in welded pressure vessels. Recently (2024), Zhang and co-researchers have studied the effects of local rapid-cooling heat treatment on residual stress of pressure vessel inner wall welding layer; they have reported that the residual stress on the surface of welding layer can be reduced to about 100 MPa and a 3 mm compressive stress welding layer can be obtained when the cold source was applied at 400 °C and the surface is cooled to 25 °C with the rate of 400 °C/h (Zhang, et al. 2024).

12.6 Worked Numerical Examples on Residual Stress and Distortion

Example 12.1 Calculating the Weld Length After the Longitudinal Shrinkage
A steel work-piece was heated to 1500 °C during fusion welding and then cooled to 25 °C. The original weld metal length was 1.5 m. Calculate the final length of the weld metal, if the co-efficient of linear expansion of the steel is 12.2 × 10^{-6}/°C.

Solution

l_o = 1.5 m, α = 12.2 × 10^{-6}/°C, T_o = 1500 °C, T = 25 °C, l = ?

By using Eq. 12.1,

$$\delta l = l_o \cdot \alpha \cdot (T_o - T) = 1.5 \times 12.2 \times 10^{-6}\ (1500 - 25) = 0.027\ \text{m}$$

$$\text{Final length} = l = l_o - \delta l = 1.5 - 0.027 = 1.473\ \text{m}$$

Example 12.2 Calculating the Longitudinal Shrinkage in Arc Welding
A current of 120 A was applied to weld a 15-mm-thick metal plate by arc welding. Calculate the longitudinal shrinkage, if the weld length is 250 mm.

Solution

$I = 120$ A, L = 250 mm, $t = 15$ mm. By using Eq. 12.2,

$$\text{Longtinudal shrinkage} = \delta_L = 3.15\times10^{-5}\left(\frac{I*L}{t}\right)$$

$$= 3.15\times10^{-5}\left(\frac{120\times250}{15}\right) = 0.063\text{mm}$$

Example 12.3 Calculating the Transverse Shrinkage in a T-Joint Weld

Calculate the transverse shrinkage in the steel weld, as shown in Fig. 12.9.

Solution

$A_w = a^2 = 3^2 = 9$ mm^2, t = 10 mm, k = 0.1. By using Eq. 12.3,

$$\text{Transverse shrinkage} = \delta_T = k\left(\frac{A_w}{t}\right) = \frac{0.1\times9}{10} = 0.09\text{mm}$$

Example 12.4 Calculating the Transverse Shrinkage in a V-Joint Weld

Calculate the transverse shrinkage in a steel V-joint with a weld width of 6 mm.

Solution

$k = 0.1$, w = 6 mm, transverse shrinkage =?

By using Eq. 12.5,

$$\text{Transverse shrinkage} = \delta_T = \text{k}(1/2\text{w}) = 0.5\times0.1\times6 = 0.3\text{mm}$$

Example 12.5 Calculating the Transverse Shrinkage in a V-Joint Fusion Weld

A fusion welding was performed for a butt joint in a 90-mm-thick metal plate. The cross-sectional area of the weld is 27 mm^2 and the root opening is 2.2 mm. Calculate the transverse shrinkage.

Solution

By using Eq. 12.6,

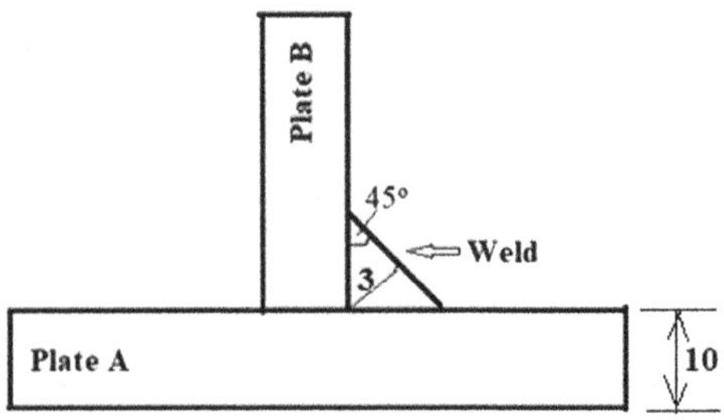

Fig. 12.9 Welding dimensions in a T-joint (in mm) for Example 12.3

$$\delta_{T(butt)} = 5.08 * \left(\frac{A_w}{t}\right) + 1.27 * d = \delta_{T(butt)} = 5.08 * \left(\frac{27}{90}\right) + (1.27 \times 2.2) = 4.32\text{mm}$$

Example 12.6 Calculating the Difference in the Residual Stresses of Weld and Base Metal
Two steel plates were butt-joined by a fusion welding process. The temperature difference in cooling was 230 °C. The Young's modulus of steel is 2.15×10^5 MPa, and its coefficient of linear expansion is 11×10^{-6} $(^oC)^{-1}$. Calculate the difference in the residual stresses of the weld and the base metal.

Solution
By using Eq. 12.10,

$$\sigma_b = \sigma_w + \text{E} \cdot \alpha \cdot \Delta T = \sigma_w + \left(2.15 \times 10^5 \text{ MPa} \times 11 \times 10^{-6} \left(C^\circ\right)^{-1} \times \left(-230^\circ\text{C}\right)\right)$$

$$\sigma_b = \sigma_w - 544\text{MPa}$$

or $$\sigma_b - \sigma_w = -544\text{MPa}$$

or $$\sigma_w - \sigma_b = 544\text{MPa}$$

It means that the residual stress in the weld metal is greater than the residual stress in the base metal by 544 MPa.

Example 12.7 Calculating the Residual Stresses in a Weld and a Base Metal
By using the data in Example 12.6, calculate the residual stresses in the weld and the base metal.

Solution
By reference to the solution of Example 12.6,

$$\sigma_b - \sigma_w = -544\text{MPa}$$

By using Eq. 12.12, we can substitute $\sigma_w = -2\,\sigma_b$ in the above equation, so:

$$\sigma_b - (-2\sigma_b) = -544\text{MPa}$$

$$\sigma_b = -181.3\text{MPa}$$

$$\sigma_w = -2\sigma_b = -2 \times -181.3 = 363\text{MPa}$$

Thus, the residual stress in the weld is 363 MPa (tensile), and the residual stress in the base metal is 181.3 MPa (compressive).

Example 12.8 Calculating the Residual Strains in a Weld and Base Metal
By using the data in Example 12.7, calculate the residual strains in the weld and the base metal.

Solution
$\sigma_w = 363$ MPa, $\sigma_b = -181.3$ MPa, E = 2.15×10^5 MPa. By using Hook's law,

$$\text{Residual strain in the weld} = \varepsilon_w = \frac{\sigma_w}{E} = \frac{363}{2.15\times 10^5} = 0.001688$$

$$\text{Residual strain in the base metal} = \varepsilon_b = \frac{\sigma_b}{E} = \frac{-181.3}{2.15\times 10^5} = -0.000843$$

Example 12.9 Calculating the Mass Penetrated in Arc Welding
A welding current of 130 A was passed for a minute in arc welding of steel. Calculate the mass penetrated. The coefficient of penetration for steel = 2.445 × 1 0^{-6} kg•A^{-1} s^{-1}.

Solution

$\alpha_N = 2.44 \times 10^{-6}$ kg A^{-1} s^{-1}, $I_w = 130$ A, t = 1 min = 60 s, m = ?

By using Eq. 12.15,

$$\text{Mass penetrated} = \text{m} = \alpha_N \cdot I_w \cdot \text{t} = 2.44\times 10^{-6}\times 130\times 60$$
$$= 19032\times 10^{-6}\,\text{kg} = 19.03\text{g}$$

Example 12.10 Calculating the Heat Input in Arc Welding for a Butt Joint
A welding current of 110 A was passed using a voltage of 30 V in SMAW welding of steel plates with a butt joint of weld cross-sectional area = 5 mm^2. Calculate the heat input.

Solution

η_o=0.7 (from Table 12.1), I_w=110A, U=30V, A_w=5mm^2, ρ=7.85×10^{-6}kg/(mm)^3.

By using Eq. 12.18,

$$\text{Heat Input} = H_i = \eta_o \frac{U\cdot\rho\cdot A_w}{\alpha_N} = \frac{0.7\times 30\times 7.85\times 10^{-6}\times 5}{2.445\times 10^{-6}} = 337\text{J/mm}$$

Example 12.11 Calculating the Heat Input in SMAW for a Fillet Weld
A voltage of 30 V was applied to weld steel plates in SMAW welding of steel plates for providing a fillet weld of leg length = 6 mm. Calculate the heat input.

Solution

z = 6 mm, η_o = 0.91, U = 30 V. By using Eqs. 12.18–12.19,

$$A_w = \frac{z^2}{2} = \frac{6^2}{2} = 18\text{mm}^2$$

$$H_i = \eta_o \frac{U\cdot\rho\cdot A_w}{\alpha_N} = \frac{0.91\times 30\times 7.85\times 10^{-6}\times 18}{2.445\times 10^{-6}} = 1577.7\text{J/mm} = 1.58\text{kJ/mm}$$

Example 12.12 Calculating the Heat Input for a GMAW Fillet Weld
A voltage of 30 V was used in GMAW welding of steel plates for a fillet weld of leg length = 6 mm. Calculate the heat input.

Solution

$A_w = 18\ \text{mm}^2$, $\eta_o = 0.68$, $U = 30$ V, $H_i = ?$

By using Eq. 12.19,

$$H_i = \eta_o \frac{U \cdot \rho \cdot A_W}{\alpha_N} = \frac{0.68 \times 30 \times 7.85 \times 10^{-6} \times 18}{2.445 \times 10^{-6}} = 1179\text{J/mm} = 1.179\text{kJ/mm}$$

Example 12.13 Calculate the Residual Strain at the Center of Gravity (CG) of a Welded Beam

A heat input of 1.5 kJ/mm was used to fabricate by arc welding a simply-supported beam. The cross-sectional view of the beam is shown in Fig. 12.10. Calculate the residual strain at the center of gravity (CG).

Solution

$H_i = 1500$ J/mm, A = 2∗(12 × 8) + (12 × 8) = 192 + 96 = 288 (mm)^2.

By using Eq. 12.13,

$$\varepsilon_G = 0.844 \times 10^{-3} \frac{H_i}{A} = 0.844 \times 10^{-3} \times \frac{1500}{288} = 4.39 \times 10^{-3}$$

Example 12.14 Predicting the Residual Stress after Ring Splitting Test of Welded Pipe

The following results were obtained in a ring splitting test of an HSAW pipe with dimensions: $\varphi 660 \times 7.1$ mm, section length = 200 mm. The circumferential opening is 440 mm, the axial misalignment is 10 mm, and the radial misalignment is nil. The

Fig. 12.10 Cross-sectional view of simply-supported beam (dimensions in mm)

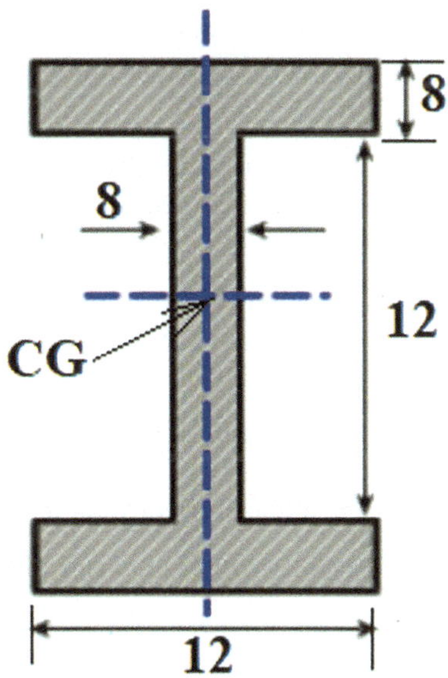

elastic modulus of the pipe material is 200,000 MPa. Predict the residual stress in the welded pipe.

Solution

D = 660 mm, D^2 = 435,600 mm^2, h = 7.1 mm, ΔL = 440 mm, Δz = 10 mm, E = 200,000 MPa,

$\Delta r = 0$, $\sigma_r = ?$

By using Eq. 12.20,

$$\sigma_r = (\mathrm{Eh}) * \sqrt{\left(\frac{\Delta L}{\pi \cdot D^2 + D \cdot \Delta L} + \frac{2\Delta r}{\pi \cdot D^2}\right)^2 + 4\left(\frac{\Delta z}{1.3 * \pi * D^2}\right)^2}$$

$$\sigma_r = (200{,}000 \times 7.1)$$

$$\times \sqrt{\left(\frac{440}{\pi \times 435600 + 660 \times 440} + \frac{2 \times 0}{\pi \cdot D^2}\right)^2 + 4\left(\frac{10}{1.3 \times \pi \times 435600}\right)^2}$$

$$\sigma_r = 1420000 \times \sqrt{\left(\frac{440}{1368477.76 + 290400}\right)^2 + 4(0.000005621)^2}$$

$$\sigma_r = 3763000 \times 10^{-4} = 376.3\mathrm{MPa}$$

The residual stress = 376.3 MPa.

Questions and Problems 12.1(MCQs). Encircle the most appropriate answers for each of the following statements:

(1). Which type of shrinkage occurs when the contracting weld metal tends to pull the plates towards the center line of the joint?

(a) Longitudinal shrinkage, (b) transverse shrinkage, (c) complex shrinkage, (d) none.

(2). The local expansion and contractions during the welding operation results in ________

(a) Shrinkage, (b) cavity, (c) distortion, (d) fracture.

(3) The stress that would exist in a body on removal of the external loads is called______.

(a) Tensile stress, (b) shear stress, (c) torsional stress, (d) residual stress.

(4). What is the main cause of stress-corrosion cracking (SCC) in welded pressure vessels?

(a) Complex residual stresses, (b) galvanic corrosion, (c) tensile stresses, (d) torsion.

(5). Which type of weld joint has the greatest thermal efficiency?

(a) SMAW butt welds, (b) SMAW fillet welds, (c) GMAW fillet welds.

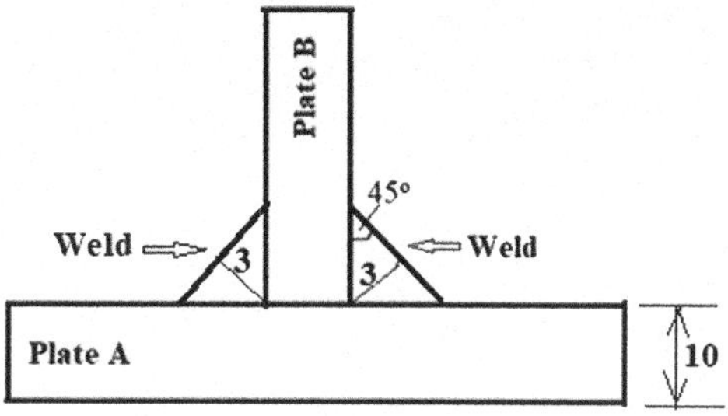

Fig. 12.11 Welding dimensions in a double T-joint (in mm) in P12.10

(6). Which type of weld joint has the least thermal efficiency?

(a) SMAW butt welds, (b) SMAW fillet welds, (c) GMAW fillet welds.

12.2. Draw sketches showing various shrinkages and distortion due to thermal stresses in welding.

12.3. Why does shrinkage/distortion occur in fusion welding?

12.4. What allowances should be made to cover shrinkage, when welding steels, for: (a) fillet weld, and (b) butt weld?

12.5. Draw a sketch showing the parameters for transverse shrinkage in a butt-joint weld.

12.6. Derive a mathematical expression for the residual stress in base metal in terms of the residual stress in weld, the modulus of elasticity, and the coefficient of linear expansion.

12.7. Why is the study of the residual distortion in a welded structure important?

12.8. Derive an expression for the residual strain at the center of gravity for steel simply- supported beams in terms of the beam cross-sectional area, thermal efficiency, voltage, the density of weld metal, the cross-sectional area of the weld, and the coefficient of penetration.

P12.9. A current of 130 A was applied to weld a 12-mm-thick metal plate by arc welding. Determine the longitudinal shrinkage, if the weld length is 300 mm.

P12.10. Calculate the transverse shrinkage in the steel weld, as shown in Fig. 12.11.

P12.11. A fusion welding was performed for a butt joint in a 70-mm-thick metal plate. The cross-sectional area of the weld is 20 mm^2 and the root opening is 2 mm. Calculate the transverse shrinkage.

P12.12. Two steel plates were butt-joined by a fusion welding process. The temperature difference in cooling was 250 °C. The Young's modulus of steel is 2.15 $\times 10^5$ MPa, and its coefficient of linear expansion is 1.1×10^{-5} $(C^o)^{-1}$.

Calculate: (a) the difference in the residual stresses of the weld and the base metal, and

(b) the residual stresses in the weld and the base metal.

P12.13. A voltage of 35 V was used in SMAW welding of steel plates for a fillet weld of leg length = 7 mm. Calculate the heat input.

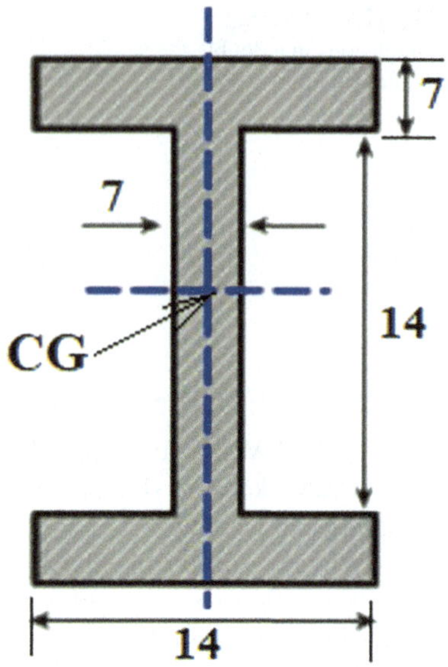

Fig. 12.12 Cross-sectional view of simply-supported beam (in mm) in P12.14

P12.14. Heat input of 1.8 kJ/mm was used to fabricate by arc welding a simply-supported beam.

The cross-sectional view of the beam is shown in Fig. 12.12. Calculate the residual strain at the center of gravity (CG).

References

Farkas, J. and Jármai, K. (2006). Special cases of the calculation of residual welding distortions. Proc. IIW Congress Stellenbosch, South Africa, March 8-10, 2006.

Farkas, J. and Jármai, K. (1997). Analysis and Optimum Design of Metal Structures. Balkema, Rotterdam-Brookfield.

Farkas, J. and Jármai, K. (1998). Analysis of some methods for reducing residual beam curvatures due to weld shrinkage. *Welding in the World*, Vol. 41, No. 4, pp. 385–398.

Khan, M.I. (2007), Welding Science and Technology, New Age Publishers, New Delhi

Kou, S. (2003). Welding Metallurgy, Wiley Inter-Science, New Jersey.

Qingren, X., Wenzhen, Z., Maosheng, Z., Yaorong, F., Lingkang, J., Feng, H. (2009). Effects of axial and radial distortion on residual stress of HSAW pipe. Proc. *ASME Pressure Vessels and Piping Conference*, Prague, Czech Republic, July 26-30, 2009.

TWI-global (2023), How much shall I allow for welding shrinkage? Internet Source: https://www.twi-global.com/technical-knowledge/faqs/faq-how-much-shall-i-allow-for-weld-shrinkage Accessed on February 4, 2023.

Xiong, Q., Li, Y., Shen, W., *et al.* (2024). Measurement and control of residual stress of high grade and large-diameter welded pipe for oil and gas transportation, *Journal of Materials Engineering & Performance*, Vol. 33, pp. 7479–7486.

Zhang, Y., Luo, Y., Jiang, W. (2024). Effect of local rapid cooling heat treatment on residual stress of pressure vessel inner wall welding layer, *Journal of Materials Engineering & Performance*, Vol. 33, pp. 7487–7499

Chapter 13
Testing of Weldments

Abstract This chapter develops readers' skills of weld testing to ensure the integrity and safety of the joint. The non-destructive testing (NDT), metallographic testing, and destructive mechanical testing of welds are explained, covering both qualitative and quantitative aspects of the tests. The NDT is discussed with reference to visual examination (VE), radiographic testing (RT), ultrasonic testing (UT), and dye penetrant testing (DPT). Metallographic inspection (both macro and microscopic examinations) of welds are illustrated with the aid of grained microstructures in various zones of weldments. Microscopic inspection covers the steps in metallographic specimen preparation (cut-off, grinding, polishing, etching, leveling, and microscopy) followed by microstructural analysis of welds with particular reference to various techniques of grain size measurement. The destructive mechanical testing of welds covers guided bend test (face bend test, root bend test, side bend test), free bend test, back bend test, nick bend test, and tensile testing; here, the parts of a fusion weld are also diagrammatically illustrated. This chapter includes 20 worked numerical examples, 13 diagrams, 16 formulas/equations, 7 exercise problems, and 8 MCQs with their answers given at the end of the book.

Keywords Non-destructive testing (NDT) · VE · RT · UT · DPT · Metallographic testing · Grain size measurement · Destructive testing · Weld metal · Heat-affected zone (HAZ) · Weld face · Weld root · Guide bend test · Nick bend test · Tensile testing of welds

Abbreviations

AWS	American Welding Society
DPT	Dye-penetrant testing
HAZ	Heat affected zone
ISO	International Standards Organization
MPT	Magnetic particle testing
NA	Numerical aperture

Z. Huda, *Welding and Joining Engineering*,
https://doi.org/10.1007/978-3-032-20462-2_13

NDT	Non-destructive testing
OM	Optical microscope
RT	Radiographic testing
UT	Ultrasonic testing
VE	Virtual examination

Nomenclature

I	Radiation intensity
d	Distance from the eradiation source
D	Radiographic density
I_0	Intensity of light incident on the film
I_t	Intensity of light transmitted through the film
L	Sound path length
θ	Refracted angle of the sound beam
n	Refractive index between the objective lens and the sample
α	Half-angle aperture
G	ASTM grain size number
F	Bending force
σ_{UTS}	Ultimate tensile strength of the bend specimen
w	Width of the bend specimen
t	Specimen's thickness

13.1 Testing/Inspection of Weldments: An Overview

The testing and inspection of welds are crucial to industries for ensuring the integrity and safety of the joint. In order to achieve this objective, a welding inspector must monitor all aspects of the welding process before, during, and after welding to ensure that the finished weldment is fit for purpose (Hughes 2009). The AWS, *American Welding Society*, has specified standards for qualifying welders and verifying weld quality across different materials and applications (AWS 2015). Additionally, the ISO, *International Standards Organization*, has developed and published the "ISO/TC 44/SC 5" standard for Testing and inspection of welds (ISO 2025). The testing of welds is broadly classified into two categories: (1) non-destructive testing (NDT), and (2) destructive testing. *Destructive testing of welds* may be further divided into two types: (a) metallographic examination, and (b) destructive mechanical testing. These techniques are explained in the following paragraphs.

Non-Destructive Testing (NDT) The NDT of welds uses various techniques to inspect welds for flaws (*e.g.* cracks, porosity, etc.) without causing damage, thereby ensuring the integrity and safety of the joint. The following NDT methods are commonly used to inspect welds: visual examination, dye penetrant testing (DPT), radiographic testing (**RT**), ultrasonic testing (UT)**,** and magnetic particle testing (MPT). The NDT allows the weldments to remain in service after inspection. The various techniques of NDT of welds are explained in Sect. 13.2.

Metallographic Inspection Metallographic inspection of welds is a destructive testing technique that is carried out at two levels: (I) macroscopic examination, and (II) microscopic examination. The macroscopic examination involves the use of low-magnification microscopes with magnifications up to 50×. Macro examination is generally done on un-mounted cross-sections through welded joints; it involves cutting and coarse/fine grinding techniques followed by etching. This technique enables us to examine the macro features of the weld joint (*e.g.* weld geometry, depth of penetration, surface imperfections, etc.). The microscopic examination is carried out by the use of optical microscopes with magnifications up to 1,000; here, the provision of a polished, optically flat surface is required (see Sect. 13.3). Microscopic examination is carried out for studies of weld imperfections and microstructural features (*e.g.* porosity, micro-cracks, grain size, etc.).

Destructive Mechanical Testing Destructive mechanical testing involves the assessment of the weld's mechanical properties by stressing it to the breaking point. The following are the techniques used in destructive mechanical testing: (a) guided bend test, (b) free bend test, (c) back bend test, (d) nick bend test, and (e) tensile test (see Sect. 13.4).

13.2 Non-Destructive Testing (NDT) of Welds

13.2.1 Visual Examination

Visual examination (VE) involves the examination of welds visually through the eyes to determine surface discontinuities (Halmshaw 1997). In most industrial settings, the VE involves the use of a magnifying glass (see Fig. 13.1). The VE should

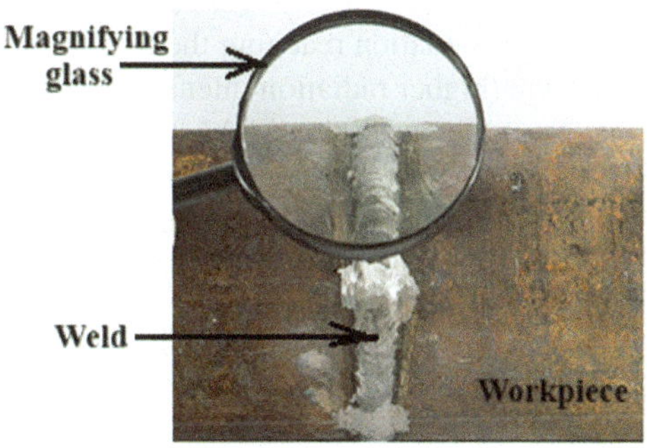

Fig. 13.1 Visual examination of a weld using a magnifying glass

be carried out in three stages. (I) pre-weld inspection, (II) in-process inspection, and (III) post-weld inspection. The pre-weld inspection enables us to detect and correct unacceptable conditions before they develop into actual welding problems. The in-process inspection during the welding operation can prevent problems in the completed weld through verification of the welding conditions and procedural requirements. Finally, the post-weld visual inspection is a relatively easy method of conducting a completed weld quality evaluation. The most common welding discontinuities found during the VE include undersized welds, undercut, overlap, surface cracking, surface porosity, under-fill, incomplete root penetration, excessive root penetration, burn through, and the like.

13.2.2 Dye Penetrant Testing (DPT)

D*ye penetrant test* (DPT) detects surface-breaking defects in the weld. This test enables us to detect visible cracks, flaws, pores, leaks, and other welding defects on the weld's surface. The DPT begins with the application of a liquid penetrant (*e.g.* a colored dye) to the surface of the weld, covering the area of interest. The penetrant, having a low surface tension, seeps into any surface openings or defects by capillarity. Then, the penetrant is allowed to dwell on the surface for a pre-determined time period. During this time, the penetrant infiltrates any defects, filling them. After the dwell period has passed, any excess penetrant is removed using solvents or emulsifiers. Finally, a developer (a suspension or white powder) is applied to the surface; the developer helps to draw out the penetrant trapped within the defects, making them visible.

13.2.3 Radiographic Testing

Radiographic testing (RT) is generally conducted using either X-rays or gamma rays. In the *RT*, the weld to be inspected is placed between the radiation source and a piece of radiation-sensitive film (see Fig. 13.2). The weld will stop some of the radiation where thicker; more dense areas will stop more of the radiation. The radiations that pass through the weld will expose the film and form a shadowgraph of the weld. The film darkness (or film density or radiographic density)will vary with the amount of radiation reaching the film through the weld; darker areas indicate more exposure (higher radiation intensity) and lighter areas indicate less exposure (lower intensity). This variation in the image darkness is used to detect the presence of any flaw, defect, or discontinuity inside the weld.

The radiation intensity is related to its distance (from the radiation source) according to inverse square law, as follows:

$$\frac{I_1}{I_2} = \frac{d_2^2}{d_1^2} \tag{13.1}$$

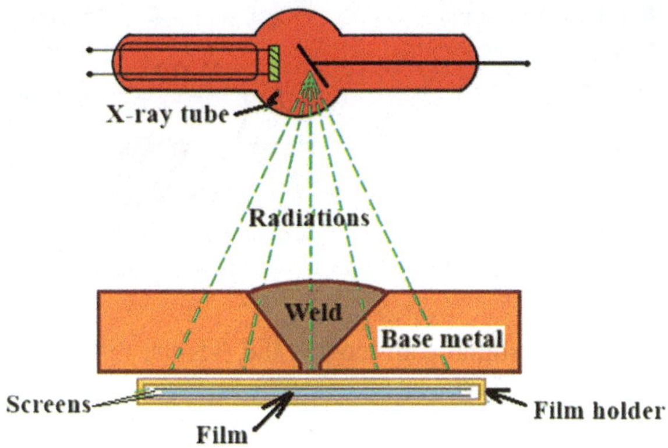

Fig. 13.2 Radiographic testing of a weld using an X-ray source

where I_1 and I_2 are the radiation intensities at distances d_1 and d_2 form the radiation source, respectively (see Example 13.1). Radiographic density (D) is determined as the logarithm of the ratio of the intensity of light incident on the film (I_0) to the intensity of light transmitted through the film (I_t), as follows:

$$D = \log\frac{I_0}{I_t} \tag{13.2}$$

Typical film densities, in RT of welds, are in the range of 1.8–2.5 (X-ray source) and 2.0–4.0 (gamma ray) (see Example 13.2).

13.2.4 Ultrasonic Testing of Welds

Ultrasonic testing (UT) of welds enables us to detect internal flaws and discontinuities in welds, such as cracks, porosity, and the like. The *UT* of welds involves the use of a transducer (probe), which receives an electric signal from the ultrasonic machine (see Fig. 13.3); the probe converts the electrical signal to a sound beam. The **sound beam travels into the test object (weld) and gets reflected at interfaces or discontinuity**; the piezoelectric material (in the probe) converts the reflected sound beam into an electrical signal that can be displayed on a liquid crystal display (LCD) screen of the machine.

The measurement of *surface distance*, in UT, is important. The *surface distance* (d_s) is the measurement along the surface of a material from the probe's index point to the point where the sound beam first exits the probe's surface after being reflected from an internal flaw. It can be determined by:

$$d_s = L^* \sin\theta \tag{13.3}$$

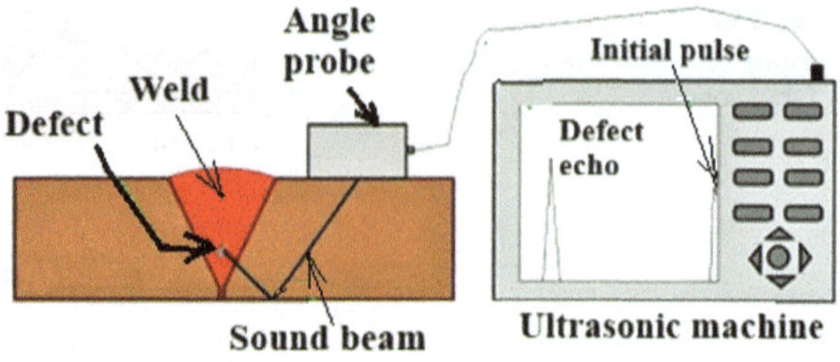

Fig. 13.3 Ultrasonic testing of a weld

where L is sound path length, and θ is the refracted angle of the sound beam (see Example 13.3).

The depth of a flaw from the top surface of the weldment can be determined by:

$$\text{Depth} = \text{L}^{*}\left(\cos\theta\right) \tag{13.4}$$

The significance of Eq. 13.4 is illustrated in Example 13.4.

13.3 Metallographic Examination of Welds

13.3.1 Macro and Microscopic Examinations: Grained Microstructure of a Weldment

Metallographic examination plays a key role in the quality assurance of weldments, both as a process control tool and as a post-mortem examination of failed components (Voorts 2011). Macroscopic testing of welds involves cutting and coarse/fine grinding techniques followed by etching. Macro-examination enables us to check weld geometry and depth of penetration; it also can detect cracks, and surface imperfections in the weld or in the heat-affected zone (HAZ).

Microscopic examination can detect weld imperfections and microstructural features (*e.g.* porosity, micro-cracks, grain size, etc.). In particular, the control of grain size (both in welds and HAZ) is critical for QC (see Fig. 13.4) (see also Chap. 3, Fig. 3.2). The measurement of grain size is discussed in Sect. 13.3.3.

13.3.2 Metallographic Specimen Preparation from a Weldment: Microstructure of HAZ

The preparation of a metallographic specimen, for microscopic examination, involves the following steps: (I) cutting, (II) rough grinding, (III) mounting, (IV) fine grinding, (V) polishing, (VI) etching, and (VII) microscopy. These steps are explained in the following paragraphs.

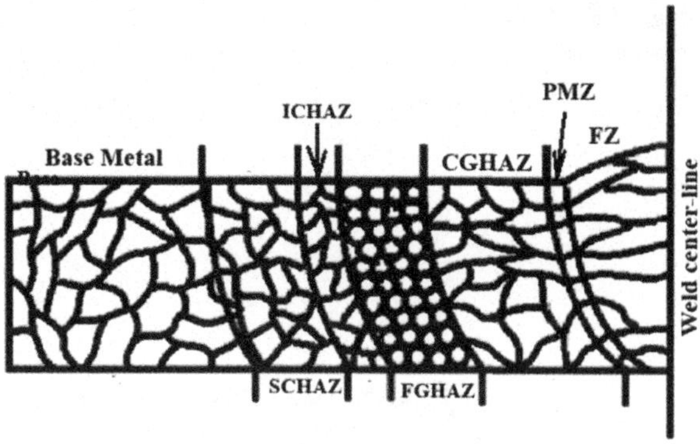

Fig. 13.4 Grained microstructure of a fusion weldment—*Schematic* (*FZ* fusion zone, *PMZ* partially melted zone, *CGHAZ* coarse-grained heat affected zone, *FGHAZ* fine grained HAZ, *ICHAZ* inter-critical HAZ, *SCHAZ* sub-critical HAZ)

Cutting-Off a Sample Firstly, a metallographic sample (a small piece) is cut out from the weldment by use of a cut-off machine. When cutting a specimen from the weldment, care must be taken to ensure that the specimen contains all the information required to investigate a feature of interest. It is important to ensure cooling by using cutting fluid during sectioning. For low deformation cutting, it is appropriate to use a low-speed diamond saw.

Rough Grinding and Mounting *Rough grinding* involves the removal of rust/surface deposits or leveling of irregular surfaces by filing or by use of a belt/disc grinder. *Mounting* of specimen provides a safe means of holding it and protecting its edges from rounding. In general, a mounting press is used to mount rigid metals, and the process is called *compression molding*.

Fine Grinding The purpose of fine grinding is to remove the surface layers damaged by cutting and to systematically abrade the metallographic specimen. *Fine grinding* involves the use of a series of abrasive papers with grits of decreasing coarseness. The fine-grinding operation may be performed either on a manual metallographic grinder or a rotating wheel grinding machine. Mounted specimens are ground with rotating discs of abrasive paper [*e.g.* wet silicon carbide (SiC)/emery paper]. The coarseness of the paper is indicated by a grit number—the number of SiC grains per square inch. For example, a 700-grit paper is finer than 600 grit paper.

Polishing *Polishing* involves the removal of the remaining scratches on the ground specimen to produce a smooth lustrous surface. *Polishing* is accomplished by using a polishing machine with rotating *polishing discs* that are covered with soft cloth impregnated with abrasive particles (alumina powder suspension or diamond paste) and an oily lubricant or water lubricant. The polished specimen reflects light like a mirror (see Fig. 13.5).

Etching Metallographic etching involves the use of etching reagents (acids/chemicals) for revealing the microstructure of the metal through selective chemical

Fig. 13.5 Polished metallographic specimen

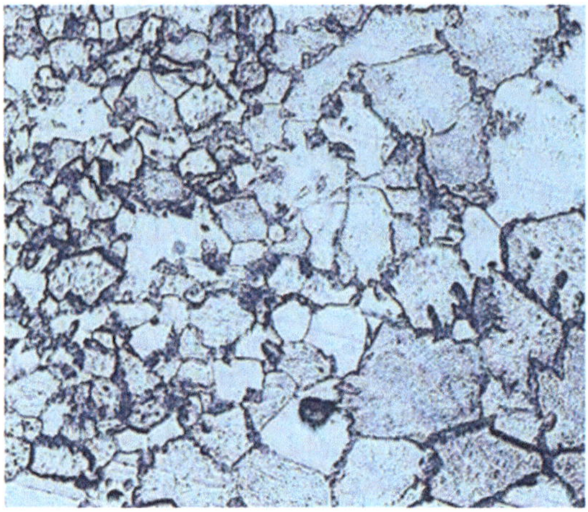

Fig. 13.6 The optical micrograph of an etched metallographic specimen showing the microstructure of HAZ of a weldment

attack. The etching reagent preferentially etches high energy sites such as grain boundaries, which results in a surface relief that enables different crystal orientations, grain boundaries, phases and precipitates to be easily distinguished. For etching carbon steels, *nital* (*e.g.* 2% nitric acid-98% alcohol) solution is used. For etching stainless steel or copper alloys, a saturated aqueous solution of ferric chloride (Fe_3Cl) containing a few drops of *HCl* is used. *Metallographic etching* is applied using a grit-free cotton bud wiped over the surface a few times; the specimen should then immediately be washed in alcohol and dried.

Leveling and Microscopy The metallographic etched specimen to be examined under a microscope should be perfectly flat and leveled; the *leveling* is accomplished by using a specimen leveling press. The leveling device presses the mounted specimen into plasticene on a microscope slide, making it leveled. The etched and leveled metallographic specimen is finally observed under an optical microscope. Figure 13.6 shows the optical micrograph for an etched metallographic specimen showing the microstructure of HAZ of a weldment.

13.3.3 *Quantitative Metallography: Microscope's Resolution and Grain Size Measurement*

The quantification of microstructure, especially grain size, in welds/materials is a vital aspect to understand the structure-property relationships in the materials (Chen et al. 2017). A welding metallurgist should, therefore, have a good knowledge of quantitative metallography, especially the resolution of a metallurgical/optical microscope and the measurement of grain size. The resolution of a microscope is the minimum resolvable distance, *i.e.* the minimum distance to resolve fine detail in the sample being observed. A metallurgical microscope's resolution depends on the numerical aperture (*NA*) of its objective lens. The *NA* of the objective is given by:

$$\mathrm{NA} = n * \sin\ \alpha \tag{13.5}$$

where n is the refractive index between the objective lens and the sample, and α is the half-angle aperture (the half inclination of the lens to the objective points) (see Example 13.5). The spatial resolution of an optical microscope (*OM*) is given by Rayleigh's equation, as follows:

$$\Delta r_0 = \frac{0.62\,\lambda}{NA} = \frac{0.62 * \lambda}{n * \sin\ \alpha} \tag{13.6}$$

where Δr_0 is the resolution of the *OM* (nm), λ is the wavelength of the light source (nm), and *NA* is the numerical aperture of the objective (see Example 13.6). Equation 13.6 suggests that for good resolution (Δr_0<200 nm), the *NA* of the microscope should be reasonably high.

Grain Size Measurement There are mainly three methods of grain size determinations: (a) *line intercept method*, (b) *ASTM* grain size number method, and (c) *Jeffries Planimetric* method. The line-intercept and ASTM grain-size methods are explained in the following paragraphs.

Line Intercept Method. The *line intercept method* (or AGI, *average grain intercept method*) is a manual technique used to quantify the grain size. Firstly, it is important to find the magnification of the micrograph (*M*) by using the following formula:

$$M = \frac{\textit{Measured length of the micron bar}}{\textit{Length shown on the bar}} \tag{13.7}$$

The significance of Eq. 13.7 is illustrated in Example 13.7. Once the magnification has been calculated, a set of randomly-positioned line segment (each of length l) are drawn on the micrograph. Then, the number of times each line segment intersects a grain boundary is counted and recorded. The av. line length intersected is determined by:

$$\overline{N_l} = \frac{l}{\overline{N}} \tag{13.8}$$

where $\overline{N_l}$ is the average line length intersected (mm), l is the length of each line segment (mm), and N is the average number of grain boundaries intersected (see Examples 13.8 and 13.9). Finally, the average grain size or grain diameter is determined by:

$$d = \frac{N_l}{Magnification} \tag{13.9}$$

where d is the average grain size of the metal (mm) (see Example 13.10).

ASTM Grain Size Number. The ASTM (American Society for Testing of Materials) grain-size number for a given photomicrograph can be determined by:

$$\mathrm{N} = 2^{G-1} \tag{13.10}$$

where N is the number of grains observed in 1-in^2 area of on a photomicrograph taken at a magnification of 100×, and G is the ASTM grain size number. It is evident in Eq. 13.10 that the ASTM grain size number increases with an increase in the number of grains per unit area. By re-arranging the terms in Eq. 13.10, we obtain:

$$\mathrm{G} = \frac{\log N}{\log 2} + 1 \tag{13.11}$$

The significance of Eq. 13.11 is illustrated in Examples 13.11 and 13.12. The ASTM grain size number can also be used to calculate the average grain size, as follows (Huda 2020):

$$d = \sqrt{\frac{(25.4)^2}{(2)^{G-1}}} \times 10 \tag{13.12}$$

where d is the average grain size of the metal/weld zone (μm), and G is the ASTM grain size number (see Example 13.13).

13.4 Destructive Mechanical Testing of Welds

13.4.1 Weld Parts

The destructive testing of welds calls for a good knowledge of the parts of a weld, especially the face and the root of the weld. Figure 13.7 illustrates the following weld parts: face, root, toe, reinforcement, etc. *Weld face* is the exposed weld surface of the side where the heat source created the weld. *Weld root* is the interface of the weld and base metal at the "bottom" of the weld (opposite the face). *Weld toe is t*he interface of the weld face and base metal. *Fusion zone* (FZ) or *weld metal* is the metal that was melted during fusion welding.

13.4.2 Guided Bend Test

13.4.2.1 Test Procedure

The guided bend test is used to determine the ductility and quality of the weld. The test also assesses the extent of fusion and degree of penetration of the weld to the base metal. This testing usually requires a jig. The required specimen for testing is machined from the already welded plate, ensuring that the thickness of the specimen is within the capacity of the jig. The specimen is placed on the die member of the jig with the weld at mid-span (see Fig. 13.8).

There are three most commonly used types of guided bend testing: (a) root bend test, (b) face bend test, and side bend test. In the *root bend test*, the specimen is placed with the root of the weld toward the gap, *i.e.* the weld root is in tension while bending; this method allows inspectors to detect internal defects. The *face bend test*

Fig. 13.7 Parts of a fusion weld. (*FZ* fusion zone, *PMZ* partially melted zone)

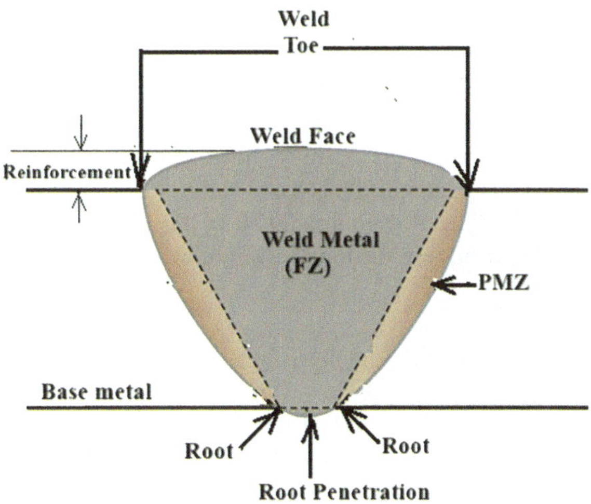

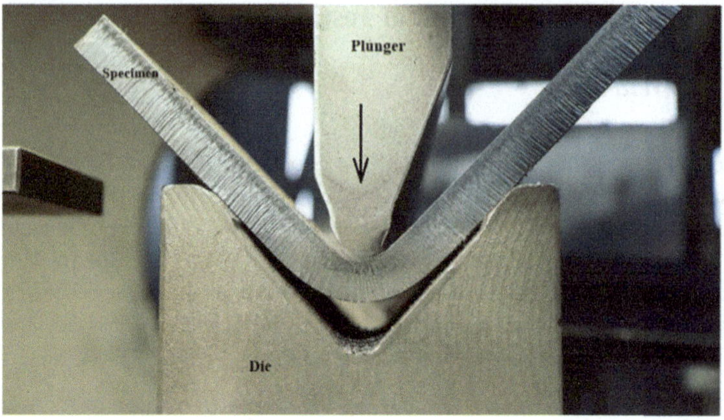

Fig. 13.8 The guided bend test in action

involves the positioning of the face of the weld directed toward the gap, *i.e.* the weld face is in tension during bending. This method enables us to detect defects that may appear on the weld surface (*e.g.* porosity, incomplete fusion, undercut, etc.). In the *side bend test*, the specimen is placed with the side showing greater discontinuity, if any, directed toward the gap. The type of test is performed on thick welded joints; it provides a complete picture of weld quality, *i.e.* it allows us to identify defects throughout the entire cross-section of the weld.

13.4.2.2 Engineering Analysis of Guided Bend Test

The bending force can be computed by:

$$\mathrm{F} = \frac{K_{bf} * \sigma_{UTS} * w * t^2}{D} \tag{13.13}$$

where F is the bending force (N), σ_{UTS} is the tensile strength of the bend specimen (MPa), w is the width of the bend specimen (mm), t is the specimen's thickness (mm), D is the die-opening dimension (mm), and $K_{bf}$is the bending factor (K_{bf}= 1.33) (see Example 13.14).

A 180° bend around radius R of the specimen induces a significant strain in the guided bend test; the induced strain can be determined by:

$$\text{Strain} = \frac{t}{2R + t} \tag{13.14}$$

where t is the thickness of the plate (see Example 13.15). According to AWS D1.1 code for root/face guided bend test, the criteria for acceptable weld quality are as follows: (a) either no crack or crack opens >3 mm in any direction on the tension face, and (b) small corner cracks <3 mm if the crack is not in weld/HAZ (see Example 13.16).

13.4.3 Free Bend Test

Face bend test involves bending a weld coupon by pushing the ends together without external support until the weld is deformed to a specified angle. Firstly, a strip or "coupon" of the welded joint is prepared by cutting it from the weldment. Then, the coupon is bent, often in a vise, by pressing the ends together until the specimen is fully bent to a specified angle. The outer surface of the bend is stretched, and any flaws in the weld will cause it to crack under the tensile strain. The specimen is examined for cracks longer than a specified size (*e.g.* 3 mm) to determine if it passes or fails the test.

13.4.4 Back Bend Test of Welds

Back bend test determines the weld quality and penetration as well as the weld's ability to withstand deformation without fracture. The test involves the bending of a weld specimen with the root of the weld on the outside (tension side). In the back-bend test, the specimens must bend to a specified angle, often 90°, without breaking.

13.4.5 Nick Bend Test

The nick bend test (also called nick break test) is a welding destructive test used to detect the internal defects (e.g. porosity, inclusion, lack of fusion, etc.) of a weld by fracturing a welded specimen. In the nick-break test, a notch is cut into the weld, and the specimen is then bent to cause a fracture along the weld's centerline (see Fig. 13.9). The test allows the broken surfaces to be inspected for defects.

13.4.6 Tensile Testing of Weldments

A tensile test can be carried out on all-weld region. However, tensile tests of butt-welded specimens are generally carried out in accordance with ISO 15614 and ASME IX specifications. These tests are generally transverse tensile tests that involve cross joint (CJ) specimens. The CJ specimen has a rectangular cross-section that is oriented across the weld so that both base metals, both heat-affected zones (HAZs) and the weld metal itself are tested (see Fig. 13.10).

Fig. 13.9 Nick bend test—*schematic*

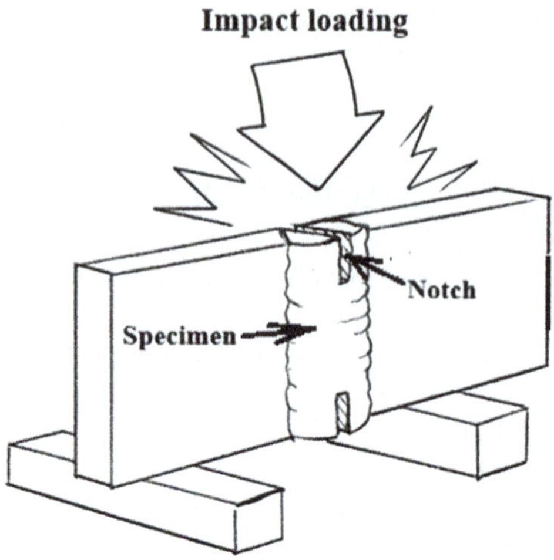

Fig. 13.10 Cross joint (CJ) specimen for tensile testing of weldment

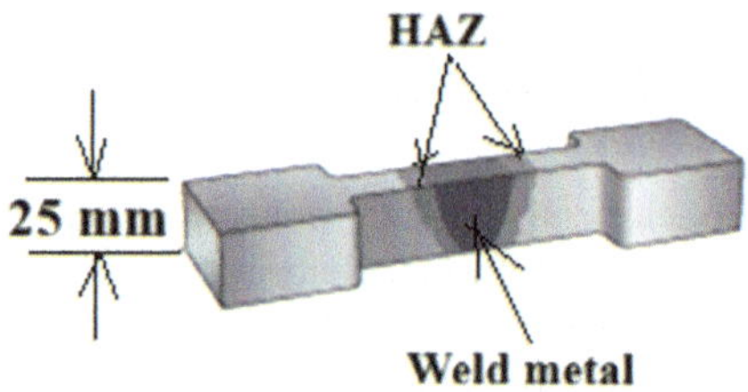

To perform a tensile strength test, a sample of weldment is loaded in tension, in a tensile testing machine, until the point of failure. The test equipment calculates the yield strength (σ_{ys}), ultimate tensile strength (σ_{UTS}), breaking strength (σ_b), % elongation, and % reduction in area (Huda 2025). The σ_{UTS} is the maximum stress withstood before failure; it can be determined by:

$$\sigma_{UTS} = \frac{F_{\max}}{A_o} \tag{13.15}$$

where F_{max} is the maximum force during the tensile test, and A_0 is the original cross-sectional area (see Examples 13.17–13.19). A weld efficiency can be determined by:

$$\eta_w = \frac{\sigma_{UTS} \text{ of the WM}}{\sigma_{UTS} \text{ of the BM}} \times 100\% \tag{13.16}$$

where η_w is the weld efficiency (see Example 13.20).

13.5 Worked Numerical Examples in Testing of Welds

Example 13.1 Determining the Radiation Intensity in Radiographic Testing (RT)
In a radiographic examination of a weld, the intensity of the received dose of radiation, at a distance of 2 cm from the radiation source, is 120 mR/h. Determine the radiation intensity at a distance of 1 m from the source.

Solution

$I_1 = 120$ mR/h, $d_1 = 2$ cm, $d_2 = 1$ m $= 100$ cm, $I_2 = ?$

By using Eq. 13.1,

$$\frac{120}{I_2} = \frac{100^2}{2^2}$$

$$\frac{120}{I_2} = 2500$$

$$I_2 = 48 \times 10^{-3} \text{ R / h}$$

The radiation intensity at a distance of 1 m from the source $= I_2 = 48$ mR/h.

Example 13.2 Determining the Radiation Density in RT
A film section, in RT of a weld, receives an incident light intensity of 800 lux. The transmitted radiation intensity is 4 lux. Determine the radiation density.

Solution

$I_0 = 800$ lux, $I_t = 4$ lux, D = ?

By using Eq. 13.2,

$$\text{Radiation density} = \text{D} = \log \frac{I_0}{I_t} = \log \frac{800}{4} = 2.3$$

Example 13.3 Determining the Surface Distance in Ultrasonic Testing (UT)
Ultrasonic testing (UT) is performed on a weld. The sound path reading on the UT machine is 35 mm, and the refracted angle of the sound beam is 60°. Determine the surface distance in the UT.

Solution

L = 35 mm, $\theta = 60°$, $d_s = ?$

By using Eq. 13.3,

$$\text{Surface distance} = d_s = \text{L} * \sin\theta = 35 \times \sin 60° = 30.3\,\text{mm}$$

Example 13.4 Determining the Depth of a Flaw from the Top Surface in UT
By using the data in Example 13.3, determine the depth of the flaw from the top surface.

Solution

By using Eq. 13.4,

$$\text{Depth} = \text{L} * \cos\theta = 35 \times \cos 60° = 17.5 \text{ mm}$$

The depth of the flaw = 17.5 mm.

Example 13.5 Determining Numerical Aperture of the Objective Lens of an OM

The refractive index between the sample and the objective lens of an optical microscope (OM) is 1.8. The half-angle aperture is 72°. Determine the numerical aperture of the lens of the *OM*.

Solution

$n = 1.8$, $\alpha = 72°$, NA = ?

By using Eq. 13.5,

$$\text{NA} = n * \sin\ \alpha = 1.8 \times \sin 72° = 1.8 \times 0.951 = 1.71$$

Example 13.6 Determining the Resolution of a Metallurgical Microscope

By using the data in Example 13.5, determine the resolution of the microscope if the visible light from the illumination system has a wavelength of 540 nm.

Solution

$\lambda = 540$ nm, $NA = 1.71$, $\Delta r_0 = ?$

By using Eq. 13.6,

$$\Delta r_0 = \frac{0.62\ \lambda}{NA} = \frac{0.62 \times 540}{1.71} = 195.78\,\text{nm}$$

The resolution of the metallurgical microscope is 195.78 nm.

Example 13.7 Determining the Magnification of an Image in an Optical Micrograph

Determine the magnification of the micrograph, as shown in Fig. 13.11.

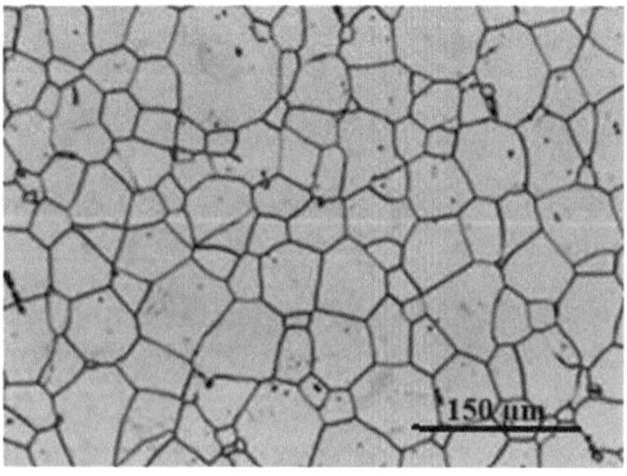

Fig. 13.11 The optical micrograph for Example 13.7

Solution

The measured length of the micron bar = 1.5 cm.

The length shown on the micron bar = 150 μm = 150×10^{-6} m = 150×10^{-4} cm

By using Eq. 13.7,

$$\text{Magnification} = M = \frac{\textit{Measured length of the micron bar}}{\textit{Length shown on the bar}}$$

$$= \frac{1.5\ cm}{150 \times 10^{-4}\ cm} = 10 \times 10 = 100$$

Example 13.8 Determining the Av. Number of Grain Boundaries Intersected in a Micrograph

By using the optical micrograph (Fig. 13.11) in Example 13.7, determine the average number of grain boundaries intersected in the micrograph.

Solution

By drawing a set of randomly-oriented line segments (each of length 42 mm) onto the micrograph, we obtain Fig. 13.12 and the data in Table 13.1.

$$\text{The average number of grain boundaries intersected} = \bar{N} = \frac{9+10+11+9+11}{5} = 10$$

Example 13.9 Determining the Average Line Length Intersected in a Micrograph

By using the data in Example 13.8, determine the average line length intersected.

Solution

$\bar{N} = 10$, $l = 42$ mm, $\bar{N_l} = ?$

By using Eq. 13.8,

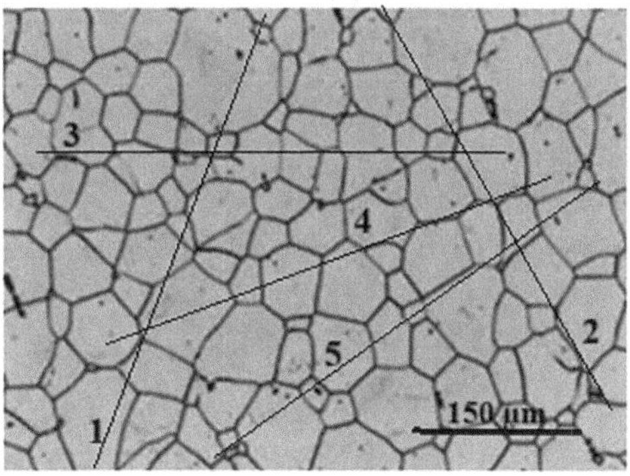

Fig. 13.12 The modified micrograph in Example 13.8

Table 13.1 Data showing the number of grain boundaries intersected in Example 13.8

Line #	1	2	3	4	5
No. of grain boundaries intersected	9	10	11	9	11

$$\text{Average line length intersected} = \overline{N_l} = \frac{l}{\overline{N}} = \frac{42\ mm}{10} = 4.2\ \text{mm}$$

Example 13.10 Determining the Average Grain Diameter in a Micrograph
By using the data in Examples 13.7–13.9, determine the average grain diameter for the microstructure shown in the micrograph (Fig. 13.11).

Solution

By using Eq. 13.9,

$$d = \frac{N_l}{Magnification} = \frac{4.2\ mm}{100} = 0.042\ \text{mm} = 42 \times 10^{-6}\ \text{m} = 42\ \mu m$$

The average grain diameter = 42 μm.

Example 13.11 Determining the Number of Grains in 1-in^2 Area in a Micrograph at 100×
By using the micrograph (Fig. 13.11) in Example 13.7, determine number of grains observed in an area of 1 in^2 on the micrograph at 100×.

Solution

We draw 1 in.^{2} area onto the micrograph, as shown in Fig. 13.13.

It is known that the magnification is 100× (from Example 13.7). By reference to Fig 13.13, the number of grains observed in an area of 1 in.^{2} on the micrograph at 100× = N = 26.

Example 13.12 Determining the ASTM Grain Size Number in a Micrograph
By using the data in Example 13.11, determine the ASTM grain size number for the micrograph.

Solution

By using Eq. 13.11,

$$G = \frac{\log N}{\log 2} + 1 = \frac{\log 26}{\log 2} + 1 = 4.7 + 1 = 5.7$$

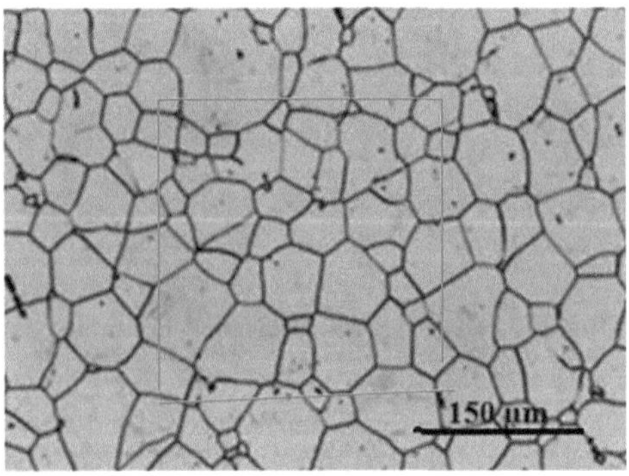

Fig. 13.13 Drawn 1 square inch area onto the micrograph in Example 13.11

The ASTM grain size number of the metal = G = 5.7.

Example 13.13 Using ASTM Grain Size Number Value to Determine the Grain Size
By using the data in Example 13.12, determine the average grain size of the metal.

Solution

$G = 5.7$, $d = ?$ µm

By using Eq. 13.12,

$$d = \sqrt{\frac{(25.4)^2}{(2)^{G-1}}} \times 10 = \sqrt{\frac{25.4^2}{2^{5.7-1}}} \times 10 = 4.98 \times 10 = 49.8\ \mu m$$

The average grain size = 49.8 µm.

Example 13.14 Determining the Bending Force in a Guided Bend Test
A face bend test was carried out on a SMAW welded mild-steel bend specimen with a thickness of 10 mm. The width of the specimen is 38 mm, and the die-opening dimension is 40 mm. Determine the bending force. Take the ultimate tensile strength of mild steel = 400 MPa.

Solution

D = 40 mm, K_{bf}= 1.33, t = 10 mm, $t^2 = 100\ mm^2$, $w = 38$ mm, $\sigma_{UTS} = 400$ MPa, F = ?

By using Eq. 13.13,

$$F = \frac{K_{bf} * \sigma_{UTS} * w * t^2}{D} = \frac{1.33 \times 400 \times 38 \times 100}{40} = 50540\,N$$

The bending force = F = 50,540 N = 50.54 kN.

Example 13.15 Determining the Strain in a Weld in a Guided Bend Test
A guided bend test was performed on a weldment (specimen) with a plate thickness of 10-mm and bend-die radius of 25 mm (bend angle = 180°). Determine the induced strain.

Solution

t = 10 mm, R = 25 mm.

By using Eq. 13.14,

$$\text{Strain} = \frac{t}{2R + t} = \frac{10}{2 \times 25 + 10} = 0.166$$

The induced strain = 0.1666 or 16.7%.

Example 13.16 Checking the Success of a Weld in a Guided Bend Test
The guided bend test, as mentioned in Example 13.15, resulted in a 2.4-mm-long crack observed on the surface of the tension side. Does the weld pass the test?

Solution

According to AWS D1.1 code, a crack (on the tension face) with a length less than 3 mm is acceptable. Since 2.4 mm < 3 mm, the weld passes the guided bend test.

Example 13.17 Determining the Ultimate Tensile Strength of a Weld in Tensile Testing

A tensile test was carried out using a CJ specimen with original cross-sectional area of 80 mm^2. The maximum load at the weld metal (WM), during tensile testing, was 26,000 N. The ultimate tensile strength (σ_{UTS}) of the base metal (BM) is 350 MPa. Determine the σ_{UTS} of the WM.

Solution

$F_{max} = 26{,}000$ N, $A_0 = 80$ mm^2.

By using Eq. 13.15,

$$\sigma_{UTS} = \frac{F_{\max}}{A_0} = \frac{26000}{80} = 325\,\text{MPa}$$

The ultimate tensile strength of the WM = σ_{UTS} = 325 MPa.

Example 13.18 Commenting on the Performance of a Weld Based on Tensile Test Data

By using the data in Example 13.17, comment on the performance of the weld.

Solution

The σ_{UTS} of the WM = 325 MPa, The σ_{UTS} of the BM = 350 MPa.

Since the σ_{UTS} of the WM is lower than that of the BM, the **WM slightly under performed** (as compared to the BM) in the tensile test.

Example 13.19 Checking the Success of a Weld in a Tensile Test

A tensile specimen containing a weldment fractures **outside the weld** in the base metal. The maximum load was 30,500 N, and the original cross-sectional area of the specimen was 85 mm^2. Comment on the performance of the weld if the σ_{UTS} of the BM is 350 MPa.

Solution

$F_{max} = 30{,}500$ N, $A_0 = 85$ mm^2.

$$\sigma_{UTS} = \frac{F_{\max}}{A_0} = \frac{30500}{85} = 358.8 \cong 359\,\text{MPa}$$

The σ_{UTS} of the WM (359 MPa) is higher than that of the BM; this is why the weldment failed outside the weld. Hence, the **weld passes** the tensile test.

Example 13.20 Determining the Efficiency of a Weld Based on Tensile Test Data

By using the data in Examples 13.17–13.18, compute the efficiency of the weld in the weldment.

Solution

The σ_{UTS} of the WM = 325 MPa, The σ_{UTS} of the BM = 350 MPa.
By using Eq. 13.16,

$$\text{Weld efficiency} = \eta_w = \frac{\sigma_{UTS}\text{ of the }WM}{\sigma_{UTS}\text{ of the }BM} \times 100\% = \frac{325}{350} \times 100\% = 92.8\%$$

Questions and Problems

13.1(MCQs). Encircle the most appropriate answers for each of the following statements.

(1). The interface of the weld face and base metal is called___________

(a) Weld face, (b) weld root, (c) weld toe, (d) weld metal.

(2). The exposed weld surface of the side where the heat source created the weld is called___

(a) Weld face, (b) weld root, (c) weld toe, (d) weld metal.

(3). The interface of the weld and base metal at the bottom of the weld is called________

(a) Weld face, (b) weld root, (c) weld toe, (d) weld metal.

(4). The metal that was melted during fusion welding is called _________

(a) Weld face, (b) weld root, (c) weld toe, (d) weld metal

(5). Which of the following test enables us to determine the grain size in the weld's HAZ?

(a) NDT, (b) metallographic testing, (c) bend test, (d) tensile test.

(6). Which test enables us to detect internal flaws and discontinuities in welds?

(a) Ultrasonic test, (b) bend test, (c) tensile test, (d) visual test

(7). Which of the following tests involves impact loading of specimen?

(a) Tensile test, (b) visual test, (c) nick bend test, (d) face bend test.

(8). Which NDT can best detect surface-breaking defects in a weld?

(a) Visual test, (b) dye-penetrant test, (c) ultrasonic test, (d) radiographic test.

13.2. List four techniques for each of the following types of weld testing:

(a) NDT, and (b) destructive mechanical tests.

13.3. Draw the sketch showing various parts of a fusion weld.
13.4. (a) Illustrate cross-joint (CJ) specimen with the aid of a sketch.
(b) Briefly explain tensile testing of welds.
13.5. Describe ultrasonic testing (UT) of welds.

Table 13.2 Data of number of grain boundaries intersected in P13.11

Line #	1	2	3	4	5
No. of grain boundaries intersected	12	10	8	11	13

13.6. Distinguish between macro- and microscopic examinations of welds.

13.7. List the steps in metallographic specimen preparation.

P13.8. A film section, in RT of a weld, receives an incident light intensity of 700 lux. The transmitted radiation intensity is 3.5 lux. Determine the radiation density.

P13.9. UT is performed on a weld. The sound path reading on the UT machine is 40 mm, and the refracted angle of the sound beam is 50°. Determine the (a) surface distance, and (b) depth of the flaw from the top surface.

P13.10. The refractive index between the sample and the objective lens of an optical microscope (OM) is 1.6. The half-angle aperture is 70°. Determine the numerical aperture of the lens.

P13.11. An optical micrograph was analyzed using line intercept method; and the data for the number of grain boundaries intersected is presented in Table 13.2. The magnification of the image in the micrograph is 200. Determine the average grain size.

P13.12. The number of grains observed in an area of 1 in^2 on a micrograph at 100× is 30. Determine the ASTM grain size number for the micrograph.

P13.13. A face bend test was carried out on a welded mild-steel (MS) bend specimen with a thickness of 10 mm. The width of the specimen is 38 mm, and the die-opening dimension is 41 mm. Determine the bending force. Take the σ_{UTS} of MS = 410 MPa.

P13.14. A welded specimen was tested using tensile testing machine. The maximum load during the test was 30,000 N, and the original cross-sectional area of the specimen was 82 mm^2. Determine the σ_{UTS} of the weld metal.

P13.15. By using the data in P13.14, determine the efficiency of the weld if the the σ_{UTS} of the base metal is 355 MPa.

References

AWS (2015). WI:2015 Welding Inspection Handbook, American Welding Society, USA

Chen, H. Yao, Y., Warner, J.A., *et al.* (2017). Grain size quantification by optical microscopy, electron backscatter diffraction, and magnetic force microscopy, *Micron,* Vol. 101, pp. 41–47

Halmshaw, R. (1997). Introduction to the Non-Destructive Testing of Welded Joints, Woodhead/ *Elsevier Publishers*, Netherlands

Huda, Z. (2020). Metallurgy for Physicists and Engineers, *CRC Press*, FL, USA

Huda, Z. (2025). Metal Casting Engineering: Design, Processes, Calculations, *Springer Nature,*Switzerland.

Hughes, S.E. (2009). A Quick Guide to Welding and Weld Inspection, Woodhead Publishing Ltd., Cambridge, UK.

ISO (2025). ISO/TC 44/SC 5 Testing and inspection of welds, *International Standards Organization*, Afnor, France. Online: https://www.iso.org/committee/48620.html. Accessed on Nov. 27, 2025.

Voort, G.F.V. (2011). Metallography of Welds, *Advanced Materials and Processes*, Vol. 169, Issue 6, pp. 19–23

Chapter 14
Welding Defects, Prevention, and Failure Analysis

Abstract This chapter presents both qualitative and quantitative aspects of welding defects to enable welding engineers to produce defect-free welds. The chapter begins with the definition and classification of welding defects, including the ISO standards. *External welding defects* (cracks, undercuts, overlaps, porosity, etc.) are explained with reference to their causes and prevention. The *cracks* are discussed with the aid of a sketch showing longitudinal cracks, transverse cracks, toe cracks, root cracks, under-bead cracks, and the like; here, the causes and prevention of *hot cracks* and *cold cracks* are also described. The discussion on *internal welding defects* emphasizes on the causes and prevention of *slag inclusions* and *incomplete penetration*. Finally, real-life examples of weld failures are presented as ***case studies on weld joint failures***; here the failures are analyzed for the Union Oil Co.'s amine-absorber tower/pressure vessel, the centrifugal pump (failed in 2022), and the *Liberty ship*. This chapter contains 9 worked numerical examples, 6 diagrams, and 8 multiple-choice questions (MCQs) with their answers given at the book's end.

Keywords Cracks · Undercuts · Overlaps · Porosity · Slag inclusion · Incomplete penetration · Weld joint failures

Abbreviations

ASTM	American Society of Testing of Materials
AWS	American Welding Society
CTOD	Crack tip opening displacement
FCAW	Flux-core arc welding
HAZ	Heat affected zone
HISC	Hydrogen-induced stress cracking
ISO	International Standards Organization
MEA	Mono-ethanol-amine
NDT	Non-destructive testing

Z. Huda, *Welding and Joining Engineering*,
https://doi.org/10.1007/978-3-032-20462-2_14

PWHT Post-welding heat treatment
SAW Submerged arc welding
SMAW Shielded metal arc welding

14.1 Welding Defects: Classification and Consequences

Welding defects refer to flaws, irregularities, and imperfections formed in a weldment; these defects adversely affect either the intended use of the weldment or its aesthetic appeal. The weld defects that adversely affect the weldment's performance are classified according to ISO-6520. On the other hand, the imperfections that do not adversely affect the weld are classified according to their acceptable limits under ISO 5817 and ISO 10042. Weld irregularities can be classified into two categories: (a) external **welding defects** and (b) internal **welding defects**.

External welding defects refer to the irregularities in the weld metal that are noticeable by either visual examination or other methods (e.g. magnetic particle testing, dye penetrant test, etc.). Typical examples of external weld defects are cracks, undercuts, overlaps, porosity, spatter, and the like. *Internal weld defects* are usually not open to the weld's surface; they can be detected by using ultrasonic testing (UT) and radiographic testing (RT). The examples of internal weld defects include slag inclusions, incomplete penetration, incomplete fusion, and the like. Any of these defects (either external or internal) are potentially disastrous as they can all give rise to high stress intensities, which may result in catastrophic failure below the design load or in the case of cyclic loading, failure after fewer load cycles than predicted (Kur and Fadol 2017).

14.2 External Welding Defects and Their Prevention

14.2.1 Cracks in a Weld

Types of Cracks and Their Causes A crack is the worst welding defect since it can rapidly grow, thereby leading to failure. Cracks are fracture-type discontinuities with harp tips and a high ratio of length and width to the displacement of the opening (AWS 2024). Cracks occur when localized stresses exceed the ultimate tensile strength of the metal. Stress concentration near discontinuities, mechanical notches, and hydrogen embrittlement can contribute to crack formation. There are a variety of types of weld cracks; these include longitudinal cracks, transverse cracks, toe cracks, under-bead cracks, weld-metal cracks (see Fig. 14.1). It is evident in Fig. 14.1 that *longitudinal cracks* are formed parallel to the weld bead, whereas *transverse cracks* appear across the width of the bead. *Crater cracks* are formed at

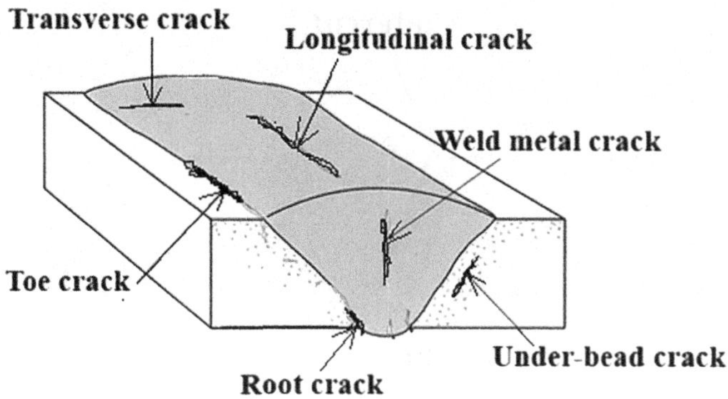

Fig. 14.1 Cracks in a weld

the end of the bead. The welding code allows a maximum cumulative crack length of 20 mm/m (see Example 14.1).

Depending on the temperature they occur, welding cracks may be either hot cracks or cold cracks. *Hot cracks* form due to thermal stresses when weld metal crystallizes as the base metals are heated above 10,000 °C. They occur when the metal contains high impurity or carbon content or when there is a disruption in heat flow during the welding process. *Cold cracks* form after the weld metal has cooled to ambient temperature. Cold cracks occur mainly due to a lack of preheating, high stresses, low temperature, high hydrogen content, and the like.

Preventing Weld Cracks The most important consideration in preventing weld cracks is to ensure the weldability of the base metals of the workpieces to be joined; additionally, welding process and filler materials should be compatible, especially their composition control by using the appropriate sulfur and carbon contents. In the case of arc welding, the use of the right welding speed and current will ensure a controlled heat input (see Sect. 14.4). In order to avoid thermal stresses, the base metal should be preheated. The use of hydrogen (H_2) as a shielding gas for ferrous metals must be avoided to prevent H_2 embrittlement/weld cracking. A reduction in the gap between weld joints also helps prevent weld cracks. It is also important to maintain the correct weld bead depth-to-width ratio in the weldment.

14.2.2 Undercut Defect

Causes *Undercut defects* are irregular recess (grooves) with notches that are formed on the base metal; they run parallel to the weldment, thereby causing a loss in thickness (see Fig. 14.2). This defect is a stress amplifier and often causes premature failure. *Undercuts* occur when the base metal melts away from the weld zone. The main cause of undercut defect is high heat-energy input, which may be either the use of excessive current or too low welding speed in arc welding (see Chap. 4,

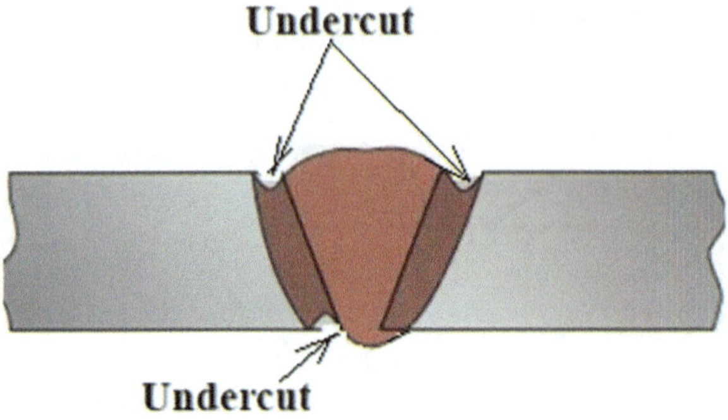

Fig. 14.2 Undercut welding defect

Fig. 14.3 Overlap weld defect

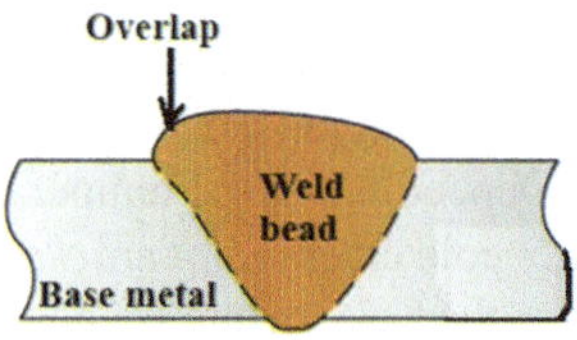

Eq. 4.2). Huda and Alostath have reported an *undercut* defect in shielded metal arc welding (SMAW) of steel using SK-260-M E6013 electrodes; this welding defect was caused by a high heat input of 1.4 kJ/mm (Huda and Alostath 2021). The welding codes, in general, allow a maximum undercut depth of 0.5 mm.

Prevention The use of the right current (amperage) helps to reduce heat input. A lower current is recommended if there are long, consistent undercuts. Alternatively, a lower voltage will also reduce heat input; this can be done by using the voltage knob's sweet spot that produces no undercuts, a smooth finish, and an even bead profile. The use of compatible filler is also necessary to avoid undercuts. The welding angle should be proper to deposit filler evenly in the weld pool. This can be done by observing each side of the bead and noticing differences. A welder must choose the correct electrode diameter (see Chap. 4, Eq. 4.1); this would help fill up the groove and deliver energy evenly. Additionally, a stable arc should be maintained; this can be adjusted using arc indicators.

14.2.3 *Overlap Defect*

Causes *Overlap* defect is the excess metal that spreads out around the weld bead; this happens when the spread-out filler metal is not properly mixed with the base metals. Typically, this defect appears as a curved shape over the weld joint (see Fig. 14.3). When the amperage or current is set too high, which leads to excess filler material in the weld pool, causing it to flow out and solidify improperly. Another cause of *overlap* is too low travel (welding) speed.

Prevention The following are the techniques to prevent *overlap* weld defect. A current settings chart should be referred to determine the optimal amperage settings for the specific welding job. If overlap still occurs, the amperage should be incrementally lowered until a clean weld is achieved. A low travel speed results in filler material accumulating in the weld pool, leading to overlap. In order to prevent *overlap*, the travel speed should be reasonably increased. Another remedy for *overlap* is to use correct electrode selection and its proper angling. Additionally, ensure proper alignment of the base metals before one begins welding. The cleaner and better-aligned the setup is, the less likely is the occurrence of *overlap*.

14.2.4 Porosity Defect

Causes *Porosity* defects occur when there is an entrapment of air or gas bubbles in the weld. The welding process often generates gases, such as steam, hydrogen, and carbon dioxide. A cross-section of porous weld bead often resembles a sponge with an accumulation of trapped air bubbles (see Fig. 14.4). The entrapped gases may be either localized or uniformly distributed in the weld. Porosity defects can weaken the weld joint, predisposing them to fatigue failures.

Prevention The following are the means to avoid porosity weld defect. Preheat the base metals before welding. Select the suitable electrode and filler material. Clean properly the base metal and prevent pollutants from entering the welding area. Use recommended welding current and travel speed. Use high-purity shielding gases. Configure the gas flow meter to the correct flow settings in the range of 22–30 cubic feet per hour.

Analysis The AWS D1.1 welding code allows a maximum of 5% porosity. The percentage porosity can be determined by:

$$\%\text{Porosity} = \frac{Total\ volume\ of\ pores}{Total\ weld\ volume} \times 100 \tag{14.1}$$

The significance of Eq. 14.1 is illustrated in Examples 14.2 and 14.3

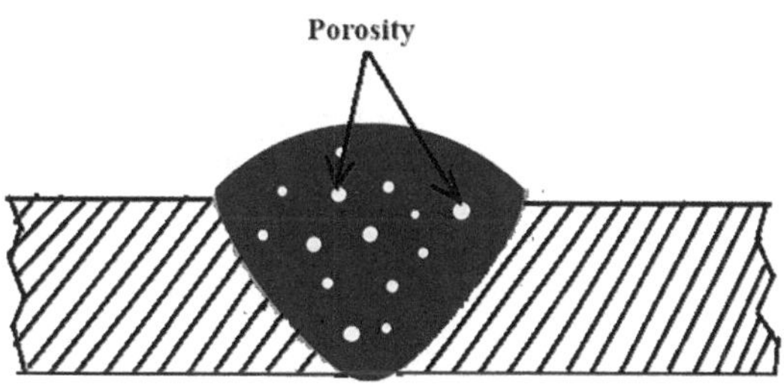

Fig. 14.4 Porosity weld defect

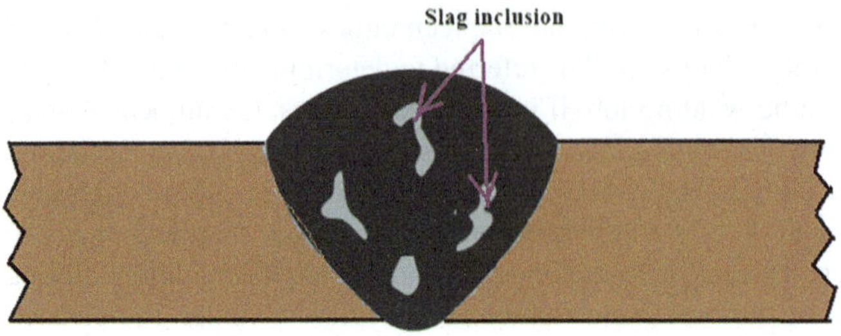

Fig. 14.5 Slag inclusion welding defect

14.3 Internal Welding Defects and Their Prevention

14.3.1 Slag Inclusion

Causes *Slag inclusions* are internal weld defects; they often appear as trapped impurities within or on the surface of the welded areas (see Fig. 14.5). Slag inclusions occur when a flux melts on the surface of the weld or within the weld region. Slags emerge in various processes, such as shielded metal arc welding (SMAW), flux-core arc welding (FCAW), and submerged arc welding (SAW). The presence of slag affects the metal's weldability, toughness, and the structural performance of the weldment. This is why most radiographic-testing standards for welds limit the maximum inclusion area to 5 mm^2 (see Example 14.4).

Prevention The following are the remedies for *slag inclusion* defect. Choose appropriate consumables (*e.g.* good quality flux, dry electrodes, etc.). Thoroughly clean the base metal between passes. Use correct welding-process parameters (*e.g.* higher amperage, steady travel speed), especially sufficient current to float slag out. Prevent rapid cooling and employ other techniques (e.g. smooth beads, adequate overlap, etc.) to ensure slag floats to the surface and is fully removed before the next layer. Redesign joints to ensure there is sufficient space for the proper use of a puddle of molten welds.

14.3.2 Incomplete Penetration

Causes *Incomplete penetration* defect refers to the incomplete filling of the weld root (Fig. 14.6); it occurs when the metal groove is too narrow and is not filled. As a result, the weld metal does not entirely spread through or get to the bottom of the weld joint. This reduces the strength of the weld joint and causes weld failure. The primary cause of incomplete penetration defect is the low heat input, thereby result-

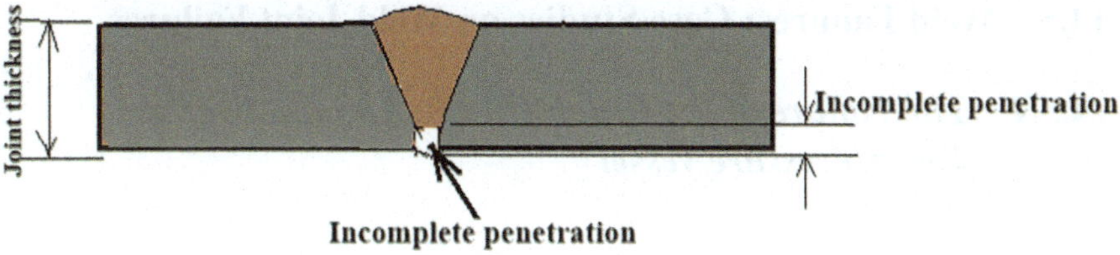

Fig. 14.6 *Incomplete penetration* defect

ing in insufficient melting of metal. The secondary causes are: improper joint alignment, incorrect positioning of the electrode, and the like.

Prevention The following are the techniques to prevent incomplete penetration defects.

Increase welding current (amperage). Reduce welding (arc travel) speed. Use the correct joint geometry and proper alignment. Ensure enough weld metal deposition. Ensure accurate positioning of electrodes.

Analysis Welding codes generally require full penetration (0% incomplete penetration). The percentage of incomplete penetration in butt weld joints can be determined by:

$$\%\text{Incomplete penetration} = \frac{\textit{Incomplete penetration}}{\textit{Thickness of weld joint}} \times 100 \qquad (14.2)$$

The significance of Eq. 14.2 is illustrated in Example 14.5.

14.4 The Role of Heat Input in Weld Quality

Heat input has been discussed in detail in Chaps. 2 and 4 (see Eq. 4.2); it plays an important role in weld quality, as both too high and too low heat inputs result in welding defects. A high heat input would cause excessive melting (spattering), distortion, and coarse-grained microstructure; the latter is caused by slow cooling rate. On the other hand, low heat input would result in incomplete penetration, insufficient fusion, and increased brittleness. It is, therefore, critical to control heat input in a welding operation (see Examples 14.6–14.9).

14.5 Weld Failures: Case Studies on Weld Joint Failures

14.5.1 The Failure of the Union Oil Co. Amine-Absorber Tower/Pressure Vessel

The weld joint failures in pressure vessels is crucial to industries, as a number of historical pressure-vessel failures/disasters have been reported (Hayes, 1996). One of such failures is the failure of the Union Oil Co.'s amine-absorber tower/pressure vessel. The Union Oil Co. refinery in Illinois, USA was seriously damaged by an explosion and fire on July 23, 1984; the explosion was caused by the ignition of a cloud of flammable gas mixture (propane + butane), which had leaked from a ruptured amine-absorber pressure vessel. The disaster resulted in the death of 17 people working at the refinery and heavy property damage. The absorber tower/vessel was a cylindrical vessel 2.6-m in diameter and of overall height of 16.8 m. The shell section consisted of six courses of 25-mm thick ASTM A516 Grade 70 steel, which were joined by full penetration submerged arc welds (SAW) in the as-welded condition. The vessel, built to ASME Section VIII, was designed to strip hydrogen sulfide (H_2S) from the propane/butane gas mixture passing through it. Mono-ethanol-amine (MEA) was fed through the tower as part of this process. The operating conditions were 1.4 MPa internal pressure at 38 °C. The environment in the tower was corrosive. Soon after the amine absorber tower entered service in 1970, hydrogen blisters were found in the lower two courses of the shell and laminations were detected in the steel. The growth of hydrogen blisters continued. The second course of the tower was replaced on site using manual shielded metal arc welding (SMAW) with no preheat or post-weld heat-treatment (PWHT) in 1974. A Monel liner to reduce corrosion was fitted in the bottom head and first course of the tower in 1976, but it did not cover the repair section.

The failure analysis showed that the tower fractured at the circumferential weld between the replacement ring and the lower course. Four large cracks in the heat affected zone (HAZ) had been detected prior to the failure, originating at the inner surface of the tower and extending almost through the wall thickness.

Causes of the Failure The pressure vessel fractured because the cracks propagated along a path that was weakened due to a repair weld joining a replacement section to the vessel. The vessel had a history of hydrogen-induced blisters and delaminations in *courses 1 and 2*; the source of hydrogen was the H_2S in the process stream. The pre-existing cracks initiated in regions of hard phase (in the microstructure) known to be susceptible to hydrogen-induced stress cracking (HISC). This hard phase formed during the repair welding of the replacement section, and at the surface, it was not tempered by subsequent weld passes or PWHT. The cracks grew through the vessel wall due to the HISC. As soon as the depth of the largest of these pre-existing cracks exceeded 90% of the wall thickness, leakage occurred because the thin ligament of steel remaining in the cracked section ruptured.

Conclusion It can be concluded that the failure occurred because the welding procedure (SMAW with no PWHT) was used when replacing a section of the vessel; the poor welding practice caused the formation of a hard phase in the HAZ of the weld. This hard region was susceptible to HISC, resulting in growth of large cracks in the vessel. Additionally, the un-cracked material in the vicinity of the existing cracks had low toughness due to hydrogen embrittlement and failed at the applied crack tip opening displacement (CTOD) in the vessel arising from the operating pressure and residual stresses associated with the weld (see Chap. 12 for techniques to reduce residual stresses).

14.5.2 Failure of Weld Joint of a Centrifugal Pump

Recently (2022), Mao and co-researchers reported failure analysis on the weld joint of a centrifugal pump diffuser for oil and gas pipeline transportation (Mao et al. 2022). The investigation of the failure involved the use of modern material characterization techniques. The results indicated that the strength and toughness of the HAZ were the highest, followed by matrix, and the weld joint had the lowest strength. Defects such as slag inclusion and residual austenite created initial conditions for the corrosion of the weld joint. Severe chloride-ion corrosion resulted in micro-cracks on the weld surface. Meanwhile, corrosion reduces the surface energy of the micro-cracks. Therefore, the concentrated driving force makes the micro-cracks with reduced surface energy spread rapidly, eventually causing the rapid failure of the diffuser weld joint. A possible remedy to this problem is to ensure a clean weld (free from slag inclusion) and good microstructural control in the weld joint with no residual austenite.

14.5.3 The Failure of Liberty Ship

The failure of the *Liberty ship* that cracked in half during World War II is attributed to a brittle (fast) fracture that resulted due to poor welding design and weld quality of the fabricated ship structure. The Liberty ship was fabricated by welding steel plates to form a continuous body with no rivet joints. Cracks began at notches or arc-weld spots and spread all the way around the hull, eventually splitting the ship in half. A possible design improvement to avoid the fracture could be to provide crack arresters by incorporating some riveted joints into the ship structure.

14.6 Worked Numerical Examples

Example 14.1 Calculating Cumulative Crack Length to Decide Weld Success/Failure
A weld is inspected using the NDT. Three cracks are detected in a meter length; the lengths of crack 1, crack 2, and crack are 9 mm, 6 mm, and 10 mm, respectively. Does the weld pass?

Solution

Crack-1 length = 9 mm, crack-2 length = 6 mm, crack-3 length = 10 mm.

Total crack length = 9 + 6 + 10 = 25 mm.

The welding code allows a maximum cumulative crack length of 20 mm/m.

Since 25 mm > 20 mm, the weld fails the NDT.

Example 14.2 Determining the Percentage of Porosity in a Weld
A weld bead has a total volume of 600 mm^3. Radiographic testing (RT) of the weld detected pores with a combined volume of 80 mm^3. Determine the percentage of porosity.

Solution

Total weld volume = 600 mm^3, total volume of pores = 80 mm^3, % porosity = ?

By using Eq. 14.1,

$$\%\text{Porosity} = \frac{\textit{Total volume of porosity}}{\textit{Total weld volume}} \times 100 = \frac{80}{600} \times 100 = 13.3$$

The porosity = 13.3%.

Example 14.3 Deciding the Success/Failure of a Weld as Per AWS Code
By using the data in Example 14.2, decide the success/failure of the weld as per AWS D1.1 code.

Solution

The AWS D1.1 welding code allows a maximum of 5% porosity. Since 13.3% > 5%, the weld fails the radiographic inspection.

Example 14.4 Determining the Area of Slag Inclusion and Deciding the Fate of the Weld
Radiographic testing of a weld revealed a slag inclusion approximated as a rectangle of length 2 mm and height 1.5 mm. (a) Determine the area of the inclusion. (b) Does the inclusion pass the NDT as per standards?

Solution

Slag-inclusion dimensions: length = 2.5 mm, height = 1.5 mm.

(a) Inclusion area = 2.5 × 1.5 = 3.75 mm^2
(b) Since 3.75 mm^2 < 5 mm^2, the weld passes the NDT.

Example 14.5 Determining the % Incomplete Penetration in a Weld and Predicting Its Fate

The thickness of a butt-weld joint is 14 mm. The NDT reveals incomplete penetration of 3 mm. (a) Determine the % incomplete penetration. (b) Does the weld pass the NDT?

Solution

Incomplete penetration = 3 mm, Weld-joint thickness = 14 mm.

By using Eq. 14.2,

$$\%\text{Incomplete penetration} = \frac{\textit{Incomplete penetration}}{\textit{Thickness of weld joint}} \times 100 = \frac{3}{14} \times 100 = 21.4$$

Incomplete penetration = 21.4%.

Since the welding code requires full penetration, the weld fails the NDT.

Example 14.6 Determining the Heat Inputs for Four Different Welding Speeds

An arc welding operation is performed with an arc voltage of 20 V and current of 200 A. The heat-transfer efficiency for the welding operation is 75%. Determine the heat input into the weld joint for the following four different welding speeds: 3 mm/s, 4 mm/s, 5 mm/s, and 6 mm/s.

Solution

U = 20 V, I = 200 A, $\eta_o = 75\%$, $H_i = ?$

Using Eq. 12.18 for $v_w = 3\ mm/s$,

$$H_i = \eta_o \frac{U * I_w}{v_w} = \frac{0.75 \times 20 \times 200}{3} = 1000\text{J / mm}$$

$$\text{For}\, v_w = 4 mm/s, H_i = \eta_o \frac{U * I_w}{v_w} = \frac{0.75 \times 20 \times 200}{4} = \frac{3000}{4} = 750\text{J / mm}$$

$$\text{For}\, v_w = 5 mm/s, H_i = \eta_o \frac{U * I_w}{v_w} = \frac{0.75 \times 20 \times 200}{5} = \frac{3000}{5} = 600\text{J / mm}$$

$$\text{For}\, v_w = 6 mm/s, H_i = \eta_o \frac{U * I_w}{v_w} = \frac{0.75 \times 20 \times 200}{6} = \frac{3000}{6} = 500\text{J / mm.}$$

Example 14.7 Commenting the Weld Qualities Based on the Heat Inputs

By using the data in Example 14.6, comment on the weld quality for the four cases.

Solution

v_w = 3 mm/s ➔ H_i = 1000 J/mm; 4 mm/s ➔ 750 J/mm; 5 mm/s ➔ 600 J/mm; 6 mm/s ➔ 500 J/mm.

The welding speed of 3 mm/s results in too high heat input of 1000 J/mm, which may result in excessive melting, burn, undercut, and distortion (warping). The welding speed of 6 mm/s results in too low heat input of 500 J/mm which may cause insufficient fusion and incomplete penetration. It is, therefore, recommended to use welding speed of either 4 mm/s or 5 mm/s.

Example 14.8 Determining the Heat Inputs Based on Thermal Efficiencies of Four Different Welding Processes

Four similar weld joints are made using four different welding processes (SMAW, GTAW, OFW, and SAW) with the following parameters: arc voltage = 22 V, current =190 A, and welding speed = 4 mm/s. The thermal efficiencies for the SMAW, GTAW, OFW, and SAW processes are 80%, 70%, 50%, and 95%, respectively. Determine the heat inputs for the four operations.

Solution

U = 22 V, I = 190 A, v_w = 4 mm/s, H_i =?

Using Eq. 12.18 for SMAW ($\eta_o = 80\%$).

$$H_i = \eta_o \frac{U * I_w}{v_w} = \frac{0.80 \times 22 \times 190}{4} = 0.8 \times 1045 = 836\,\mathrm{J/mm}$$

By Eq. 12.18 for GTAW ($\eta_o = 70\%$).

$$H_i = \eta_o \frac{U * I_w}{v_w} = \frac{0.7 \times 22 \times 190}{4} = 0.7 \times 1045 = 731.5\,\mathrm{J/mm}$$

By Eq. 12.18 for OFW ($\eta_o = 50\%$).

$$H_i = \eta_o \frac{U * I_W}{v_W} = \frac{0.5 \times 22 \times 190}{4} = 0.5 \times 1045 = 522.5\,\mathrm{J/mm}$$

By Eq. 12.18 for SAW ($\eta_o = 95\%$)

$$H_i = \eta_o \frac{U * I_w}{v_w} = \frac{095 \times 22 \times 190}{4} = 0.95 \times 1045 = 992.75\,\mathrm{J/mm}$$

Example 14.9 Commenting the Weld Qualities Resulting from Four Welding Processes

By using the data in Example 14.8, comment on the practical aspects for each of the four cases.

Solution

For SMAW, H_i= 836 J/mm; GTAW: H_i = 731.5 J/mm; OFW: H_i= 522 J/mm, SAW: H_i= 992 J/mm.

The SMAW process generally induces residual stresses, so a PWHT is required for QC. The SAW welding process (H_i = 992 J/mm) has the highest value of heat input, so this process might be suitable for welding thick plates and obtaining deeper penetration. On the other hand, OFW has too low heat input of 522 J/mm, which would cause incomplete fusion and penetration defect. However, the GTAW welding process has a reasonably low input of 731 J/mm, so it is suitable for welding thin plates.

Questions and Problems

14.1(MCQs). Encircle the most appropriate answers for each of the following statements:

(1) The weld defects that affect the weldment's performance are classified according to ______:

(a) ISO 10042, (b) ISO 6520, (c) ISO 5817. (d) ISO 9000.

(2) The weld defects that affect the weldment's aesthetic appeal are classified according to ______:

(a) ISO 10042, (b) ISO 6520, (c) ISO 9000.

(3). Which one of the following welding processes has the greatest thermal efficiency?

(a) SMAW, (b) GTAW, (c) OFW, (d) SAW.

(4). Which one of the following welding processes has the lowest thermal efficiency?

(a) SMAW, (b) GTAW, (c) OFW, (d) SAW.

(5). Which welding defect would occur owing to mechanical notches and H_2 embrittlement?

(a) Cracks, (b) incomplete penetration, (c) slag inclusion, (d) undercut.

(6). Which welding defect is most likely to occur when a flux melts within the weld region?

(a) Cracks, (b) incomplete penetration, (c) slag inclusion, (d) undercut.

(7) Which welding defect is most likely to occur due to low heat input in the weld region?

(a) Cracks, (b) incomplete penetration, (c) slag inclusion, (d) undercut.

(8). What is the maximum allowable limit of porosity defect, in accordance with AWS D1.1?

(a) 20%, (b) 15%, (c) 5%, (d) 1%.

14.2. List at least three welding defects classified under: external defects, and internal defects.

14.3. Draw a sketch showing various types of cracks in a weldment.

14.4. Describe the ways to prevent cracks in welds.

14.5. (a) Draw a sketch showing *undercut* welding defect.
(b) What are the causes of undercut defect?

14.6. (a) Why are slag inclusions formed in a weld?
(b) How can the inclusions defect be prevented?

14.7. (a) Briefly describe the failure of the Union Oil Co. amine-absorber tower/ pressure vessel.
(b) What were the causes and mechanism of the weld failure?

14.8. Mao and co-researchers (2022) reported failure analysis on the weld joint of a centrifugal pump diffuser for oil and gas pipeline transportation. Highlight the causes and prevention of the weld defect.

P14.9. An arc welding operation is performed with an arc voltage of 20 V and current of 190 A. The heat-transfer efficiency for the welding operation is 70%. (a) Determine the heat input into the weld joint for the following four different welding speeds: 4 mm/s, 5 mm/s, 6 mm/s, and 7 mm/s. (b) Comment on the weld qualities for the four cases.

P14.10. A weld is inspected using the NDT. The following three cracks were detected in a meter length; the lengths of crack 1, crack 2, and crack are 8 mm, 5 mm, and 13 mm, respectively. Does the weld pass the test?

P14.11. Four similar weld joints are made using four different welding processes (SMAW, GTAW, OFW, and SAW) with the following parameters: arc voltage = 20 V, current = 200 A, and welding speed = 5 mm/s. The thermal efficiencies for the SMAW, GTAW, OFW, and SAW processes are 80%, 70%, 50%, and 95%, respectively. (a) Determine the heat inputs for the four operations. (b) Comment on the practical aspects for each of the four cases.

P14.12. The thickness of a butt-weld joint is 20 mm. The NDT reveals incomplete penetration of 2 mm. (a) Determine the % incomplete penetration. (b) Does the weld pass the NDT?

P14.13. A weld bead has a total volume of 700 mm^3. Radiographic testing (RT) of the weld detected pores with a combined volume of 65 mm^3. (a) Determine the percentage of porosity. (b) Decide the success/failure of the weld as per AWS D1.1 code.

P14.14. Radiographic testing of a weld revealed a slag inclusion approximated as a rectangle of length 3 mm and height 2 mm. (a) Determine the area of the inclusion. (b) Does the inclusion pass the NDT as per standards?

References

AWS (2024). Understanding Weld Cracking, *American Welding Society*, USA https://www.aws.org/magazines-and-media/inspection-trends/2024/may/

Hayes, B. (1996). Six case histories of pressure vessel failures, Engineering Failure Analysis, Vol. 3, Issue 3, Pages 157-170.

Huda, Z. and Alostath, M. (2021). Effects of process parameters on SMAW weld quality using SK-260-M E6013 electrodes, *Journal of Technological Science and Engineering*, Vol. 2, No. 1, pp. 17–20.

Kur, J.E. and Fadol, N. (2017). Defects in welding and preventive actions. *LAP LAMBERT Academic Publishing*, Republic of Moldova

Mao, L., Gan, L., Li, W., Zhnag, P. (2022), Failure analysis on weld joint of centrifugal pump diffuser for oil and gas pipeline transportation, *Engineering Failure Analysis*, Vol. 140, 106620 https://doi.org/10.1016/j.engfailanal.2022.106620

Answers to MCQs and Even-Numbered Problems

Chapter 1

1.5(MCQs). (1) b, (2) a, (3) d, (4) c, (5) b, (6) d.

Chapter 2

2.9(MCQs). (1) b, (2) a, (3) d, (4) c, (5) c, (6) a, (7) c.

P-2.10.

(a) Liq. + δ-ferrite; Comp. of Liq. = Fe-0.3% C, Comp. of δ = Fe-0.05% C

(b) $M_L = 0.60$, $M_\delta = 0.40$.

P-2.12. PD = 47.77 W/mm^2.

P-2.14. (a) 10.656 J/mm^3, (b) 1.0123 MJ, (c) 20.25 kW, (d) $ 0.05628.

Chapter 3

3.10.(MCQs). (1) b, (2) d. (3) a, (4) a, (5) d, (6) b, (7) d.

P-3.12.

(a) 1503°C, (b) Liq. + δ-ferrite, (c) Liq. = Fe-0.45% C, δ = Fe-0.075% C, (d) M_L = 0.28, M_δ = 0.72.

P-3.14. *PDAS* = 4.4 μm

P-3.16. T_o = 1756.3 °C

P-3.18. (a) Cr Equiv. = 19.2, (b) Ni Equiv. = 13.5

Chapter 4

4.1.(MCQs). (1) b, (2) e, (3) a, (4) d, (5) b, (6) a, (7) g, (8) b, (9) a, (10) d.

P-4.8. H_i = 0.5 kJ/mm

P-4.10. v_w = 4.9 mm/s

P-4.12. $H_i^{pc-GTAW}$ = 0.605 kJ/mm

P-4.14. Volume rate of metal welded = 8241 mm^3/min.

Chapter 5

5.1(MCQs). (1) b, (2) d, (3) a, (4) c, (5) a, (6) b.

P5.6. Q = 3840 J

P5.8. F = 2 kN

P5.10. I_s = 9600 A

P5.12. F = 1290 N, Pitch = 1 mm.

Z. Huda, *Welding and Joining Engineering*,
https://doi.org/10.1007/978-3-032-20462-2

Chapter 6

6.1(MCQs). (1) d, (2) a, (3) d, (4) d, (5) b, (6) a, (7) b, (8) d.
P6.6. Heat energy = 5.322 MJ.
P6.8. Mass of iron = 698.12 grams

Chapter 7

7.1(MCQs). (1) c, (2) d, (3) a, (4) b, (5) a, (6) c, (7) b, (8) a.
P7.8. p = 170.3 MPa,
P7.10. No. of cycles = 7500
P7.12. Bonding strength = 2.22 MPa

Chapter 8

8.1.(MCQs). (1) c, (2) a, (3) c, (4) d, (5) b, (6) b.
P8.6. Rate of heat exchange = 14,140 W = 14.14 kW
P8.8. $\dot{Q}_{FSW}$ = 10,778.63 W = 10.78 kW
P8.10. Feed rate = 130 mm/min.
P8.12. I = 43.2 kg•m^2

Chapter 9

P9.10. L_l = 4.9 mm
P9.12. γ_{sl} = 0.327 N/m
P9.14. r_w = 1.157 mm
9.17(MCQs). (1) b, (2) c, (3) a, (4) b, (5) a

Chapter 10

10.1(MCQs). (1) b, (2) d, (3) a, (4) c, (5) c, (6) c, (7) b, (8) c
P10.10. E = 80 kJ
P10.12. P_i = 1.59 MPa
P10.14. E = 1.44×10^{-14}
P10.16. d = 0.0263 mm

Chapter 11

11.6(MCQs). (1) a, (2) a, (3) b, (4) d, (5) a
P11.8. h = 2.82 mm
P11.10. Throat area = 125 $(mm)^2$.
P11.12. F = 81.98 kN.
P11.14. A = 317.5 $(mm)^2$

Chapter 12

12.1(MCQs). (1) b, (2) c, (3) d, (4) a, (5) b, (6) c
P12.9. 0.1024 mm
P12.11. 3.99 mm
P12.13. 2.505 kJ/mm.

Chapter 13

13.1(MCQs). (1) c, (2) a, (3) b, (4) d, (5) b, (6) a, (7) c, (8) b
P13.8. D = 2.3
P13.10. NA = 1.5
P13.12. G = 5.9
P13.14. $\sigma_{UTS} \cong$ 366 MPa

Chapter 14

14.1(MCQs). (1) b, (2) a, (3) d, (4) c, (5) a, (6) c, (7) b, (8) c
P14.10. Total crack length = 26 mm. The weld fails the NDT.
P14.12. % Incomplete penetration = 10
P14.14. Inclusion area = 6 mm^2. The weld fails the NDT.

Index

Z. Huda, *Welding and Joining Engineering*,
https://doi.org/10.1007/978-3-032-20462-2

The manufacturer's authorised representative in the EU is Springer Nature Customer Service Centre GmbH, Europaplatz 3, 69115 Heidelberg, Germany. If you have any concerns regarding our products, please contact ProductSafety@springernature.com

Printed and bound by CPI Group (UK) Ltd, Croydon, CR0 4YY
09/09/2026
02213710-0002